Complex Systems Design & Management

Daniel Krob · Lefei Li · Junchen Yao ·
Hongjun Zhang · Xinguo Zhang
Editors

Complex Systems Design & Management

Proceedings of the 4th International
Conference on Complex Systems Design &
Management Asia and of the 12th Conference
on Complex Systems Design & Management
CSD&M 2021

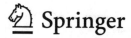 Springer

Editors
Daniel Krob
CESAMES
Paris, France

Junchen Yao
Chinese Society for Aeronautics
Beijing, China

Xinguo Zhang
Tsinhua University
Beijing, China

Lefei Li
Tsinhua University
Beijing, China

Hongjun Zhang
CSSC Systems Engineering Research
Institute
Beijing, China

ISBN 978-3-030-73541-8   ISBN 978-3-030-73539-5   (eBook)
https://doi.org/10.1007/978-3-030-73539-5

This Springer imprint is published by the registered company Springer Nature Switzerland AG
The registered company address is: Gewerbestrasse 11, 6330 Cham, Switzerland

Conference Organization

Conference Chairs

General Co-chairs

Prof. Daniel Krob, Institute Professor at Ecole Polytechnique, President of CESAMES, INCOSE Fellow, France

Dr. Xinguo Zhang, Distinguished Professor, Director of the Complex Systems Engineering Research Center, Tsinghua University, INCOSE ESEP, President of INCOSE Beijing Chapter, People's Republic of China

Organizing Committee Chair

Junchen Yao, Secretary-General, CSAA, People's Republic of China

Program Committee Co-chairs

Prof. Lefei Li, Associate Professor, Deputy Head of the Department of Industrial Engineering, Tsinghua University, People's Republic of China (academic co-chair)

Hongjun Zhang, Member of Standing Committee of Science and Technology Committee, China State Shipbuilding Corporation Limited, People's Republic of China (industrial co-chair)

Program Committee

The program committee consists of 28 members (15 academics and 13 industrialists) of high international visibility. Their expertise spectrum covers all of the conference topics.

Academic Members

Co-chair

Prof. Lefei Li, Associate Professor, Deputy Head of the Department of Industrial Engineering, Tsinghua University, People's Republic of China (academic co-chair)

Members

Michel-Alexandre Cardin, Associate Professor, Imperial College London, UK
Olivier De Weck, Professor of Aeronautics, Astronautics and Engineering Systems, Massachusetts Institute of Technology, USA
Yanghe Feng, Associate Professor, National University of Defense Technology, People's Republic of China
Xinghai Gao, Professor, Beihang University, People's Republic of China
Ruo-Xi Guan, Ph.D., Director Assistant of Complex Systems Engineering Research Centre, Tsinghua University, People's Republic of China
Mengyu Guo, Assistant Professor, Tsinghua University, People's Republic of China
Omar Hammami, Professor, Department of Systems Engineering, ENSTA ParisTech, France
Peter Jackson, Professor and Head of Pillar, Singapore University of Technology and Design (SUTD), Singapore
John Koo, Director of Cyber-Physics Systems, Hong Kong Applied Science and Technology Research Institute, People's Republic of China
Yisheng Lv, Associate Professor, Chinese Academy of Sciences, People's Republic of China
Antoine Rauzy, Professor, Norwegian University of Science and Technology, Norway
Chen Wang, Associate Professor, Tsinghua University, People's Republic of China
Yingxun Wang, Professor, Beihang University, People's Republic of China
Xiaolei Xie, Associate Professor, Tsinghua University, People's Republic of China

Industrial Members

Co-Chair

Hongjun Zhang, Member of Standing Committee of Science and Technology Committee, China State Shipbuilding Corporation Limited, People's Republic of China (industrial co-chair)

Members

Alain Dauron, Expert Leader Systems Engineering, Renault-Nissan, France
Marco Ferrogalini, Vice President, Head of Modelling and Simulations, Airbus, France
Shudong Hou, Chief Expert, Huawei, People's Republic of China
Robert Ong, Technical Director, No Magic, Singapore
Yongjun Qie, Technical Director, Aviation Industry Corporation of China Ltd., People's Republic of China
Paul Schreinemakers, EMEA Sector Director, INCOSE, The Netherlands
Rainer Stark, Director, Fraunhoffer Institute, Germany
Zhixiao Sun, Vice Director, Aviation Industry Corporation of China Ltd., People's Republic of China
Yuejie Wen, Senior Engineer, China Academy of Space Technology, People's Republic of China

Chenguang Xing, Director of Systems Engineering Institute, Aviation Industry Corporation of China Ltd., People's Republic of China
Victor Yee, Business Development Manager, IBM, People's Republic of China
John Zhang, Chief Digital Design Scientist, Huawei, People's Republic of China

Organizing Committee

The organizing committee consists of 17 academic and industrial members of high international visibility. The organizing committee is in charge of defining the program of the conference and of identifying the keynote speakers. The organizing committee also has to ensure the good functioning and organization of the event (communication, sponsoring, …).

Chair

Junchen Yao, Secretary-General, CSAA, People's Republic of China

Members

Chuangye Chang, Research Associate Professor, Beihang University, People's Republic of China
Gilles Fleury, Director Paris Campus—Sciences-Po, Former Director of Centrale Beijing (Beihang University), France
Minyan Gong, Director, CESAMES China, People's Republic of China
Xiangjun Jin, Deputy Secretary-General, Chinese Society of Naval Architects and Marine Engineers (CSNAME), People's Republic of China
Serge Landry, Director, INCOSE Asia-Pacific, INCOSE Asia-Pacific, Singapore
Dong Liu, Managing Director, Accenture Greater China Technology Innovation, People's Republic of China
Mingliang Liu, Vice Secretary-General, Chinese Institute of Electronics, People's Republic of China
Tong Long Liu, President, Nortek, People's Republic of China
Xiao Liu, Professor and Deputy Director of International Cooperation, Shangai Jiao Tong University, People's Republic of China
Kerry Lunney, President of INCOSE, Country Engineering Director at Thales, Australia
Yiran Wang, VP and Secretary-General, Chinese Society of Astronautics (CSA), People's Republic of China
(Mrs.) Jinrui Ye, Deputy Secretary-General, Chinese Society for Composite Materials, People's Republic of China
Ce Yu, Director of International Cooperation, CSAA, People's Republic of China
Jianfu Yu, Secretary-General, Chinese Nuclear Society (CNS), People's Republic of China
Xiaohu Yu, Vice Chairman and Secretary-General, China Ordnance Society (COS), People's Republic of China

Jian Zhang, Vice Secretary-General, China Instrument and Control Society (CIS), People's Republic of China

Invited Speakers

Plenary Sessions

Thierry Chevalier, Head of Digital Design Manufacturing Thrust at Airbus, France
Olivier De Weck, Professor of Aeronautics, Astronautics and Engineering Systems, Massachusetts Institute of Technology, USA
Kerry Lunney, President of INCOSE, Country Engineering Director at Thales, Australia
Antoine Rauzy, Professor, Norwegian University of Science and Technology, Norway
Fanli Zhou, CEO of Suzhou Tongyuan Software and Control Information Technology Co. Ltd., People's Republic of China
Wenfeng Zhang, Research Professor at the Institute of Aerospace System Engineering, People's Republic of China

"INCOSE Beijing Summit" Sessions

Daniel Krob, Institute Professor at Ecole Polytechnique, President of CESAMES, INCOSE Fellow, France
Serge Landry, Director INCOSE Asia-Pacific, INCOSE Asia-Pacific, Singapore
Pao Chuen Lui, Emeritus Professor NUS and Adviser to the National Research Foundation, Prime Minister's Office and INCOSE Fellow, Singapore
Xinguo Zhang, Distinguished Professor, Director of Complex Systems Engineering Research Center, Tsinghua University, China INCOSE ESEP, President of INCOSE Beijing Chapter, People's Republic of China

NB. At the time of publication, we are still waiting for the confirmation of some invited speakers; you will find all of them on the conference Web site.

Preface

Introduction

This volume contains the proceedings of the 4th Asia-Pacific conference on Complex Systems Design & Management (CSD&M Asia 2021) and of the 12th international conference on Complex Systems Design & Management (CSD&M 2021) which are two international series of conferences on systems architecting, modeling and engineering that merged this year (see the two conference Web sites www.2020. csdm-asia.net or www.csdm-asia.cn for more details).

Hosted by the Chinese Society of Aeronautics and Astronautics (CSAA) and organized by the Center of Excellence on Systems Architecture, Management, Economy and Strategy (CESAMES) with the support of its Chinese branch, CESAMES China, the 4th CSDM Asia and 12th CSDM edition was held in Beijing for two days.

The conference also benefited from the sponsorship and technical and financial support of many organizations such as APSYS, AVIC, AVIC China Aeronautical Radio Electronics Research Institute, AVIC Xian Flight Automatic Control Research Institute, Beihang University, China Instrument and Control Society, China Ordnance Society, China State Shipbuilding Corporation Limited, Chinese Institute of Electronics, Chinese Nuclear Society, Chinese Society for Composite Materials, Chinese Society of Astronautics, Chinese Society of Naval Architects and Marine Engineers, Dassault Systèmes, Eclipse Capella Consortium and OBEO, INCOSE, INCOSE Asia-Oceania Sector, MBSE Consulting, PGM Technology and Tsinghua University. Our sincere thanks, therefore, to all of them.

Many other academic, governmental and industrial organizations were involved in the CSD&M 2021 program and organizing committees. We would like to sincerely thank all their members who helped a lot through their participation and contribution during the conference preparation.

Why a CSD&M Conference?

Mastering complex systems requires an integrated understanding of industrial practices as well as sophisticated theoretical techniques and tools. This explains the creation of an annual *go-between* forum—which did not exist before—alternating between Europe and Asia and jointly dedicated to academic researchers and governmental and industrial actors working on complex industrial systems architecting, modeling and engineering. Facilitating their *meeting* was actually for us a *sine qua non* condition in order to nurture and develop in Europe and Asia the science of systems which is currently emerging and developing worldwide.

The purpose of the "Complex Systems Design & Management" (CSD&M) conference is exactly to be such a forum. Its aim is to progressively be *the* European and Asian academic-industrial conference of reference in the field of complex industrial systems architecting, modeling and engineering, which is a quite ambitious objective.

The last 11 CSD&M conferences—which were all held from 2010 to 2020 in Paris (France)—and four CSD&M Asia conferences—which were all held from 2014 to 2018 in Singapore—were the first steps in this direction. Last year, participants were again 310 to attend our 2-day CSD&M 2020 conference—exceptionally managed online—which proves that the interest in systems architecting, modeling and engineering does not fade. In 2021, a key point was the merge of our two previously independent European and Asia-Pacific streams, resulting in a unique CSD&M series that will alternate between Europe and Asia from now on.

Our Core Academic—Industrial Dimension

To make the CSD&M conference a convergence point between the academic, governmental and industrial communities working in complex industrial systems, we based our organization on a principle of *parity* between academics and industrialists (see the conference organization sections in the next pages). This principle was implemented as follows:

- program committee consisted of 50% academics and 50% industrialists, and
- invited speakers came in a balanced way from numerous professional environments.

The set of activities of the conference followed the same principle. They indeed consist of a mixture of research seminars and industrial experience sharing, academic articles and industrial presentations, software tools and training offers presentations, etc. The conference topics cover the most recent trends in the emerging field of complex systems sciences and practices from both an academic and industrial perspective, including the main *industrial domains* (aeronautics and aerospace, defense and security, energy and environment, high tech and electronics, software and communication, transportation), *scientific and technical topics* (systems fundamentals, systems architecting, modeling and engineering, systems metrics and quality, systems safety, systems integration, systems verification and validation, model-based systems engineering and simulation tools) and types of systems (transportation systems, embedded systems, energy production systems, communication systems, software and information systems, systems of systems).

CESAM Community

The CSD&M series of conferences are organized under the guidance of CESAM Community (see cesam.community/en), managed by the Center of Excellence on Systems Architecture, Management, Economy and Strategy (CESAMES; see cesames.net and cesames.cn).

CESAM Community aims in organizing the sharing of good practices in systems architecting and model-based systems engineering (MBSE) and certifying the level of knowledge and proficiency in this field through the CESAM certification.

The CESAM systems architecting and model-based systems engineering (MBSE) certification are especially currently the most disseminated professional certification in the world in this domain through more than 1000 real complex system development projects on which it was operationally deployed and around 10,000 engineers who were trained on the CESAM framework at international level.

The CSD&M 2021 Edition

The CSD&M 2021 edition received 89 submitted papers, out of which the program committee selected 33 regular papers to be published as full papers in the conference proceedings. The program committee also selected 20 papers for a collective presentation during the poster workshop of the conference.

Each submission was assigned to at least two program committee members, who carefully reviewed it, in many cases with the help of external referees. These reviews were discussed by the program committee during an online meeting that took place by October 23, 2020, and was managed through the EasyChair conference system.

We also chose several outstanding speakers with great scientific and industrial expertise who gave a series of invited talks covering the complete spectrum of the conference during the two days of CSD&M 2021. The conference was organized this year around a common topic: "Digital Transformation in Complex Systems Engineering." Each day proposed various invited keynote speakers' presentations on this topic and a "à la carte" program consisting in accepted paper presentations managed in different sessions.

Furthermore, we had "poster workshops," to encourage presentation and discussion on interesting, but "not-yet-polished," ideas. Finally, CSD&M 2021 also offered booths presenting the last state-of-the-art engineering and technological tools to the conference participants.

Paris, France   Daniel Krob
Beijing, China   Lefei Li
Beijing, China   Junchen Yao
Beijing, China   Hongjun Zhang
Beijing, China   Xinguo Zhang
January 2021

Acknowledgements

We would like to thank all members of the program and organizing committees for their time, efforts and contributions to make CSD&M 2021 a top-quality conference. Special thanks also go to the CESAM Community, CESAMES and CSAA teams who permanently and efficiently managed the administration, communication and logistics of the CSD&M 2021 conference (for more details, see cesam.community/en, cesames.net, cesames.cn and www.csaa.org.cn).

The organizers of the conference are grateful to the following partners without whom the CSD&M 2021 event would not exist:

- **Founding partner**

  - CESAM Community managed by CESAMES (Center of Excellence on Systems Architecture, Management, Economy and Strategy)

- **Industrial and institutional partners**

  - The French company APSYS—法国APSYS公司
  - The Chinese company AVIC—中国航空工业集团
  - The AVIC China Aeronautical Radio Electronics Research Institute—航空工业无线电电子研究所
  - The AVIC Xian Flight Automatic Control Research Institute—航空工业西安飞行自动控制研究所
  - Beihang university—北京航空航天大学
  - The China Instrument and Control Society—中国仪器仪表学会
  - The China Ordnance Society—中国兵工学会
  - The China State Shipbuilding Corporation Limited—中国船舶集团有限公司,
  - The Chinese Institute of Electronics—中国电子学会
  - The Chinese Nuclear Society—中国核学会
  - The Chinese Society for Composite Materials—中国复合材料学会
  - The Chinese Society of Astronautics—中国宇航学会
  - The Chinese Society of Naval Architects and Marine Engineers—中国造船工程学会
  - The French Group Dassault Systèmes—达索集团

- The Eclipse Capella Consortium, managed by the French company OBEO—法国OBEO公司
- The International Council on Systems Engineering (INCOSE)—国际系统工程协会
- The International Council on Systems Engineering (INCOSE) Asia-Oceania Sector—国际系统工程协会亚太区
- The Hong Kong company MBSE Consulting—香港MBSE咨询公司
- The Chinese company PGM Technology—中国仆勾山科技有限公司
- Tsinghua University, Department of Industrial Engineering—清华大学工业工程系

At the time of publication, we are still waiting for the confirmation of other partners; you will find all of them on the conference Web sites.

Contents

Contents

Posters

Regular Papers

A Deep Learning and Ontology Based Framework for Textual Requirements Analysis and Conceptual Model Generation

Yongjun Qie, Huanhuan Shen, and Aishan Liu

1 Introduction

With the increasing complexity of systems, model-based systems engineering (MBSE) has been widely adopted in the system development process to reduce expense and improve efficiency leading to a paradigm shift from document to model. Analyzing requirements correctly and creating subsequent conceptual models effectively is a key part of the system engineering process, thus playing a critical role in the whole-life development of systems [1, 2]. However, for the textual requirements processing and model generation process, most of the current strategies mainly based on human efforts, which inevitably introduce errors and mistakes [3, 4]. Thus, a question emerges: is it possible for us to automatically analyze textual requirements and generate conceptual models through artificial intelligence technology? It is obvious that, with such an automatic process, the efficiency of system development will be greatly improved.

Deep learning is the most rapid-developing research direction in the field of artificial intelligence in recent years, and it is highly concerned by the academia and industry. Deep learning is a general term for machine learning algorithms based on feature self-learning and deep neural networks. Recent years have witnessed great success in many applications from different fields including speech recognition, computer vision, and natural language processing.

Y. Qie
Tsinghua University, Haidian District, Beijing, China
e-mail: xiyj19@mail.tsinghua.edu.cn

H. Shen · A. Liu (✉)
AVIC-Digital, No.7 Jingshun Road, Beijing, China
e-mail: talentedlas@126.com

H. Shen
e-mail: shenhh@avic-digital.com

© The Author(s), under exclusive license to Springer Nature Switzerland AG 2021
D. Krob et al. (eds.), *Complex Systems Design & Management*,
https://doi.org/10.1007/978-3-030-73539-5_1

In addition to that, as one of the important directions of artificial intelligence, natural language processing (NLP) has shown promising performance when studying the effective communication between humans and machines through natural language. In recent years, NLP has achieved great breakthroughs in word segmentation, part-of-speech tagging and syntactic analysis showing many commercial success cases and industrial practice cases. Nowadays, using deep learning technology to solve natural language processing problems is a research hotspot in artificial intelligence. Meanwhile, ontology, a formal, explicit specification of a conceptualization, is used in many knowledge related fields to improve machine inference and reasoning abilities.

This paper proposed a framework to assist the model-based systems engineering process through natural language processing and ontology. With the help of artificial intelligence, it is much more convenient for engineers to complete the most time-consuming and effort-consuming tasks (i.e., textural requirements analyzing and conceptual model generation). Thus, as we discussed before, the efficiency and quality of system development can be improved.

In the development of aeronautical systems, engineers are required to conduct requirements analysis. With top-level needs or requirements captured from stakeholders, engineers are required to understand, analyze, and process those textual requirements, then, in turn, generate formalized requirements and conceptual models.

In our deep learning and ontology-based framework, textual requirements inputs are firstly parsed and analyzed using natural language processing technology. Then, with deep convolutional neural networks, key concepts and relationships in textural requirements are captured and extracted. In addition, we introduce domain ontology which could be beneficial to verify the completeness and consistency of the extracted information. With the information above, conceptual models, such as a block definition diagram, are automatically generated by using external APIs.

In order to evaluate our proposed framework, we further take the Air Traffic Management (ATM) system as a case study for research and experiment. All ontology, corpus, and test sets come from the ATM domain.

2   Proposed Approach

Figure 1 illustrates the overall architecture of the automatic model generator from the textual requirement. It mainly consists of 3 parts: textual requirements analysis, properties verification, and model generation.

Textual Requirements Analysis. In order to generate models automatically, the key point is to analyze the nature language described textual requirement. Only by 'understanding' the semantics of described requirement, could the computer extract the containing information, and build corresponding models consequently. The semantic information contains users concerned concepts, attributes of concepts, the dynamic and static relationship between these concepts, etc.

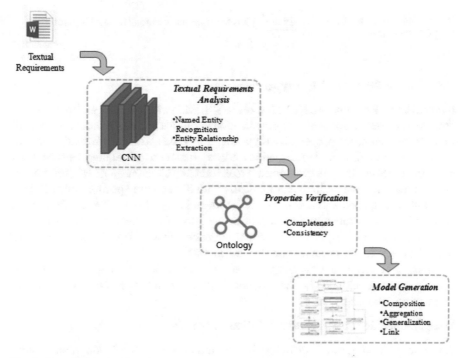

Fig. 1 Overall framework

Properties Verification. The semantic information extracted from textual requirements is coarse-grained, in which repetition and inconsistency may somewhat occur causing bad modeling results. Based on the observation, we further introduce a properties verification process, where domain ontology is used to check and verify the extracted properties, e.g., the consistency of concepts attributes, missing of important attributes of concepts, etc.

Model Generation. This module is the final step of the whole progress. We need to integrate the information extracted from sentences at the first step, and then generate corresponding models (e.g. class diagrams, sequential diagrams) in the software tool.

2.1 Textual Requirements Analysis

To generate conceptual models automatically, it is required to analyze and decompose the textual requirements into elements automatically. To solve the problem, we mainly divide the process into two parts: domain-specific named entity recognition and domain-specific entity relation extraction.

Fig. 2 Entities in natural language sentences

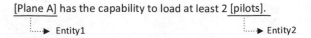

[Plane A] has the capability to load at least 2 [pilots].

......► Entity1   ......► Entity2

Domain-specific Named Entity Recognition.

Named Entity Recognition (NER) is a key task in NLP, which is designed to recognize the names of people, places, and organizations within a text. As a fundamental skill in the natural language process, NER plays a really crucial role in information extraction, information filter, and information retrieval. In our textual requirements analysis scenario, NER could be very important to extract key elements, e.g., subject, object, etc., from a requirement item automatically with the domain-specific information.

Intuitively, named entities described by users in a particular domain for certain include important concepts that would be used to build conceptual models in the future. As shown in Fig. 2, in this textual sentence below, 'Plane A' and 'pilots' are named entities in this domain, and they are annotated.

Thus, it is significant to recognize and extract named entities in a particular domain, which could also be the foundation step of domain-specific entity relation extraction in the next section.

Domain-specific Named Entity Relation Extraction.

An entity is a thing or set of things in the natural world. A relation is an explicit or latent semantic connection between pairs of entities. The goal of entity relation extraction is to find semantic relations between entities in a text. Entity-relationship extraction has been widely adopted in fields including information extraction, expert system, information retrieval [5, 6].

In this paper, we mainly focus on binary relation extraction. In other words, each sentence only contains two entities. The definition can be found below:

$$R(e1, \; Ri, \; e2)Ri \; S,$$

where e1 is the first entity whereas e2 is the second entity. Ri represents the relation between them, which is constrained in a set S.

To solve the problem, in this paper, we formally treat entity extraction as a supervised classification task in machine learning, which can be described below.

$$f(x) = \begin{cases} 1, \text{ entity pair in x has particular semantic relation} \\ 0, \text{ entity pair in x doesn't have particular relation} \end{cases}$$

Besides, x is a sentence containing a pair of entities, f(x) is a classifier trained by supervised methods to classify the input sentence.

Different from the previous work [7–9], this paper utilized deep convolutional neural network (CNN) based techniques [10–15] as the basic model to classify the input sentence into different labels, i.e., different relationship types. The CNN is trained with plenty of corpus and manually designed features.

Figure 4 illustrates the architecture of the deep convolutional neural network for relation extraction. It mainly consists of two parts: lexical-level feature abstraction and sentence-level feature abstraction.

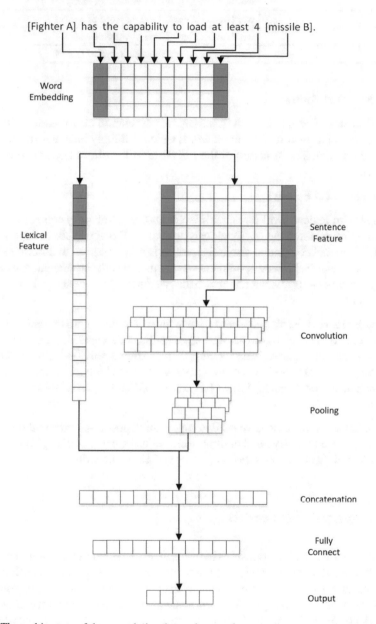

Fig. 4 The architecture of the convolutional neural network

Table 1 Lexical level features

Feature name	Meaning
T1	word embedding of *entity1*
T2	word embedding of *entity2*
T3	word embedding of the left word of *entity1*
T4	word embedding of the right word of *entity1*
T5	word embedding of the left word of *entity2*
T6	word embedding of the right word of *entity2*

(a)Lexical-level Feature.

The lexical-level feature plays an important role in relation classification. This kind of feature contains information about words which definitely indicates the pattern of a specific relation. The structure of the lexical-level feature can be found in Fig. 2 (Table 1).

(b)Sentence-level Feature.

Although with lexical-level features, it is not enough to classify sentence relationships with limited information. Obviously, lexical-level feature fails to capture long-distant feature and compound semantics, which is only designed to describe similarities between words. However, humans usually understand the semantics of natural language through long-distant features. Thus, we further take sentence-level features into account.

Context Feature. Distributed Theory has pointed out that words coming from the similar context tend to have the similar semantic meanings. So, we use a 'slide window' to capture context features. Supposing a sliding window with window-size m, we form a context window with each m/2 words on the left and right side of word x in a sentence. For instance, $WS = \{..., xt - 1, xt, xt + 1, ...\}$, besides the size of WS is m.

Position Feature. Moreover, we need to supply more positional information for each word of distance to entity1 and entity2. Here we use a pair (Left, Right) to indicate the word-level distance to entity1 and entity2 of a specific word.

2.2 Properties Verification

The named entities and relationships between them extracted by convolutional neural networks are still coarse-grained, and there may exist many problems such as repetition and inconsistency. For example, textual requirements may miss some key factors and aspects of the system (i.e., not complete), the extracted information may contain some inconsistency and violation between each other (i.e., not consistent and conforming). In order to further generate fine-grained requirements leading to better modeling performance, these situations need to be considered and these problems need to be solved.

Ontology is a formal, explicit specification of conceptualization. Knowledge exists in every single filed, which might be explicit and structural. Ontology is used in many knowledge related fields, for example, knowledge representation and storage. When a new requirement comes forth, an ontology can be a 'domain expert' with great knowledge, who can help engineers to verify requirement iteratively so that models could be built more easily.

Based on the observation, we thus introduce a properties verification process, where domain ontology is used to check and verify the extracted properties from the previous step.

Given the extracted entity and relationship, the ontology is utilized to check and verify the completeness and consistency of requirements, i.e., whether the set of requirements contains everything pertinent to the definition of system or system element is specified and whether the set of requirements are not contradictory nor duplicated.

2.3 Model Generation

After extracting the entities and classifying their semantic relationships, we need to further map the semantic relations to relations in SysML. In addition, the relations between entities coming from textual sentences might be duplicated and are required to be washed. In this paper, we just discuss the Block Definition Diagram in SysML. We map the semantic relations extracted from the previous step to relations in the block definition diagram as shown in Table 2.

Specifically, relations like composition and aggregation are classified more accurate than generalization. The reason for this phenomenon might be that generalization is more 'abstract' and may lie in deep semantics of a sentence. How to promote the accuracy of classification of generalization would be a part of our work in the future.

Table 2 Semantic relation to model relation

Semantic relation	Model relation
Component-Whole	Composition
Member-Collection, Content-Container	Aggregation
General-Special	Generalization
Others	Link

3 Experiment and Evaluation

Air traffic control is an important specialty and research direction in the field of civil aviation. The Air Traffic Management (ATM) system applies communication and navigation technology to monitor and control aircraft flight activities and to ensure flight safety and flight order. ATM is critical for maintaining the efficient, safe and orderly operation of transportation systems.

This article takes the ATM system as a case study. Using the framework proposed, requirement documents of Air Traffic Management system can be analyzed automatically.

3.1 Air Traffic Management

Air Traffic Management is an aviation system encompassing all systems that assist aircraft to depart from an aerodrome, transit airspace, and land at a destination aerodrome. Apparently, ATM is a system of systems (SoS), in which systems include air traffic control, air traffic safety electronics personnel, aeronautical meteorology, air navigation system, air space management, air traffic service, Air Traffic Flow Management, etc., interacts with each other. As an SoS, however, ATM contains several challenges and problems to be solved.

Challenges and Problems.

- The growing complexity of the ATM system. More and more integrated functionalities for different services in the system, e.g., more planes, passenger services, collision avoidance, etc.
- The growing complexity of systems within the ATM system. There are different factors increase the complexity of the systems in ATM. Technical complexity growth by increased electronic and software-driven functionality of the subsystems. Increased performance requirements lead to more sophisticated and precise control of the subsystems.
- Strong uncertainties. With such big and huge capacities, ATM contains a lot of uncertainties, e.g., meteorology uncertainty, prediction uncertainty, environmental uncertainty, etc. (Fig. 5).

3.2 Experimental Settings

The training set we use in this paper is SemEval-2010 Task 8: Multi-Way Classification of Semantic Relations Between Paris of Nominals [16], which is designed specifically for relation classification. The main purpose of this task is to classify the sentence into a specific relation that has been defined in the dataset with the two entities annotated. The relations contained in this dataset have been displayed in Table

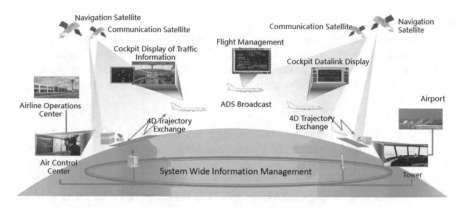

Fig. 5 Air traffic management system

Table 3 SemEval-2010 Task 8: multi-way classification of semantic relations between paris of nominals corpus

Relation	Number	Percentage (%)	Meaning
Cause-Effect	1331	12.42	Cause and effect
Component-Whole	1253	11.69	Component and whole
Content-Container	732	6.83	Component and whole, more about physical relation
Entity-Destination	1137	10.61	Entity and its destination
Entity-Origin	974	9.09	Entity and its origin
Instrument-Agency	660	6.16	agency
Member-Collection	923	8.61	Component and whole
Message-Topic	895	8.35	Message and topic
Product-Producer	948	8.85	Product and producer
Other	1864	17.39	others

3. Meanwhile, this paper adds 500 extra sentences of Special-General to indicate the generalization relation which could be beneficial to the model generation process in the next step.

3.3 Implementation Details

Entity Recognition
We choose stanfordNLP to do natural language processing tasks including parse, POS, syntax analysis, and entity recognition. Besides, we use a crawler to crawl plenty of domain-specific words.

Entity Relation Extraction

We use Sena for word embedding [17]. As for the implementation of the deep convolutional neural network, we use Tensorflow [18]. For our deep neural network, we set the window size to 3, embedding dimension to 50, and distance dimension to 5. We train our model using SGD optimizers with the learning rate as 0.1 [19].

Domain Ontology

We build our knowledge base with more than 100 different entities from different levels. Since ATM is a system of system, we build most entities at system level including airport, airplane, air control center, navigation satellite, etc., and we further create subsystem level entities such as flight management systems. In this paper, we use Web Ontology Language (OWL) to build our ontology and their relationships.

Model Generation

The model generation module is implemented in Java, in which input is relations between entities extracted in the previous step. This module analyzes input data firstly and converts them into the $R(e_1, R_i, e_2)$ styled triple. After that, processes like a filter, classification, combination are proposed. Finally, we call APIs in Rhapsody to create models dynamically.

This module mainly contains three layers. The first layer is the data layer, which is responsible to store and load data in databases (MySQL, Json, XML). Then, an operation layer is designed for data analysis. Finally, the UI layer is used for visualization.

3.4   Evaluation and Discussion

The dataset we use is SemEval-2010 Task 8: Multi-Way Classification of Semantic Relations Between Paris of Nominals. Besides, the sentences in this corpus are described under some general situations, which are not suitable for the information in ATM. Thus, in order to improve the generalization ability of our model, we further import domain-specific corpus and train the deep neural network under that. Moreover, we try to modify and update some of the words in each sentence to fit the ATM scenario. After this modification, we see a significant improvement in performance.

Future Work

What we generate in this paper is static models like class diagrams. Though the static model shows a good representation of the static structure of concepts, it still lacks for the description of dynamic behaviors, business flow, and capabilities. So, as for a better description of system features, dynamic model generation should be considered in the future.

4 Conclusion

This article proposed a deep learning and ontology-based framework for textual requirement analysis and conceptual model generation. The framework includes three modules, textual requirements analysis, properties verification, and automatic model generation, and our framework achieves the automatic generation from textual requirements to conceptual models. Further, we use the Air Traffic Management (ATM) system as a case study. From the experimental results, the framework is able to automatically analyze textual requirements and generate conceptual models.

References

1. Hause, M.C.: SOS for SoS: A new paradigm for system of systems modeling. In: 2014 IEEE Aerospace Conference. IEEE, pp. 1–12 (2014)
2. Jackson, S.: Systems Engineering for Commercial Aircraft: A Domain-Specific Adaptation. Ashgate Publishing, Ltd. (2015)
3. Walden, D.D., et al. :Systems Engineering Handbook: A Guide for System Life Cycle Processes and Activities. John Wiley & Sons (2015)
4. Kossiakoff, A., et al.: Systems Engineering Principles and Practice, vol. 83. John Wiley & Sons (2011)
5. Kapurch, S.J.: NASA Systems Engineering Handbook. Diane Publishing (2010)
6. Bach, N., Badaskar S.: A review of relation extraction. In: Literature review for Language and Statistics, II2 (2007)
7. Miller, S., et al.: A novel use of statistical parsing to extract information from text. In: Proceedings of the 1st North American chapter of the Association for Computational Linguistics conference. Association for Computational Linguistics (2000)
8. Kambhatla, N.: Combining lexical, syntactic, and semantic features with maximum entropy models for extracting relations. In: Proceedings of the ACL 2004 on Interactive poster and demonstration sessions. Association for Computational Linguistics (2004)
9. Zhou, G.D., et al.: Exploring various knowledge in relation extraction. In: Proceedings of the 43rd annual meeting on association for computational linguistics. Association for Computational Linguistics (2005)
10. Jiang, J., Zhai, C.X.: A systematic exploration of the feature space for relation extraction. In: Human Language Technologies 2007: The Conference of the North American Chapter of the Association for Computational Linguistics; Proceedings of the Main Conference (2007)
11. Zeng, D., et al.: Relation classification via convolutional deep neural network. In: Proceedings of COLING 2014, the 25th International Conference on Computational Linguistics: Technical Papers (2014)
12. Kim, Y.: Convolutional neural networks for sentence classification. arXiv preprint arXiv:1408. 5882 (2014)
13. Sahu, S.K., et al.: Relation extraction from clinical texts using domain invariant convolutional neural network. arXiv preprint arXiv:1606.09370 (2016)
14. Yin, W., Schütze, H.: Multichannel variable-size convolution for sentence classification. arXiv preprint arXiv:1603.04513 (2016)
15. Kalchbrenner, N., Grefenstette, E., Blunsom, P.: A convolutional neural network for modelling sentences. arXiv preprint arXiv:1404.2188 (2014)
16. Hendrickx, I., et al.: Semeval-2010 task 8: Multi-way classification of semantic relations between pairs of nominal. In: Proceedings of the Workshop on Semantic Evaluations: Recent Achievements and Future Directions. Association for Computational Linguistics (2009)

17. Turian, J.W., Ratinov, L., Bengio, Y.: Word representations: a simple and general method for semi-supervised learning. In: Proceedings of the 48th annual meeting of the association for computational linguistics. Association for Computational Linguistics (2010)
18. Abadi, M., et al.: TensorFlow: A System for Large-Scale Machine Learning. OSDI, vol. 16 (2016)
19. Srivastava, N., et al.: Dropout: a simple way to prevent neural networks from overfitting. J. Mach. Learn. Res. **15**(1), 1929–1958 (2014)

A Design of Commercial Aircraft Health Management Using N-F-R-P Process

Shuo Chang, Jian Tang, Yi Wang, Zhaobing Wang, and Shimeng Cui

1 Introduction

After decades of maturity system engineering now has entered into a golden era, used extensively by all engineering fields especially aerospace industry. And there are several system engineering handbooks published and constantly renewed by organization like INCOSE [1] and NASA [2] as well as guideline of applying system engineering in different field such as SAE ARP4754A. Chinese aviation industry doesn't want to be left behind in this big trend. With the ambition to become a market breaker, system engineering methodology is well adopted by Commercial Aircraft Corporation of China Ltd. (COMAC). For COMAC, the motivation comes from requirements of airworthiness authorities and experiences collected in half century. In COMAC's main project C919, CR929 aircraft development, system engineering is heavily emphasized hoping to catch up with international competitor. To show determination of applying system engineering in aircraft project and to guide company activity in system engineering, COMAC system engineering manual [3] is published. A X model/N-F-R-P design process which is need-function-requirement-physical is presented.

S. Chang (✉) · J. Tang · Y. Wang · Z. Wang · S. Cui
Beijing Aircraft Technology Research Institute, Commercial Aircraft Corporation of China (COMAC), Beijing 102211, People's Republic of China
e-mail: changshuo@comac.cc

© The Author(s), under exclusive license to Springer Nature Switzerland AG 2021
D. Krob et al. (eds.), *Complex Systems Design & Management*,
https://doi.org/10.1007/978-3-030-73539-5_2

2 System Engineering Design Processes and Commercial Aircraft Health Management

2.1 Different System Engineering Design Process

There are different top down system engineering design processes, namely Need-Requirement-Architecture-Physics (NRAP) process in INCOSE system engineering handbook/ISO IEC IEEE 15,288 and Need-Function-Requirement-Physics (NFRP) process in COMAC System Engineering Manual. Both processes start from stake holder needs, then translate needs to formal requirements. The comparison of these two processes is whether a functional structure formed before requirements capture. The choosing of different processes should depend on the project type. NRFP is more fit for engineering projects that seek for new and innovated solution such as defense projects since the function are formed after the requirement capture process. These projects focus on the form of function which might include new and innovated solution and it might cause some iteration between function and requirement process. On the contrary, NFRP process is fit for inherited function/technology project which more focused on precise capture of requirements in a speedy way resulted from a formed top-level function structure. This makes the requirement capture process more targeted, purposeful and efficient which will reduce iteration in requirements validation.

For the reasons stated above COMAC commercial aircraft project is more fit for NFRP process, so the process is proposed in COMAC system engineering manual with emphasis on forward design which is advocated by ARP4754A [4].

Different forward design processes using either function analysis or architecture design to transform stake holder needs to real hardware/software. System architecture design is used in ISO IEC IEEE 15288 [5] technical process and INCOSE system engineering handbook which has fall into highest level of implementation. For system engineering professional, it is better to include of function analysis, which is proposed in ARP4754A as it is more focus on behavior level.

2.2 Commercial Aircraft Health Management

Commercial aircraft health management function which has onboard part and off board part [6] is a relative new compare to other functions in an aircraft. Yet it expands rapidly with pressures coming from airline operation and the evolution of enabling technologies. Such new system development and function integration is a good chance to apply N-F-R-P design process in defining the system functions, architecture and capture system requirements. But until now most of the research work done in this aera focused on the technology evolution [7]. Integrated vehicle health management (IVHM) is an emerging concept which try to integrate the former separated health management related equipment and products. IVHM cuts across all aspects of system design and implementation throughout the system life cycle, this necessitates the adoption of a systems engineering view of IVHM design [8], 9. System engineering method could better illustrate the word integrate in IVHM concept. As a main advocator of system engineering by APR 4754A and IVHM by HM-1 health management standard group [10], SAE plays key role in introducing system engineering to IVHM design [11, 12].

3 NFRP Process to Analyze Commercial Aircraft Health Management

This chapter applies NFRP process on designing health management system for commercial aircraft. With this top down design process, the design can be independent from previous hardware centric method decided by supplier hardware offer.

3.1 Operation Scenario and Stake Holder Need Capture

Health management system which has multiple inside and outside stake holders and connected to multiple onboard and offboard systems should apply NFRP process to capture needs and requirements from all kinds of stake holders and design suitable functions and architecture/hardware to meet these needs and requirements.

Stake holders of health management function are identified from the business flow of aircraft project and operation scenario analysis. From up, middle and downstream of business flow stake holders are identified and shown in Fig. 1.

To make sure the completeness stake holders are also identified from operational scenario the steps are shown below (Fig. 2).

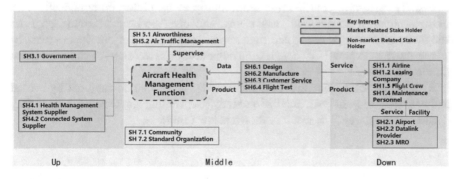

Fig. 1 Stake holders identification from business flow

Fig. 2 Scenario analysis process

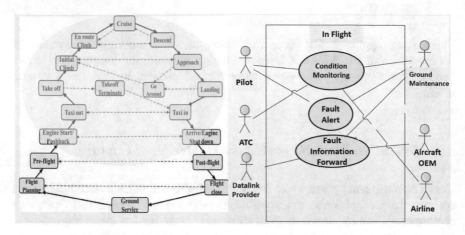

Fig. 3 In flight scenario

Three working scenarios of health management function: in flight, maintenance and test, data analysis are identified with the participants shown together (Figs. 3, 4 and 5).

From the result, one can observed design and implementation of IVHM systems necessarily integrates contributions from multiple disciplines, communicates with multiple physical and logical subsystems, and involves human interaction from designers, owners, operators, and maintainers. IVHM system developers must be keenly aware of IVHM dependencies and influences on system design and subsequent implementation [11].

After stake holder identification, stake holder needs can be captured by market analysis, scenario analysis, regulation study, coworking, government and society consideration survey. The result is shown below (Table 1).

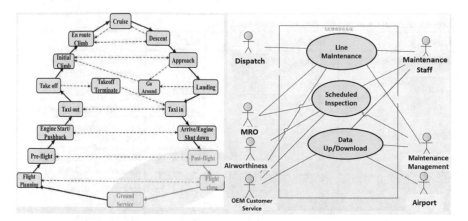

Fig. 4 Maintenance

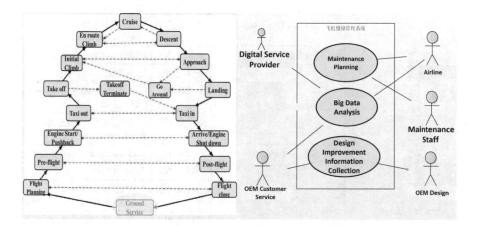

Fig. 5 Use of data

3.2 Function Analysis and Requirement Capture

3.2.1 Subfunctions

Sub functions are designed from stake holder needs:

- Support testing
- Monitor aircraft status
- Recording
- Diagnosis
- Prognostics
- Fault information forward

Table 1 Health management function stake holder needs

Category	Stake Holder	No	Needs
SH1. Customer	SH1.1 Airlines	SH1.1_N1	Extend aircraft life or increase residual value
		SH1.1_N2	Reduce operational cost
		SH1.1_N3	Raise operational safety
		SH1.1_N4	Increase aircraft availability
		SH1.1_N5	Increase aircraft and system reliability
		SH1.1_N6	Reduce unscheduled maintenance
		SH1.1_N7	Support maintenance upgrade to condition-based maintenance and predictive maintenance
	SH1.2 Leasing Company	SH1.2_N1	Provide aircraft health monitoring function
	SH1.3 Pilot	SH1.3_N1	Understand aircraft status
		SH1.3_N2	Easy to use
		SH1.3_N3	Reduce false/duplicate and unnecessary alarm
		SH1.3_N4	Maximize first time alarm to failure time
	SH1.4 Maintenance Personnel	SH1.4_N1	Reduce Maintenance Task
		SH1.4_N2	Easy to use
		SH1.4_N3	Reduce maintenance support equipment and staff
		SH1.4_N4	Automatic and fast fault locating
		SH1.4_N5	Maximize fault coverage rate
		SH1.4_N6	Easy access to data
		SH1.4_N7	Collecting onboard system data as much as possible
		SH1.4_N8	Track parts remaining useful life
SH2.Operation Support	SH2.1 Airport	SH2.1_N1	Less investment on infrastructure to match with health management system
		SH2.1_N2	Generic system for data transfer for different types of aircraft

(continued)

Table 1 (continued)

Category	Stake Holder	No	Needs
	SH2.2 Data Link Provider	SH2.2_N1	Provide high speed datalink to air-ground data transfer
	SH2.3 MRO	SH2.3_N1	Take part in the condition/predictive based maintenance mode led by health management
SH3. Government	SH3.1 Government	SH3.1_N1	Aircraft tracking when accident happen
SH4.Supplier	SH4.1 Health Management System Suppliers	SH4.1_N1	Less investment to development includes certification
		SH4.1_N2	Good revenue performance
		SH4.1_N3	Enough time for development and manufacture
		SH4.1_N4	Use inherited technology from previous product as much as possible
		SH4.1_N5	New developed technology can become generic technology used in other projects
	SH4.2 Connected system suppliers	SH4.2_N1	Interface definition easy to apply
		SH4.2_N2	Help to rise system reliability
SH5. Supervision	SH5.1 Airworthiness authority	SH5.1_N1	Meet airworthiness regulation
	SH5.2 Air traffic management	SH5.2_N1	Proper data sharing for surveillance
SH6. Aircraft OEM	SH6.1 Strategic Planning	SH6.1_N1	Master advanced commercial aircraft health management technology
		SH6.1_N2	Increase competitiveness of product
	SH6.2 Design	SH6.2_N1	Reasonable task sharing and purchase/self-development strategy
	SH6.4 Customer Service	SH6.4_N1	Reduce customer service task

(continued)

Table 1 (continued)

Category	Stake Holder	No	Needs
		SH6.4_N2	Support customer service centered business model and revenue generation
	SH6.5 Flight test	SH6.5_N1	Support aircraft test in data collection, storage, fault detection, isolation, analyze and product change
		SH6.5_N2	Provide evidence of compliance for airworthiness
SH7 Public	SH7.1 Community	SH7.1_N1	Reduce carbon emission
	SH7.2 Standard organization	SH7.2_N1	Use of publicize standard

- Data transfer
- Health condition evaluation
- Fault alert
- Report generation.

Some stake holder needs SH5.1_N1, SH4.1_N2, SH4.1_N3, SH4.1_N4, SH6.1_N1, SH6.2_N1, SH7.2_N1 are fulfilled by project management activity. The tracing of stake holder needs to sub functions are shown below (Table 2).

3.3 Requirements Capture and Allocation

The functional analysis and requirement capture process is introduced in [13]. It is done with activity diagram and functional interface diagram. Another type of diagram, N^2 diagram for interaction within subfunctions is added here. Functional requirements are catching at every activity block, interface port and interaction block. Limited by length of paper here just show the result table (Tables 3 and 4).

The requirements allocation matrix to system onboard is given below. Requirements are allocating to different ATA chapters which goes to different design teams; a further decomposition should be given to subsystems and physical components in real aircraft design working together with system suppliers (Table 5).

3.4 Physical Architecture

A detailed commercial aviation IVHM design is proposed in separate papers [6]. Complementary to the onboard avionic system the health management system should

Table 2 Stake holder needs to sub functions

Need	Subfunction
SH1.1N2 SH1.4_N2 SH1.4_N3 SH4.2_N1 SH6.5_N1 SH2.3_N1	SF1 Support Testing
SH1.1N2 SH1.1_N3 SH1.1_N6 SH1.2_N1 SH1.3_N1 SH1.3_N2 SH1.4_N5 SH3.1_N1 SH4.2_N1 SH5.2_N1 SH6.5_N1	SF2 Monitoring Aircraft Status
SH1.1_N5 SH1.4_N7 SH3.1_N1 SH4.2_N1 SH6.2_N2 SH6.5_N1 SH6.5_N2	SF3 Recording
SH1.1N2 SH1.1_N3 SH1.1_N4 SH1.3_N3 SH1.4_N1 SH1.4_N2 SH1.4_N3 SH1.4_N4 SH1.4_N5 SH6.4_N1 SH6.5_N1 SH2.3_N1	SF4 Diagnosis
SH1.1N1 SH1.1N2 SH1.1_N4 SH1.1_N6 SH1.1_N7 SH1.3_N4 SH1.4_N8 SH4.1_N1 SH4.1_N5 SH6.1_N2 SH6.4_N1 SH7.1_N1 SH2.3_N1	SF5 Prognostics
SH1.1N2 SH1.1_N4 SH2.2_N1 SH5.2_N1 SH2.3_N1	SF6 Fault Information Forward
SH1.4_N6 SH2.1_N1 SH2.1_N2	SF7 Data Transfer
SH1.1N1 SH1.1_N5 SH1.1_N6 SH1.1_N7 SH1.2_N1 SH1.3_N1 SH1.3_N2 SH1.4_N3 SH4.1_N5 SH6.1_N2 SH6.4_N1 SH2.3_N1	SF8 Health Condition Evaluation
SH1.1_N3 SH1.3_N1 SH1.3_N2 SH1.3_N3 SH1.4_N5 SH6.5_N1	SF9 Fault Alert
SH1.3_N2 SH1.4_N2 SH4.2_N2	SF10 Report Generation

Table 3 Sub functions relation matrix

	SF1	SF2	SF3	SF4	SF5	SF6	SF7	SF8	SF9	SF10
SF1										
SF2										
SF3										
SF4										
SF5										
SF6										
SF7										
SF8										
SF9										
SF10										

include data transfer link and ground health management system. Ground health management system include ground monitor control, intelligent diagnosis, maintenance expert and remote customer/expert terminal modules. For fault that cannot be solved by onboard health management system, fault condition data can be transferred to ground monitor and diagnosis center by air-ground communication system. With

Table 4 Functional requirements

Sub-Function	No	Requirement
Support Testing	Req 1	The health management system shall provide LRU level BITE information
	Req 2	The health management system shall monitor the communication between LRU and report to onboard maintenance system
Monitoring Aircraft Status	Req 3	The health management system shall monitor basic aircraft information include: type, registration number, airline, IATA code, ICAO code, MSN number, engine number, APU number and all hardware/software version information
	Req 4	The health management system shall monitor flight information
	Req 5	The health management system shall monitor time information
	Req 6	The health management system shall monitor operational information include: aircraft, subsystem and component operational hour, cycle information
	Req 7	The health management system shall monitor OOOI and POS message (OOOI message aircraft taxi out/take off/landing/taxi in information POS message aircraft position information)
	Req 8	The health management system shall use ARINC664, ARINC629 and ARINC429 to communicate with other system
	Req 9	The health management system shall send aircraft position information every 1 min
Recording	Req 10	The health management system shall provide fault information recording ability, including all system data around fault time and maintenance history
Diagnosis	Req 11	The health management system shall provide fast and precise fault diagnosis ability
	Req 12	The health management system shall keep fault diagnosis and abnormal monitor result to 2 root cause at most
Fault Information Forward	Req 13	The health management system shall transfer the fault information to ground support as soon as it happens when the aircraft is in air
Prognostics	Req 14	The health management system shall be able to monitor system and component trends
	Req 15	The health management system shall be able to predict system and component fault

(continued)

Table 4 (continued)

Sub-Function	No	Requirement
Data Transfer	Req 16	The health management system shall have high-speed air-ground data transfer capability
	Req 17	The health management system shall provide wifi/4G/5G data transfer capability when on ground
	Req 18	The health management system shall provide ethernet port and data transfer capability
Health Condition Evaluation	Req 19	The health management system shall be able to evaluate health condition of all interested systems
	Req 20	The health management system shall be able to show condition evaluation result on flight deck and maintenance equipment
Fault Alert	Req 21	The health management system shall provide immediate fault alert to flight deck display
	Req 22	The health management system shall reduce false alarm and duplication
	Req 23	The health management system fault alert shall cover xx percent of RUL level fault
Report Generation	Req 24	The health management system shall be able generate report of onboard maintenance system and other related health management information
	Req 25	The health management system shall provide onboard printing equipment

the help of intelligent diagnosis system or remote expert, faults can be localized and dealt with in an instant manner. Recent years there are trends of the ground part of health management grow into a business level digital system platform such as Airbus company's Skywise system discussed in [14].

The next step is to build a hardware in loop simulation for the proposed architecture like described in [15].

4   Conclusion and Consideration

This paper shows a design of commercial aircraft health management system by NFRP process. In real engineering environment this kind of design need to put into the whole aircraft design which means some input of the task in the paper will not done separately but rather coming from higher level design and analysis. The design is mostly from technical point of view that does not take business aspects like supply chain and production into consideration. In commercial aircraft project these aspects would post huge limit to hardware/software design.

Table 5 Sub functions relation matrix

No	ATA45 Onboard Maintenance System			ATA46 Information System		
	5–19 CMS/Aircraft General	45 Central Maintenance System (CMS)	70–89 Power Plant	10 Airplane General Information Systems	20 Flight Deck Information Systems	30 Maintenance Information Systems
Req 1						
Req 2						
Req 3						
Req 4						
Req 5						
Req 6						
Req 7						
Req 8						
Req 9						
Req 10						
Req 11						
Req 12						
Req 13						
Req 14						
Req 15						
Req 16						
Req 17						
Req 18						
Req 19						

(continued)

Table 5 (continued)

No	ATA45 Onboard Maintenance System			ATA46 Information System		
	5–19 CMS/Aircraft General	45 Central Maintenance System (CMS)	70–89 Power Plant	10 Airplane General Information Systems	20 Flight Deck Information Systems	30 Maintenance Information Systems
Req 20						
Req 21						
Req 22						
Req 23						
Req 24						
Req 25						

Currently, there are many revolutionary commercial aircraft projects and researches going on include supersonic, hybrid/electric propulsion, blended wing body (BWB) which could make huge influence on the next generation aircraft product. Big change and innovation need to be tamed by system engineering process like NFRP to control the project direction, to reduce the risk, to make development smooth and to achieve project success commercially. So, more work is expected to be done by not only the aircraft OEM design team but also all participants and departments from the whole supply chain.

Reference

1. INCOSE: Systems Engineering Handbook: A Guide for System Life Cycle Processes and Activities, version 4.0. Hoboken, NJ, USA: John Wiley and Sons, Inc (2015)
2. National Aeronautics and Space Administration: NASA System Engineering Handbook, NASA SP-2016–6105 Rev2
3. He, D., Zhao, Y., Guo, B., Qian, Z.: COMAC System Engineering Manual. Shanghai Jiaotong University Press, July 2019
4. Society of Automotive Engineer: ARP 4754A Guidelines for Development of Civil Aircraft and Systems (2010)
5. International Standard Organizatioin: ISO/IEC/IEEE 15288: 2015 Systems and Software Engineering—System Life Cycle Processes
6. Shuo, C., Yi, W.: Integrated vehicle health management technology and its applications in commercial aviation. In: 2017 International Conference on Sensing, Diagnostics, Prognostics, and Control (SDPC), Shanghai, pp. 740–745 (2017)
7. Jha, A., Sahay, G., Sivaramasastry, A.: Framework and Platform for Next Generation Aircraft Health Management System, SAE Technical Paper 2017-01-2126 (2017)

8. Jennions, I.K.: Integrated Vehicle Health Management: Perspectives on an Emerging Field. SAE International (2011)
9. Johnson, S.B., Gormley, T.J., Kessler, S.S., Mott, C.D., Patterson-Hine, A., Reichard, K.M., Scandura, P.A., Jr.: System Health Management: With Aerospace Applications. John Wiley & Sons, Ltd (2011)
10. Chang, S., Gao, L., Wang, Y.: A review of integrated vehicle health management and prognostics and health management standards. In: 2018 International Conference on Sensing, Diagnostics, Prognostics, and Control (SDPC), Xi'an, China, pp. 476–481 (2018)
11. Society of Automotive Engineer: ARP 6407 IVHM Design Guidelines (2019)
12. Society of Automotive Engineer: ARP6883 Guidelines for Writing IVHM Requirements for Aerospace Systems (2019)
13. Chang, S., Wang, Y.: Civil aircraft IVHM system analysis using model based system engineering. In: 2017 ICRSE, Beijing, pp. 1–5 (2017)
14. Chang, S., Wang, Z., Wang, Y., Tang, J., Jiang, X.: Enabling technologies and platforms to aid digitalization of commercial aviation support, maintenance and health management. In: 2019 International Conference on Quality, Reliability, Risk, Maintenance, and Safety Engineering (QR2MSE), Zhangjiajie, China, pp. 926–932 (2019)
15. Chang, S., Wang, Y., Gao, L., Yang, Y.: An aircraft level fault simulation platform to promote integrated vehicle health management technology. In: CSAA/IET International Conference on Aircraft Utility Systems (AUS 2018), Guiyang, pp. 195–199 (2018)

A MBSE-Based Development Life Cycle for Reconnaissance, Early-Warning, and Intelligence Equipment System-Of-Systems

Weiwei Zheng, Yanli Shen, Taoshun Xiao, and Ruiyuan Kong

1 Introduction

System of Systems Engineering (SoSE) includes the technical specification and management process [1–5]. The technical specification follows the idea of the decomposition-integration system theory and progressive and orderly development steps [1]. The management process includes the technical management process and the project management process. The development of an engineering system is essentially the process of establishing an engineering-system model. At the technical process level, it mainly involves in the construction, analysis, optimization, and verification of the system model. On the other hand, it includes the planning, organization, leadership, and control of the system-modeling work in the management process level [2]. Therefore, the SoSE's organizational-management technology should essentially include two levels, i.e., both the system-modeling technology and the organizational-management technology of the system-modeling work, where the system-modeling technology includes the modeling languages, the modeling ideas, and modeling tools [3, 4].

The MBSE method is an important direction in the field of the SoSE [6–12]. It supports system-engineering activities throughout the entire system-development life cycle by exploiting the formal modeling, and exploits object-oriented, graphical, and visual system-modeling languages to describe the underlying system elements, thus forming an integrated, concrete and visual system-architecture model layer by layer. The MBSE method still follows the idea of the decomposition-integration system theory and progressive and orderly development steps [13, 14]. Its core is to adopt formal, graphical, and related modeling languages as well as corresponding modeling tools, and to fully exploit the advantages of the computer and information technology

W. Zheng (✉) · Y. Shen · T. Xiao · R. Kong
China Academy of Electronics and Information Technology, No. 11, Shuangyuan Road, High-Tech Park, Shijingshan District, Beijing, China
e-mail: weiweizheng_lucky@163.com

to modeling (including analysis, optimization, and simulation), which lays a more solid foundation for both the system implementation and the field verification and thus improves the efficiency of the entire development process [10, 15, 16].

In this paper, we propose a MBSE-based Development Life Cycle (MDLC) for Reconnaissance, Early-warning, and Intelligence Equipment System-of-Systems (REIESoS). We used model-based systems engineering technologies to refine the traditional V model, considering the characteristics of the joint REIESoS and the complexity of the System of Systems. In addition, we present both the process method of the scheme design and the requirement of the scheme-design model for the MBSE-based REIESoS. Our proposed MDLC for REIESoS provides the model-based construction, analysis, optimization, and verification at each stage of the development life cycle, allowing us to reduce the risk of the repeated design or development and thus improving the efficiency of the entire development process significantly.

The remainder of the paper is organized as follows. We begin in Sect. 2 by presenting the MBSE-based REIESoS used throughout this paper. Then, in Sect. 3, we introduce our scheme design for MBSE-based REIESoS, including the process methods for the scheme design as well as requirements for scheme-design models. Finally, we conclude this paper and present future work in Sect. 4.

2 MBSE-Based REIESoS

The MBSE-based reconnaissance, early-warning, and intelligence equipment system-of-systems adopt the model expression method to describe such activities as the requirement, design, analysis, verification, and confirmation in the entire life cycle of the system, including both the model-based technical process and the model-based management process.

2.1 The Model-Based Technical Process

The model-based technical process is a cyclic and iterative process [17–19]. Following the system life cycle, this process is a series of technical processes from the user- requirement identification to the system development, operation, and handling. Each process has its defined process and includes several activities. In this paper, basically following the V model [1, 20, 21] and taking into account the characteristics of the joint Reconnaissance, Early-warning, and Intelligence Equipment System-of-Systems (REIESoS) all well as the complexity of the System of Systems (SoS), we refine the V model on the basis of incorporating the ideas of the SoSE to form a model-based technical-process model, as shown in Fig. 1. We remark that the technical process in Fig. 1 is model-driven and includes four stages, i.e., the requirement

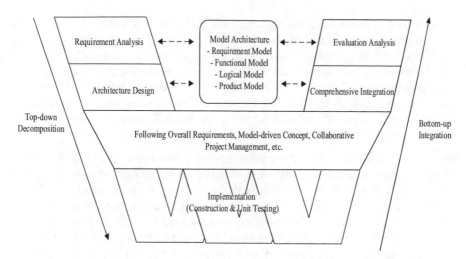

Fig. 1 The model for the model-based systems engineering technology process

analysis, the architecture design, the comprehensive integration, and the evaluation analysis.

Requirement Analysis. The demands or requirements for the SoS construction is obtained from relevant parties [22]. Also, the environmental constraints refined are transformed into system requirements, thus identifying system requirements and corresponding stakeholders. Based on the above, the system demands or requirements and the functional architecture are further defined. Based on the MBSE's requirement analysis, the above activities are performed, thus obtaining such requirement-analysis models and function-analysis models for the system as requirement diagrams, requirement tables, use case diagrams, black-box activity diagrams, black-box sequence diagrams, black-box internal block diagrams, black-box state diagrams, which further provides requirement bases for designing the architecture subsequently.

Architecture Design. The architecture-design scheme is achieved by converting the functional architecture from both the system-requirement analysis and the system-function analysis stakeholders expect, i.e., according to the existing technical conditions, the functional architecture is mapped into the physical architecture, thus forming the specific architecture-design scheme. The key for this mapping is the analysis and design of the system architecture. As a result, the design of the MBSE-based scheme provides such architecture-analysis models and architecture-design models for the system as block-definition diagrams, white-box activity diagrams, white-box sequence diagram, white-box internal block diagrams, and white-box state diagrams.

Comprehensive Integration. The comprehensive integration for the SoS targets the realization of overall SOS capabilities. This integration is cross-domain, cross-level, and cross-time and is the core of the SoS construction. The main problem that the integration solves is how to organically integrate a large number of existing systems, and to allow them to cooperate with each other and coordinate actions in order to construct the SoS with specific capabilities, which finally forms an overall efficiency that meets predetermined demands or requirements. As a result, the MBSE-based comprehensive integration provides such comprehensive-integration models as parameter diagrams, internal block diagrams. Also, these comprehensive-integration models can be verified whether the demands or requirements are met by comprehensively evaluating such influencing factors as the SoS and its various functions, performance, security, and reliability.

Evaluation Analysis. A comprehensive capability evaluation and analysis for the SoS is conducted, confirming that the SoS meets the initial capability requirements and the SoS-capability delivery is achieved. The development of the MBSE-REIESoS exploits the model to transmit information throughout the development cycle, and establishes the connection relationship between the model and the SoS requirements. Also, the model evaluation and analysis can be made at each stage of the development cycle, confirming the compliance of the SoS design and requirements and verifying whether the requirements meet the original requirements of the stakeholders. At the initial design stage of the development cycle, the act of verifying system-function behavior, interface, and control-logic integration can greatly improve the early design maturity of the system, and may reduce the risk for repeated redesign caused by the later verification, thus decreasing the system-development time, labor and material costs significantly.

2.2 The Model-Based Management Process

The model-based management process is a comprehensive management project. Through the comprehensive management against various technical means, such a management process provides the guarantee for the accurate definition of SoS-capability goals as well as the correct delivery of final capabilities. The development of the MBSE-REIESoS is essentially the process of establishing engineering-system models. At the technical process level, it includes the construction, analysis, optimization, and verification for system models. On the other hand, it also provides the management technical support for the implementation of the above system-modeling work.

2.3 Model Requirements at Each Stage

To ensure that the implementation and application for the technology process of the proposed MBSE-REIESoS, the requirements should be imposed on the model formed at each stage as follows.

Hierarchy. The model can be characterized by using the hierarchical representation or multiple views from different levels. It can be applied to various processes of the product design, e.g., the parameter matching and the scheme analysis in the design phase, the parameter optimization and the design verification in the detailed design phase, and the virtual experiments in the test phase. The model is continuously refined and enriched and the reuse of the model is ensured as to the utmost along with the development phase. As a result, the model is transited from the simple to the complex. This transition meets the characteristics of the development-process needs from the system to functional units and from requirements to the specific implementation.

Verifiability. The model itself is a formal description for the design and has the characteristics of no ambiguity. The system can be evaluated at various stages of the development by running the model. In particular, there are different requirements for testing at different stages of the development, e.g., the requirements-based testing, which typically occurs during the integration and evaluation phase, focuses on testing functions from a requirement perspective. Also, the model is fully tested at the design stage.

Convertibility. Errors caused by manually converting models into codes can be avoided to the greatest extent, using automatic methods. The generated code can be used to build a rapid prototype of the product, and to perform a comprehensive function and performance test against the product by using semi-physical simulation before the final implementation. On the other hand, for the algorithm model corresponding to the software part of the product, product-level codes can be generated and is used in the final actual product. In addition, it is required to ensure that documents can be automatically generated through the model to reduce the workload of document preparation.

Traceability. It is convenient to control the change of the development process as a result of the model-based integration for the entire development process. The product consistency during the development can be easily ensured by resorting to checking and testing methods against models. In essence, requirements can be linked to models or designs, and these models are linked to generated codes and documentations. As such, the consistency for design changes can be easily implemented.

3   The Scheme Design for MBSE-Based REIESoS

The scheme design of the MBSE-REIESoS is to convert the functional architecture from both the system-requirement analysis and the system-function analysis stakeholders expect, i.e., according to the existing technical conditions, the functional architecture is mapped into the physical architecture, thus forming the specific architecture-design scheme. In the INCOSE Systems Engineering Handbook, the system architecture is defined as "the arrangement of elements and subsystems and their function allocation in order to meet system requirements" [16, 23, 24]. The

US Department of Defense Architecture Framework (DoDAF) defines system architecture as "the structure of components, their relationships, and the principles and guidelines that govern their design and evolution over time" [10]. In other words, the system architecture encompasses both the design and the description for the system, and is an evolutionary process. As such, the system architecture can be a process or a description [25].

During the scheme design of the MBSE-REIESoS, the architecture that can meet the functional and non-functional requirements is designed according to the system-function analysis model, and such system-design models as system architecture analysis models, system architecture design models, and system architecture detailed design models can also be established. The design for the architecture includes both the architecture design and the architecture analysis. The former achieves the design scheme for each system in the SoS, assigns SoS functions to systems, and simultaneously forms logical interfaces between internal systems. The latter verifies whether the architecture meets the requirements by evaluating such influencing factors as the SoS and its various functions, performance, security, and reliability.

3.1   The Process Method for the Scheme Design

The MBSE-based scheme design is a "white-box" analysis and modeling process. The black-box activity diagram model obtained in the requirement analysis is opened in accordance with the selected architecture. The performance and functions of the reconnaissance, early-warning, and intelligence equipment SoS are assigned to the system. Then, the white-box activity diagram model can be obtained and the white-box sequence diagram model based on the design of white-box activity diagram can also be established. According to the white-box sequence diagram, the interactions and interfaces between systems is identified, the white-box state diagram model can be established. Then, the model can be executed to complete the verification, and then arrives in the next iterative level. The iteration finishes until the definition of the entire design scheme of the MBSE-REIESoS is completed.

Architecture Design. The architecture design firstly needs to establish the system-architecture design model, i.e., to define the system architecture and its constituent structures through block-definition diagrams and internal-block diagrams, and to achieve the allocation for the functional requirements and non-functional requirements of the system through white-box activity diagrams. Based on the above, detailed design models for the system architecture can be established. White-box sequence diagrams are driven according to white box activity diagrams, and then internal ports and interfaces in internal-block diagrams of system-architecture structures are defined. Logical-interface control files and software and hardware requirement specifications can be generated by defining the state behavior of blocks in white-box state machine diagrams.

After the top-level customer requirements or needs are determined, final responded functions of products are need to be analyzed so as to meet design requirements, and product functions and logical architectures are established to achieve the allocation of requirements for downstream functions and logics. Then, the definition of the system architecture is completed while further starting to the design of subsystems and parts. At this stage, the system functions and logical models are completed. On the other hand, the requirements are further decomposed and re-allocated. It is necessary to carry out the overall scheme design on its design platform, and to analyze product requirements based on product functional specifications, thus establishing functional views, logical views, correlations between requirements, functions, and logic, as shown in Fig. 2.

The functional design and system RFLP architecture definition mainly solves the following problems: (1) Unified Architecture Model. requirements, functions, and logic are based on a unified architecture model. the traditional document-based product description form is completely changed, and the MBSE system architecture description form is implemented. (2) System-Interface Definition. The system interface control is clearly implemented through the definition of requirements and functions to the system-logical architecture, the physical-product allocation, and the data-relationship logic between systems, further laying the key interface foundation for the downstream design of subsystems, components, and even the supplier cooperation.

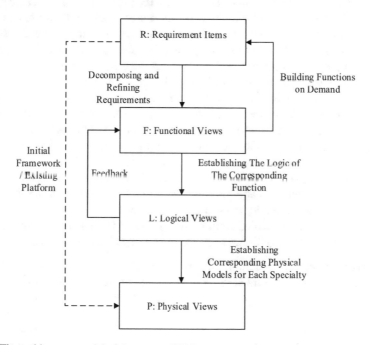

Fig. 2 The architecture model of the system RFLP

The purpose of the physical-architecture definition is to create a physical and concrete scheme or solution that adapts to the logical architecture and meets or weighs requirements of the system. The physical architecture describes physical components of a system and their interconnections. It also describes the allocation from logical components to physical components, since each physical component performs a function of a system. A functional architecture can correspond to multiple physical architectures.

The physical-architecture definition is for the system to define an alternative physical architecture. The system elements that can perform the functions of the logical architecture must be identified. The interfaces that can perform the input–output flow and control flow must also be specified. When the possible elements are identified, design attributes are necessary to be allocated in the logical architecture. These attributes are derived from system requirements. The division and allocation are activities that break down, collect, or separate functions to facilitate the identification of feasible system elements that support these functions. These system elements either already exist and can be reused, or they can be developed and technically implemented.

The physical structure of the system is decomposed. Then, the results of the functional architecture are mapping or assigned to the various elements of the physical architecture. To establish a physical organization, it needs to (1) determine physical elements or divisions; (2) decompose the functional architecture onto these physical elements; and (3) establish the interface between the physical elements of the system.

The model is passed to the next level. The ultimate purpose of the system design is to (1) obtain the system architecture that meets requirements and a reasonable balance and (2) to clearly put forward the design requirements for the next-level product. Under the model-based system-design SoS, a large amount of design information and design requirements are reflected in the model. Based on the above, the next-level products continue to carry out the detailed product design. Then, the models and documents that should be transferred to the next level of products include:

(1) The product's Port/Interface model files that represent function allocation results and logical interface requirements of the product, and is the key input for the detailed design of the next-level product.

(2) Use case diagrams, sequence diagrams, and internal block diagrams associated with the product's functions and interface definitions are used to clarify the context of the functions undertaken by the product at this level.

(3) System test cases formed from activity diagrams and sequence diagrams are used as test inputs for the product acceptance and confirmation at the next level.

Architecture Analysis. System-architecture analysis models are established at this stage. The system architecture scheme is determined and block-definition diagrams for the scheme are defined. According to the evaluation-index architecture that the system determines, evaluation results of selected schemes are determined by exploiting such system engineering methods as the association matrix method or the analytic hierarchy process. Then, parameter diagrams of block definition diagrams are established, and the final scheme is determined based on the comprehensive utility value.

The system-architecture analysis is an objective quantitative evaluation for the system in order to generate the derived engineering data and select the most effective system architecture. No scheme can have the best performance, the highest quality, and the lowest cost at the same time. In the engineering-design process, each technology choice or decision must be evaluated to determine whether the system meets the requirements. The system analysis is a rigorous method for making technical decisions. We conduct trade-off studies through system analysis, and point out that the system analysis includes modeling and simulation, cost analysis, technical-risk analysis, and effectiveness analysis.

(1) To determine analysis goals.

The first step in the system analysis is to clearly define goals of the analysis. The goals include (1) predicting certain aspects of the system, e.g., the system performance, reliability, quality, or cost; (2) optimizing the design through the sensitivity analysis; (3) evaluating and selecting among alternative design methods; (4) verifying the design; and (5) supporting the technical planning, e.g., the cost estimation and the risk analysis.

(2) To establish the analysis model.

To achieve the analysis goal, the analysis model is defined on the basis of the system architecture model. The analysis model can be a deterministic model and a probabilistic model or random model. It is mainly used to determine the estimated value and to simulate the real operation of the system through equations. In MBSE, to achieve the integration of both the system architecture model and the analysis model, part of the analysis model is built using a descriptive modeling language, and the other part (e.g., solving part) generally needs to be built and executed in other tools.

The context that uses a descriptive modeling language to build and analyze is built directly on top of the system-architecture model. The system-architecture model provides data that the analysis methods (e.g., the scoring analysis) need to be used and the analysis context needs to be estimated. The execution of the analysis model requires the specific external analysis or simulation tools, and the system-description model is integrated with the analysis models in different domains through integration tools.

The structural analysis is to develop the physical architecture of a system and to define the interfaces between internal and external components. At each level of the hierarchy, the system is represented by several views. The hierarchy of components is a basic systems-engineering abstraction to simplify the analysis. The decomposition must ensure that each system component is captured. The system is broken down into subsystems through layers, and then the subsystems are broken down into lower-level components.

3.2 Requirements for Scheme-Design Models

To realize the model-based system-wide integration test, the architecture-design process focuses on the logical decomposition for requirements in the scheme design of the MBSE-REIESoS, defining block definition diagrams and white box activity diagrams, as shown in Fig. 3. The architecture analysis focuses on the design for the scheme, providing white-box sequence diagrams, white-box internal block diagrams, and white-box transition diagrams, as shown in Fig. 4. These diagrams are defined by resorting to the syntax and semantics of the system-modeling language, which is not only convenient for humans to read, but also easy for computers to understand and process [12, 18, 25]. The benefit is that iteratively applying the process from the top to the bottom at different levels of the system, and the applying process can go deep

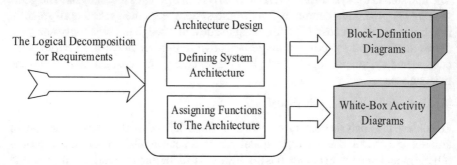

Fig. 3 Models used in each stage of the MBSE-based architecture design

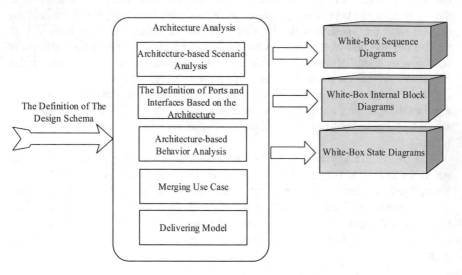

Fig. 4 All models of MBSE-based architecture analysis at each stage

into the bottom of the system, and the graphical language describing the bottommost elements is integrated to established a complete system-architecture model.

4 Conclusion

In this paper our main goal was to show a MBSE-based development life cycle for reconnaissance, early-warning, and intelligence equipment System-of-Systems. We have also shown that the traditional V model can be further refined with model-based systems engineering technologies, taking into account the characteristics of the joint REIESoS as well as the complexity of the System of Systems. Our proposed MBSE-based development life cycle is general, since the development of the entire equipment System-of-Systems can be supported, and efficient, since model-based construction, analysis, optimization, and verification at each stage of the development life cycle are executed to reduce the risk of the repeated design or development. Currently, we are investigating potential benefits for V-model refinements that may contribute to the further understanding of the proposed MBSE-based development life cycle. We leave further investigations to future work.

Acknowledgements The author would like to thank the anonymous reviewers for their invaluable suggestions which have been incorporated to improve the quality of the paper.

References

1. Jamshidi, M.: System-of-systems engineering—a definition. IEEE SMC **2005**, 10–12 (2005)
2. Northrop, L.: Ultra-Large-Scale Systems, The Software Challenge of the Future, Software Engineering Institute, Carnegie Mellon University, June 2006
3. LSCITS—Large-Scale Complex IT Systems—An EPSRC Funded Program (2007)
4. Sommerville, I.: Software Engineering: Challenges for the 21st Century, ICCBSS 2008, Madrid
5. International Council on Systems Engineering, INCOSE Systems Engineering Handbook, v.3.2, Seattle, WA, USA (2010)
6. Mazeika, D., Morkevicius, A., Aleksandraviciene, A.: MBSE driven approach for defining problem domain. In: 2016 11th System of Systems Engineering Conference (SoSE) (2016)
7. Estefan, J.A.: Survey of Model-Based Systems Engineering (MBSE) Methodologies. In: INCOSE MBSE Initiative, May 23, 2008
8. Morkevicius, A., Bisikirskiene, L., Jankevicius, N.: We choose MBSE: what's next? Proc. Sixth Int. Conf. Complex Syst. Des. Manage. CSD&M **2015**, 331–335 (2015)
9. Pearce, P., Hause, M.: ISO-15288, OOSEM and Model-Based Submarine Design (2008)
10. Department of Defence: DoD Architecture Framework Version 2.0 Volume 2: Architectural Data and Models Architect's Guide (2009)
11. Goknila, A., Kurtevb, I., Bergc, K.V.D.: Generation and validation of traces between requirements and architecture based on formal trace semantics. J. Syst. Softw. **88**, February 2014
12. Sergent, T.L., Guennec, A.L.: Data-based system engineering: ICDs management with SysML. ERTS2 (2014)

13. Gregory, J., Berthoud, L., Tryfonas, T., Prezzavento, A., et al.: Investigating the flexibility of the MBSE approach to the biomass mission. IEEE Trans. Syst. Man Cybern.: Syst. 1–9 (2020)
14. London, B., Miotto, P.: Model-based requirement generation. In: Proc. IEEE Aerosp. Conf., pp. 1–8 (2014)
15. Spangelo, S.C., et al.: Model based systems engineering (MBSE) applied to radio aurora explorer (RAX) CubeSat mission operational scenarios. In: Proc. IEEE Aerosp. Conf., pp. 1–18 (2013)
16. INCOSE: Systems Engineering Handbook: A Guide for System Life Cycle Processes and Activities, version 4.0. Hoboken, NJ, USA: John Wiley and Sons, Inc, ISBN: 978-1-118-99940-0 (2015)
17. Lindblad, L.: Data-driven systems engineering: turning MBSE into industrial reality. In: Proc. Int. Syst. Concurrent Eng. Space Appl. Workshop (SECESA), Glasgow, UK (2018)
18. Mordecai, Y., Orhof, O., Dori, D.: Model-based interoperability engineering in systems-of-systems and civil aviation. IEEE Trans. Syst. Man Cybern. Syst. **48**(4), 637–648 (2018)
19. Wibben, D.R., Furfaro, R.: Model-based systems engineering approach for the development of the science processing and operations center of the NASA OSIRIS-REx asteroid sample return mission. Acta Astronaut. **115**, 147–159 (2015)
20. Gough, K.M., Phojanamongkolkij, N.: Employing model-based systems engineering (MBSE) on a NASA aeronautics research project: a case study. In: Proc. Aviation Technol. Integr. Oper. Conf., Atlanta, GA, USA, pp. 3361–3374 (2018)
21. Cloutier, R., Sauser, B., Bone, M., Taylor, A.: Transitioning systems thinking to model-based systems engineering: Systemigrams to SysML models. In: IEEE Trans. Syst. Man, Cybern. Syst., vol. 45, no. 4, pp. 662–674 (2015)
22. Kaslow, D.: Developing a CubeSat MBSE reference model—Interim status #3. In: Proc. IEEE Aerosp. Conf., pp. 1–16 (2017)
23. Munk, P., Nordmann, A.: Model-based safety assessment with SysML and component fault trees: application and lessons learned. In: Software and Systems Modeling. Springer, Berlin (2020)
24. Adler, R., Domis, D., Höfig, K., Kemmann, S., Kuhn, T., Schwinn, J.P., Trapp, M.: Integration of component fault trees into the UML. In: Dingel, J., Solberg, A. (eds.) Models in Software Engineering, pp. 312–327. Springer, New York (2011)
25. Clegg, K., Li, M., Stamp, D., Grigg, A., McDermid, J.: A SysML profile for fault trees—linking safety models to system design. In: Romanovsky, A., Troubitsyna, E., Bitsch, F. (eds.) Computer Safety, Reliability, and Security, pp. 85–93. Springer, New York (2019)

A Model-Based Aircraft Function Analysis Method: From Operational Scenarios to Functions

Yuchen Zhang and Raphael Faudou

1 Introduction

In this paper, a model-based function analysis method is introduced. The main goal of the method is to work as a supplement to the existing function analysis method, improving the shortcomings of IDEF0, such as lack of guidance for top level function identification, lack of early validation with stakeholders because of the implement-specific functions decomposition. The method in this article learns from scenario analysis in software engineering, and provide a model-based method for the elicitation of implement-independent high level functions from operational scenario.

This paper is organized as follows. Section 1 describes the background and current problems. Section 2 introduces function-behavior transformation, the basic principle used in the method. Section 3 describe the method, and in Sect. 4 an application of the method on a UAV is presented.

1.1 The Need for Functional Analysis

Function analysis is one of the most essential processes in aircraft development and is required by applicable standards, such as ISO/IEC/IEEE 15,288 [1] and ARP 4754A [2]. According to the R-F-L-P development process [3], function analysis is

Y. Zhang (✉)
Aviation Industry Development Research Center of China, No. 14 Xiaoguan Dongli, Chaoyang District, Beijing, China
e-mail: zhangyc082@avic.com

R. Faudou
Samares-Enginnering, 31700 Blagnac, France
e-mail: raphael.faudou@Samares-Engineering.com

the foundation stone of following development, including logical design and physical design.

Especially, function analysis is essential for new aircraft concept design. It provides a better understanding of new aircraft and serves as root point for further development. In recent years, as new technologies mature, new aircrafts, such as unmanned aircrafts and electric aircrafts, are becoming potential trends [4]. Compared with traditional aircrafts, functions of these new aircrafts will change significantly, thus function analysis will play an important role in the concept design of these new aircrafts.

1.2   The Need for Model-Based Method

For the past ten years, model-based methods, such as model based systems engineering (MBSE), have been proved useful to manage complexity and improve effectiveness in complex system development [5].

The function analysis method introduced in this article is implemented with models, so that the analysis results are formalized, structured, reusable, and can be simulated. All the analyses in this article are based on SysML, which is a widely used modeling language for complex system design. The method proposed in this article can also be implemented by UAFP or Capella.

1.3   Functional Analysis Through Operational Scenarios

The existing most widely used function analysis method is IDEF0. IDEF0 has been proved a feasible and effective function analysis method. It considers function as a transformation of inputs to outputs and influenced by control and mechanism [6]. IDEF0 method starts with identified top-level functions, and decomposes functions to lower level sub-functions iteratively.

However, some surveys and studies have been done on existing function analysis method, and the result shows that some problems still exist [7]. First, no guidance are provided for the identification of top level functions. For traditional aircrafts, top level functions are already defined by precedent aircrafts and can be reused. However, for new aircrafts that are still in concept proof phase, like unmanned or electric aircrafts, top level functions are unclear and need to be re-defined. Second, the function decomposition in IDEF0 is implement-specific, so some respondents point out that functions obtained with IDEF0 can hardly be used to communicate with stakeholders for validation and also limit some design space of solution. To deal with these problems, some aircraft manufacturers, such as Bombardier [7], begin to pursue new function analysis method.

In response to the need of improvement, this article proposes a method as a supplement to IDEF0, trying to solve the top level functions identification and the

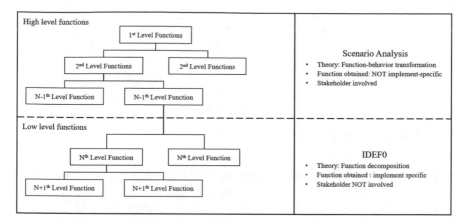

Fig. 1 Introduce scenario analysis as supplement to IDEF0

implement-specific functions problems. This article refers to the scenario analysis method in software engineering, analyzing functions through operational scenarios (Fig. 1).

For high-level functions, this article suggests a function analysis method using operational scenarios, as detailed described in Sect. 3. Through this method, top level functions can be defined from operational behaviors in operational scenarios, and functions coming from scenarios are easy to understand in validation with stakeholders. For low level functions, IDEF0 can be used to obtain much detailed sub-functions, pushing the further development and implementation.

2   Basic Theory

The method proposed in this article suggests using operational scenarios to analyze functions. The reason functions can be analyzed through operational scenarios is that operational behaviors, the basic elements of operational scenarios, are closely related to functions and transformable to functions.

The function-behavior transformation [8] is a significant process in engineering design, responsible for transforming between posited functions and expected behaviors. Numerous studies [9–11] have been done on this topic, and function-behavior transformation has been acknowledged as the foundation of much design work.

Functions and behaviors are different concepts, but are closely related. According to the definition in ISO/IEC/IEEE 24765 [12]: function is "a transformation of inputs to outputs by means of certain mechanisms and subject to certain controls", while behavior is "an observable activity of a system in terms of effects on the environment arising from internal or external stimulus".

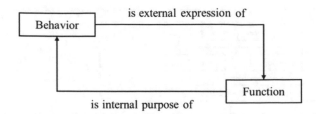

Fig. 2 Basic idea in function-behavior transformation

In function-behavior transformation studies, these two concepts are further explained [11]. Behavior, as a physical world concept, is "the manner or way in which an entity behaves". Behavior is an objective concept and is observable in physical world. Functions, as an intermediate concept between intentional world and physical world, is "a desired action on physical world aiming at transforming it from problematic state to satisfactory state". Function is a subjective concept that represents human's purpose or desire (Fig. 2).

Function and behavior are two sides of one item. For manmade products, functions are internal purposes of behaviors, while behaviors are external expression of functions. This theory provides the basis to use behaviors, the basic element in scenarios, to analyze functions.

3   Model-Based Aircraft Function Analysis Method

The model-based aircraft function analysis method in this article contains two parts. The first step is operational scenario analysis, including the definition, decomposition and extension of operational scenarios. Operational scenarios are described with detailed operational behaviors and their relationships. After that, the second step is function analysis. Functions are derived from operational behaviors in operational scenarios through function-behavior transformation.

3.1   Operational Scenario Analysis

Scenario analysis is a commonly used requirements elicitation and analysis technique in software engineering. In scenario analysis, product being developed is put into its assuming operational scenarios, analyzing its intended behaviors, finally capturing its requirements [13, 14]. Since functions can be defined from behaviors in scenarios, as stated in Sect. 2, scenario analysis can also be a useful technique for function analysis.

For aircraft, operational scenario is a sequence of observable behaviors which describe the operation of aircraft in its operational circumstance. Operational scenario

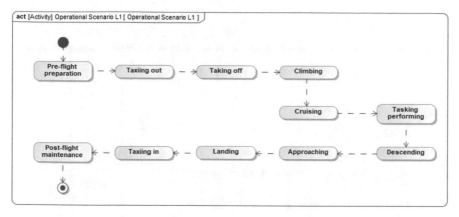

Fig. 3 Operational scenario example (UAV case level 1)

is an intuitive and direct way to describe what an aircraft is expected to do, so stakeholders who are not technical experts can understand or describe it. Compared with the technology-specific and implement-specific functions obtained from IDEF0, the intuitive operational scenarios make the discussion and early validation with stakeholders possible.

Figure 3 is a simplest aircraft operational scenario of UAV, shown in SysML Activity Diagram. It demonstrates a common high level scenario of aircraft flight.

3.2 Decomposition of Operational Scenario

As the same as functions, operational behaviors have different levels. As we want to obtain much more details about the aircraft, operational scenarios and operational behaviors need to be further decomposed.

Figure 4 shows an example of scenario decomposition. The operational behavior "taking off" in Fig. 3 is decomposed and a lower level operational scenario is created to describe the "taking off" behavior in detail.

Since all the operational scenarios are built in SysML models, they are supported to be simulated. Figure 5 shows an example of the simulation of SysML model. If necessary, three dimensional visualization models can also be built as supplement. The dynamic simulation supported by model-based techniques makes it much easier to perform validation with stakeholders and also helps discover logical errors or mistakes in early design.

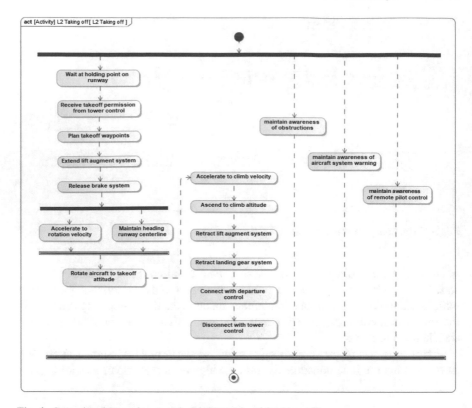

Fig. 4 Operational scenario example (UAV case level 2 taking off)

3.3 Extension of Operational Scenario

In the previous sections, definition and decomposition of operational scenario are introduced. Now the question is how many scenarios are needed to define and how can the completeness be ensured. Since functions are derived from scenarios in this method, if scenarios are not completely defined, the functions certainly are not complete.

To answer this question, this article summarizes aircraft operational scenarios into multiple dimensions, and establish a multi-dimensional matrix, to ensure the completeness of operational scenarios.

In this article, five dimensions are used:

(1) Stakeholder dimension: consider scenarios from perspective of stakeholders. E.g. task performing scenario and emergency evacuation scenario from pilot's perspective, traffic control scenario from air traffic controller's perspective, pre-flight inspection and post-flight maintenance scenario from ground crew's perspective.

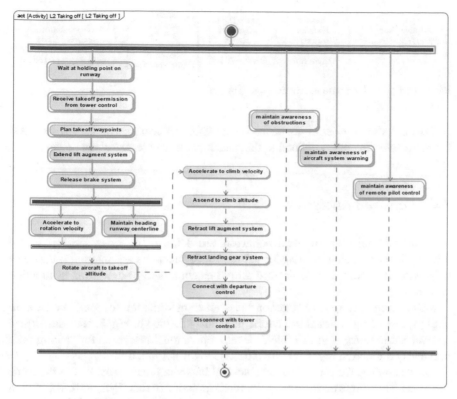

Fig. 5 Operational scenario simulation example (UAV case level 2 taking off)

(2) Mission dimension: consider scenarios from missions of flights. E.g. for commercial aircraft, missions include airline operation, flight test and etc. For military aircrafts, missions include combat, training, exercise, transition and etc.

(3) Time dimension: consider scenarios from the perspective of phases of operations. E.g. pre-flight, departure (taxiing out, taking off, climbing), enroute, arrival (descending, landing, taxiing in), post-flight.

(4) Environment dimension: consider scenarios from perspective of environment conditions of aircraft operations. E.g. climate (tropical, dry, continental, highland, polar), weather (rainy, snowy, stormy, hailing, foggy, icy), gravity, radiation, corrosion, airport condition.

(5) State dimension: consider scenarios from perspective of aircraft normal or abnormal states. E.g. lost control, lost communication, one engine failure, two engines failure, hydraulic system failure, electric system failure (Fig. 6).

$$STKH \times M \times T \times ENV \times S$$

Fig. 6 Operational scenario matrix

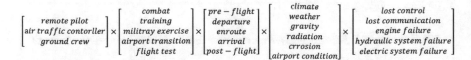

Fig. 7 Operational scenario matrix example (UAV case)

Figure 7 shows an example of the operational scenario matrix defined for UAV. From the five dimension defined in the matrix, multiple scenarios can be defined.

3.4 Function Analysis

After operational scenarios and behaviors are defined, functions can be derived through function-behavior transformation. Since functions are internal purpose of behaviors, functions can be derived from behaviors and re-described in function statement.

Below is the example of function-behavior transformation for the UAV case. In Table 1, level 1 operational behaviors, previously defined in Fig. 3, are transformed to level 1 functions. And in Table 2, level 2 operational behaviors for "taking off", previously defined in Fig. 4, are transformed to level 2 function.

As we can see from tables above, several behaviors may transform to the same functions. This is true because behaviors are expression of functions, and one function can have different expression forms. In this case, the function elicited from behaviors may be duplicated, so the functions need to be re-organized. Table 4 shows an organized example of function list.

Table 1 Transformation from behaviors to functions (UAV case level 1)

Level 1 Operational Behavior	Level 1 Functions
Pre-flight preparation	Accept inspection, accept preparation
Taxiing out	Plan on ground waypoints, control on ground movement
Taking off	Plan in air waypoints, control in air movement
Climbing	
Cruising	
Task performing	
Descending	Plan in air waypoints, control in air movement
Approaching	
Landing	
Taxing in	Plan on ground waypoints, control on ground movement
Post-flight maintenance	Accept inspection, accept maintenance

Table 2 Transformation from behaviors to functions (UAV case level 2 taking off)

Level 2 Behavior	Level 2 Functions
Wait at holding point on runway	Control on ground stop braking
Receive takeoff penrmission from tower control	Provide communication (with tower control)
Plan takeoff waypoints	Plan on ground waypoints, plan in air way points
Extend lift augment system	Configure aerodynamic configuration
Release brake system	Control on ground stop braking
Accelerate to rotation velocity	Control on ground speed (accelerate)
Maintain heading runway centerline	Control on ground direction (yaw)
Rotate aircraft to takeoff attitude	Control in air direction (pitch)
Accelerate to climb velocity	Control in air speed (accelerate)
Ascend to climb altitude	Control attitude and in air direction (yaw, pitch, roll)
Retract lift augment system	Configure aerodynamic configuration
Retract landing gear system	Configure on ground configuration
Connect with departure control	Provide communication (with departure control)
Disconnect with tower control	Provide communication (with tower control)
Maintain awareness of obstructions	Provide external awareness
Maintain awareness of aircraft system warning	Monitor aircraft status
Maintain awareness of remote pilot control	Provide communication (with remote pilot), allow remote pilot take over control

4   Application Results

In order to verify the usability and effectiveness of the method, this article applied the method in the development of a certain type of UAV.

Operational scenario matrix was firstly defined, as shown in Fig. 6, to ensure the completeness of scenarios. More than 40 operational scenarios were selected from the matrix as the combination of the five dimensions. An example of the defined operational scenario is shown in Table 3.

Table 3 Operational Scenario defined with matrix (UAV case)

Scenario Example	
Stakeholder dimension	Remote pilot, approach control, tower control
Mission dimension	Training
Time dimension	Arrival (landing)
Environment dimension	X airport, snowy, icy,
State dimension	One engine failure

Table 4 Function list section (UAV case)

Level 1 Functions	Level 2 Functions
Control on ground movement	Control on ground velocity (accelerate, decelerate)
	Control on ground direction (yaw)
	Control on ground stop braking
	Configure on ground configuration (wheel)
Control in air movement	Control in air velocity (accelerate, decelerate)
	Control in air direction (yaw, pitch,roll)
	Control aerodynamic configuration (wing, movable surface)
Provide communication	Provide communication with remote pilot
	Provide communication with air traffic control (tower, approach, area)
Provide navigation	Determine aircraft location
	Determine aircraft altitude
	Determine aircraft speed
	Determine aircraft direction
Provide unmanned control	Plan on ground waypoints
	Plan in air waypoints
	Allow remote pilot take over control
Provide environment control	Provide ice protection
	Provide rain protection
Provide situational awareness	Monitor aircraft status
	Provide external awareness
Provide power	Provide hydraulic power
	Provide electric power
Provide carriage	Provide cargo space
	Provide cargo loading/unloading

For each operation scenario, its operational behaviors were described in SysML model, and were further decomposed to more detailed lower level behaviors. An example can be seen in Fig. 4.

After the definition and decomposition of operational scenarios, different level functions were derived from operational behaviors through function-behavior transformation, shown in Tables 1 and 2. Finally, the obtained functions were re-organized to a formalized structure, example shown in Table 4.

Following the method described in this article, a structured function list was obtained, which initially confirmed the usability and effectiveness of the method.

5 Conclusion

In this article, a model-based aircraft function analysis method is proposed. It introduces sequence analysis into function analysis and applies function-behavior transformation as principle, deriving functions from operational behaviors in operational scenarios. According to the application result on UAV case, a set of implement-independent high level functions are defined, and early validation with stakeholders are supported with the simulation of model-based scenarios created. The application result initially confirms the usability and effectiveness of the method and it can serve as a supplement to the existing IDEF0 method for high level functions analysis.

References

1. ISO, IEC: ISO/IEC 15288: Systems and Software Engineering—System Life Cycle Processes. ISO, IEC (2008): 24748-1
2. ARP4754A, S. A. E.: Guidelines for development of civil aircraft and systems. SAE International (2010)
3. Ambroisine, Thierry. Mastering increasing product complexity with Collaborative Systems Engineering and PLM. Embedded World Conference (2013)
4. Bulent Sarlioglu Casey T Morris 2015 More electric aircraft: review, challenges, and opportunities for commercial transport aircraft IEEE Trans. Transp. Electrification 1 1 54 64
5. Walden, D.D., et al.: Systems engineering handbook: a guide for system life cycle processes and activities. Wiley (2015)
6. Force, US Air: ICAM architecture part II-volume IV, function modelling manual (IDEF0). US Air Force (1993)
7. Esdras, G., Liscouët-Hanke, S.: Development of core functions for aircraft conceptual design: methodology and results. Bombardier Product Development Engineering, Aerospace, Montréal QC, vol 10 (2015)
8. John S Gero 1990 Design prototypes: a knowledge representation schema for design AI Mag. 11 4 26 26
9. Iwasaki, Y., et al.: Causal functional representation language with behavior-based semantics. Appl. Artif. Intelligence Int. J. 9(1), 5–31 (1995)
10. Umeda, Y., et al.: Supporting conceptual design based on the function-behavior-state modeler. Ai Edam 10(4), 275–288 (1996)
11. Chen, Y., et al.: Towards a scientific model of function-behavior transformation. In: DS 68-2: Proceedings of the 18th International Conference on Engineering Design (ICED 11), Impacting Society through Engineering Design, vol. 2: Design Theory and Research Methodology, Lyngby/Copenhagen, Denmark, 15.-19.08. 2011 (2011)
12. IEEE Standards Association: ISO/IEC/IEEE 24765: 2010 Systems and software engineering-Vocabulary. ISO/IEC/IEEE 24765: 2010 25021. Institute of Electrical and Electronics Engineers, Inc (2010)
13. A Sutcliffe 1998 Scenario-based requirements analysis Requirements Eng. 3 1 48 65
14. Sutcliffe, A.: Scenario-based requirements engineering. In: Proceedings. 11th IEEE International Requirements Engineering Conference, IEEE (2003)

A Model-Based Requirements Analysis Method for Avionics System Architecture

Cong Chen, Bingfei Li, and Jieshi Shen

1 Introduction

With the advancement of military revolution recent years, the characteristics of SoS confrontation and informatization are becoming more and more obvious, the size of weapon equipment system is getting bigger and bigger, the connections between systems become increasingly complex, and the demands for interconnection, interworking and interoperation are becoming higher and higher. In weapon equipment system, the absence of any component system will cause the overall system efficiency to drop sharply, or even failure. As an important part of the modern military equipment system, aviation weapon equipment should be able to accomplish flexible combat tasks at different strategic and tactical levels under the condition of integrated joint operations. As the most important system of aviation weapon equipment, avionics system should carry out task and capability analysis from the equipment system level, analyze systematic requirements of avionics system, and then guide the architecture design [1, 2].

Therefore, this article employs the systems engineering thought and architecture design method of SoS (System of Systems) [3, 4], combining with the practice of aviation weapon equipment requirements demonstration, proposes the model-based systematic requirements analysis and simulation method of avionics system, as well as the architecture model co-simulation environment design scheme [5]. The method and scheme can effectively support the avionics systematic requirements analysis and the top-level design, and has important theoretical value and practical significance.

C. Chen (✉) · B. Li · J. Shen
Key Laboratory of Avionics System Integration, China Aeronautical Radio Electronics Research Institute, Shanghai, China
e-mail: cissy0104@126.com

© The Author(s), under exclusive license to Springer Nature Switzerland AG 2021 53
D. Krob et al. (eds.), *Complex Systems Design & Management*,
https://doi.org/10.1007/978-3-030-73539-5_5

2 The Challenge of Aviation Weapon Equipment Top-Level Demonstration Technology

Carrying out the top-level demonstration of aviation weapon equipment is the need to realize the SoS confrontation of aviation weapon equipment, and also the inevitable choice to realize the leapfrog of aviation weapon equipment from tracking development to independent innovation. In recent years, with the continuous enrichment and expansion of the connotation of aviation weapon equipment demonstration, significant changes have taken place in the scope, field and method of demonstration.

(1) Demonstration scope.

Due to historical reasons, the domestic aviation weapon equipment demonstration was once mainly limited to the specific foreign product model technical demonstration. However, in recent years, with the improvement of the combat environment, combat missions and own capabilities, a major change in the top-level demonstration of aviation weapon equipment is to expand from single model demonstration to SoS demonstration. The system positioning of the model and its contribution to the overall combat capability of the system should be clarified in the top-level.

(2) Demonstration field.

Military problem is an important field in aviation weapon equipment demonstration, military scenarios through the whole process of requirements demonstration, SoS design, product conceptual design and effectiveness evaluation. Operational view is an important tool to guide system design and product concept design, operational process, operational sequence and combat information interaction relationship is the critical evidence to determine the product function, performance and technical indicators. Therefore, military literacy has become a necessary quality for the top-level argumentative personnel of aviation weapon equipment, and relevant technologies such as operational plan formulation and operational attempt description have been integrated into the top-level demonstration technology system of aviation weapon equipment.

(3) Demonstration method.

Aviation weapon equipment demonstration has the characteristics of the commonness of product, also have their own personality characteristics, absorbing the domestic and foreign products in time the demonstration of the new progress in common theory, constantly summarize their own practice of new experience, and constantly enrich the perfect technology, is the necessary way of weapon equipment demonstration technology innovation. In recent years, the continuous development of foreign SoS engineering, systems engineering, architecture design and other technologies has provided new ideas for the top-level demonstration of China's aviation weapon equipment. Therefore, we should base on the national conditions to achieve digestion and re-innovation, and explore new methods, new processes and new tools to adapt to China's equipment demonstration.

3 Architecture-Based Demonstration of Top-Level Requirements for Aviation Weapon Equipment

3.1 Weapon Equipment System and System of Systems Engineering

SoS is known as "system of systems". In the information age, with the rapid development and widespread application of high and new technologies in the information field such as computer communication, intelligent object association and cloud computing, the original independent system individuals can connect each other into a "system of systems" existence form through the "link" built by "information", such as the networked weapon equipment system, as shown in Fig. 1. Weapon equipment system usually refers to "the uncertainty environment, in order to complete a particular task or mission, by a lot of features are independent of each other, operating on the strong interaction of multiple independent weapon equipment system, under the condition of certain constraints, according to a certain pattern or way of a higher level of system, including the lack of any a subsystem can cause degeneration of overall operational effectiveness, even failure".

Fig. 1 General form of SoS and Weapons SoS

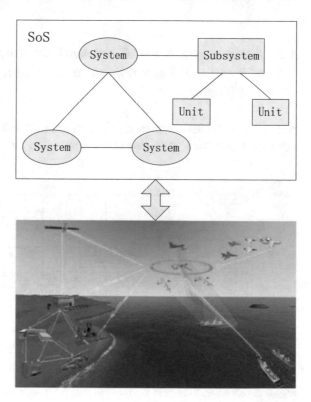

In order to deal with a series of new problems in the process of planning, design, analysis, organization and integration of SoS, such as interconnection, interworking, interoperation, complexity, oriented emergence, development and evolution, the concept of "SoS engineering" came into being. According to the definition in the literature, SoS engineering is the theory, method and technique of designing, developing and integrating complex large systems to accomplish specific tasks and achieve desired results, and to achieve capabilities, missions or desired results.

The core idea of SoS engineering includes two aspects: capability-based systematic requirements development and architecture design method. Capability-based systematic requirements development refers to the transformation from "threatbased" to "capability-based" in the demonstration of equipment demand, from the traditional single platform and single model demonstration to the demonstration of SoS, to the improvement of the overall combat capability. Architecture design method is a kind of model construction of requirements, system behavior patterns and relevant technical specifications and standards from different perspectives of operations, systems and technologies, and the introduction of computer modeling and simulation technology to support the realization of the top-level design process of complex large systems. At present, the defense department architecture framework released by the United States, Britain and other countries has been widely used in China, which can effectively guide the top-level design and demonstration of the equipment system.

3.2 Architecture-Based Framework for the Demonstration of Top-Level Requirements for Aviation Weapon Equipment

According to the design concept of JCIDS and the DoDAF [6], the aviation weapon equipment systematic requirements demonstration framework is shown in Fig. 2.

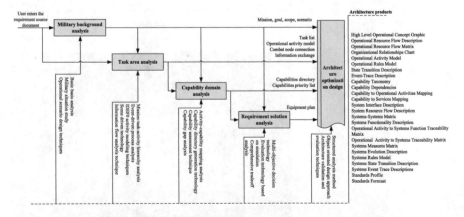

Fig. 2 Aviation weapon equipment requirements demonstration framework based on architecture

(1) Military background analysis.

Clarify the missions and tasks required by the SoS under the conditions of future integrated joint operations, analyze the requirements of future missions and missions on system capabilities, identify the capability level of the existing system, construct a conceptual system to meet the missions, revise the existing operational norms, and determine the mission, scope, objectives and intentions of the SoS requirements project.

(2) Task area analysis.

Analyzes the various typical combat styles in the future, excavates the common joint combat task list, and gives the measurement index of the capability needed to execute the task, determines the organizational structure relationship and information exchange relationship of the operational unit of the implementation task, so as to clearly understand the combat task field involved in the system.

(3) Capability domain analysis.

Analyze the capability of the SoS, generate the capability inventory, analyze the gap of the SoS capability, determine the key performance parameters of the capability, and establish the priority list of capability requirements.

(4) Requirement solution analysis.

Explore reasonable plans for the construction of weapon equipment systems to meet the needs of capability areas, analyze and evaluate alternative plans, identify key projects and key system functions of system construction, and identify and recommend the best plans.

(5) Architecture optimization design.

Architecture optimization design includes three stages: architecture design, architecture validation and evaluation, and architecture optimization. Architecture design is mainly to establish a variety of views of system architecture, from the global height, different perspectives, different dimensions to discuss the system construction ideas and technical routes; The task of architecture validation is to determine whether the architecture solution meets the system requirements; Architecture evaluation provides a scientific basis for decision makers to choose the best solution from a variety of architecture candidates. Architecture optimization design refers to the improvement design of system architecture based on the Suggestions of architecture verification and evaluation.

4 Systematic Requirements Analysis and Simulation of Avionics System Based on Model

It can be seen from the previous analysis that the top-level demonstration of aviation weapon equipment has expanded from single model demonstration to system demonstration, and any demonstration of equipment or system should be combined with the system positioning of the model and its contribution to the overall combat capability of the SoS. As the most important system of aviation weapon equipment, avionics system should carry out task and capability analysis from the level of equipment system, analyze systematic requirements of avionics system, and then guide the architecture design of avionics system. According to the architecture-based top-level demand demonstration framework of aviation weapon equipment described above, this paper proposes a model-based systematic requirements analysis process for avionics system based on the systematic requirements demonstration practice of avionics system, as shown in Fig. 3.

(1) The combat mission of aviation weapon equipment is the input of model-based systematic requirements analysis, design typical combat scenarios to determine specific combat goals that are required in specific battlefield environments and in combat phases, including the deployment, action, combat factors, processes, and allegations of combat forces, etc. Using the 3D modeling simulation tool to simulate the operation, to quickly analyze the complex operational tasks, provide the easy to understand the icon and the analysis results of the written form, to determine the organizational structure relationship and the information interaction of the operational mission unit.

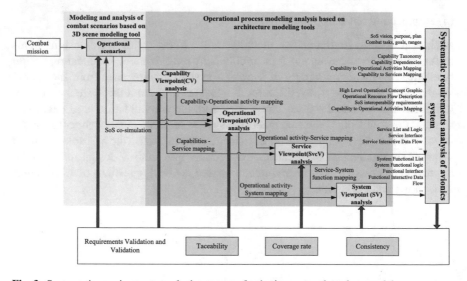

Fig. 3 Systematic requirements analysis process of avionics system based on model

(2) Based on the operational scenario analysis, guided by the DoDAF, using the Rhapsody tool to analyze the operational logic process from the capability view, operational view, service view, system view, etc. And establish a series of architecture view products to describe combat mission, combat demand, system service requirements and system function requirements.

Capability viewpoint: Analyze what capabilities avionics system should have and what capabilities it needs under SoS combat conditions, take high-level strategic planning and background as input, and take capability gap, demand and progressive capability development sequence as output, so as to provide high-level requirements for the subsequent avionics architecture design process.

Operational viewpoint: Modeling and analyzing the operational activities, operational elements, information exchange and attributes of avionics system under systematic combat, and revealing the requirements of avionics system in terms of cooperative combat capability and interoperability.

Service viewpoint: Describes the various services supporting avionics system missions and their interrelationships, relates avionics system service resources to operational requirements and capability requirements, and describes the design of service-oriented avionics architecture solutions.

System viewpoint: Analyze the systems, functions, and connectivity that support combat and capability requirements. For avionics systems under SoS combat, the system view not only needs to describe the internal structure and information interaction of avionics system, but also needs to describe the cross-linking between avionics systems and other combat platform systems. The system view is the intermediate link connecting the combat capability and design requirements of avionics system.

(3) Based on the analysis of typical operational scenarios and operational processes, the capability inventory design method based on capability-task-service-function iterative simulation and analysis is formed through architecture modeling and simulation. On each link of the model to undertake a link model, each iteration process verification coverage and consistency between different perspective model, accurate to refine the combat mission to system requirements, at the same time, different angle of view for avionics system architecture model and the avionics system requirements traceability, to ensure that each of the requirements and have traceability system architecture model, architecture model can cover all the system requirements.

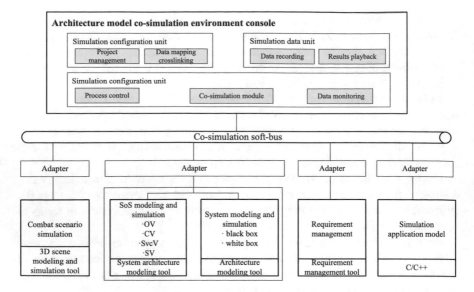

Fig. 4 Architecture modeling and simulation environment design

5 Architecture Models Co-simulation Environment Design

5.1 General Scheme

Based on the requirements analysis process of avionics system and system engineering application process, this environment integrates the modeling and simulation tools in a generalized way, and defines the interface standard of generalized development in a "soft bus" way, which is open, universal and extensible.

The environment is composed of console, interface adapter, simulation computer and so on. The console includes simulation project management module, data mapping crosslinking module, simulation data recording module, simulation result playback module, simulation process control module, co-simulation module and simulation data monitoring module. The system design scheme is shown in Fig. 4.

5.2 Working Process

The architecture modeling and simulation environment supports co-simulation validation between different tools. The working principle of system configuration is shown in Fig. 5.

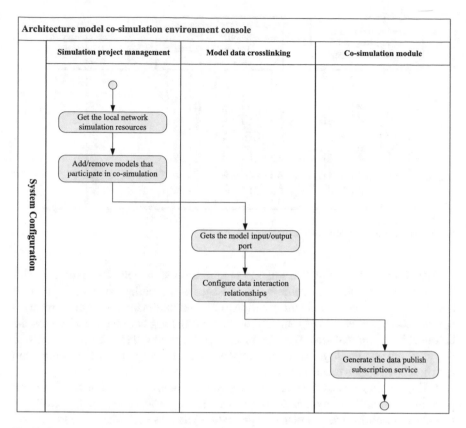

Fig. 5 System configuration and work principle

5.3 Application Scenarios

Taking this environment for an aircraft avionics system requirements analysis as an example, 3D scene modeling and simulation tools, system architecture modeling tools and C/C++ are used for joint simulation. The three simulation models interact with the simulation data soft bus and control scene operation and synchronization through the console. The data interaction relationship in the entire application scenario is shown in Fig. 6.

The system architecture modeling tool platform runs the logical model of battlefield behavior and operational behavior of aircraft platform and avionics system. It will release some combat command and state migration events to the simulation data soft bus, and at the same time, it will receive the event messages of the battlefield and some attributes of the system (including coordinates, speed, damage degree, etc.), convert these information into event messages in the model, and drive the behavior logic model to advance synchronously.

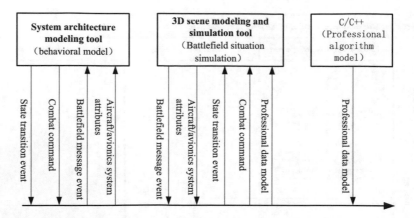

Fig. 6 Application scenario data interaction diagram

The 3D scene modeling and simulation tool runs the battlefield situation simulation model, which will send some event messages of the battlefield, aircraft platform and avionics system attribute information to the simulation data soft bus. It receives combat command and state transfer events from the logical model and drives the battlefield change and state transfer of the system model through these events. It will also receive the flight track and dynamic data from the algorithm model and synchronize the data to each entity.

C/C++ platform runs a variety of algorithm models, such as aircraft mission planning model, flight dynamics model, etc. These professional multidisciplinary models can publish the calculated flight path, dynamics and other data to the soft bus.

6   Brief Summary

This paper draws on the thought of SoS engineering and architecture design, combines the demonstration practice of aviation weapon equipment requirements, and carries out avionics system requirements analysis from the equipment system level. It builds the avionics systematic requirements analysis based on the model and simulation framework, design the system architecture model of co-simulation environment, implements the avionics system architecture model transfer between different layers, the accumulation of design information, and the convenience of design knowledge reuse, can effectively support the avionics systematic demonstration and top-level design, has important theoretical value and practical significance.

References

1. Pu, X.B.: Modern avionics system and integration, pp. 55–89. Aviation Industry Press, Beijing (2013)
2. Li, Q., Yan, J.: State of art and development trends of top-level demonstration technology for aviation weapon equipment. Acta Aeronautica ET Astronautica Sinica **37**(1), 1–16 (2016)
3. Hu, X.F., Zhang, B.: SoS complexity and SoS engineering. J. China Acad. Electronics Information Technol. **06**(5), 446–450 (2011)
4. Tan, Y.J., Zhao, Q.S.: A research on system-of-systems and system-of-systems engineering. J. China Acad. Electronics Information Technol. **06**(5), 441–445 (2011)
5. Zhao, Q.S.: Methods and techniques for system engineering and architecture modeling, pp. 125–130. Aviation Industry Press, Beijing (2013)
6. DoD Architecture Framework Working Group: DoD Architecture Framework Version 2.0 Volume 1: Introduction, Overview and Concepts. U.S.: Department of Defense (2009)

A Parallel Verification Approach for New Technology Application in Complex Radar System

Wei Yao, Hao Ming, and Lei Zhiyong

1 Introduction

Radar is the key surveillance and early warning sensor in the war environment. The trends of radar technique including intelligence, multi-functional and distribute, are making radar system more and more complex. In the development of complex radar system, Model-base system engineering (MBSE) are used to manage the complexity in the development process and ensure the delivered system will meet all needs & requirements [1, 2]. Contrasted with traditional document-based SE approach, MBSE make the formalized application of modeling to support system requirements, design analysis verification, and validation activities beginning in the conceptual design phase and continuing throughout development and later life cycle phases [3]. The applied models include structural, thermal, mechanism, electrical, software, simulation modules, etc. In this paper, we mainly discuss the approach to solve key technical problems or apply new technology in radar system, such as artifact intelligence (AI) algorithm, to fulfill the new-presented and difficult requirements [4].

Some techniques or approaches have been applied in MBSE to solve the emergency in SE process or handle the avoidance of high technical risks. Currently, system level modeling and analysis has been utilized for the requirement analysis and the design of top-level architecture [2, 5], but the rapid system model is generally limited to the evaluation of simple parametric equations or simplification of some complex radar signal process module, such as anti-jamming module, automatic target recognition module and so on. Without a high degree of detail, it is difficult to properly

W. Yao (✉) · H. Ming · L. Zhiyong
Nanjing Research Institute of Electronics Technology, Nanjing 210013, China

H. Ming
e-mail: mg.hao@163.com

L. Zhiyong
e-mail: zhylie@163.com

evaluate the design quality to perform the excellent trade-offs between performance, risk and cost. And the system level models may not been applied in the subsequent SE process, such as design definition process.

On the other hand, domain engineers/experts may use a wide variety of sophisticated models or analysis tools to analyze and design the system. But the detail baseline of these models cannot be verified and fixed until the special experiments or tests are executed. The risk of some key technical issues is difficult to be released in time on account of the long verification period of key technology. What's more, these detailed model are not connected to the system level model, the uniform algorithm baseline is difficult to build throughout the whole life cycle stages.

In this paper, a novel parallel verification approach for new technology application in complex radar system is proposed to improve the development process efficiently, to build a uniform algorithm baseline continuous renewal throughout the virtual verification to engineering realization, to provide a public platform for the domain algorithm experts in the tackle and verification of key technical problems. Section 2 discusses the typical MBSE method used in the radar industry currently, and the defects of current MBSE method when the new technology is applied in radar are analysis. In Sect. 3, the parallel verification approach that was employed in the MBSE process of radar development is discussed and the development and application of the parallel verification system are presented. And the experience of applying the parallel verification approach to several radar development projects is discussed. Section 4 conclusions the paper and discusses the future related works.

2   Related Analysis on Current MBSE Method

Nowadays, along with enhancement of radar performance requirement and acceleration of lead time, the main problems emerging in the new technology application activities are as follows:

(1)   In the perspective of SE, during the development process, most attention are paid in the decomposition of requirement baseline and capabilities baseline, and the hardware & software are developed according to the baselines [2] (typically life cycle of NASA is shown in Fig. 1). But for high-risk key technical tasks, risk avoidance is insufficient, the extensive analysis and testing is not enough.

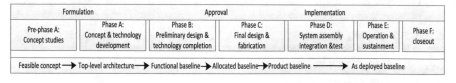

Fig. 1 NASA life cycle model

Optimize item / Role in SE	Algorithm performance	Computing speed
Algorithm expert	√	×
System engineer	√	√
Software developer	×	√

Fig. 2 Different SE roles focus on the optimization of different performance of technical solution

(2) Refer to the application requirement of product, a new technical solution should have a very high technology ready level (TRL). Simulation can only release the theoretically technical risks for key technology, but the technical maturity requirements cannot be satisfied just by simulation.

(3) In the Vee model of system life cycle, the demonstration and analysis of the key technical solution, the algorithm developer and the software engineering may involve several professional teams in different domains. For different roles, they focus on the different performance of the technical solution (shown in Fig. 2). A uniform and efficient approach is needed to solve the conflict in different domain experts and throughout the virtual verification to engineering realization.

Currently, the typical MBSE method applied in radar industry is shown in Fig. 3, there are three main model-based iterations/recursions in the life cycle. First of all, based on the typical warfare scenarios, virtual parametric simulation derives the capability requirement baseline [6, 7]. Then the functional simulation and verification confirm the design of architecture, and in the simulation process, the key performance indicators (KPIs), function, and interface are allocated and discussed. Thirdly, in the development and implementation stage, the system/sub-system models are decomposed and developed top-down, integrated and verified down-top [8, 9].

As the key sensor for warning detection and surveillance in the war field, radar has to cope with more and more complex environment and scenarios. And more and more new technologies are needed and required by radar in the 5G and AI era. If the new technology can be applied in the product, its TRL must be high enough, which means the technology should be verified by the physical prototype in the real field experiments [2, 4]. But the needed new technologies may often have low TRLs, and have to be verified and optimized on the prototype of sub-system/system. This may affect the development period. Because the change and iteration of new technology applications cannot be solidified in the early concept and design stages, this may also influence the schematic trade-offs. In order to release the high technical and schedule risks, we proposed a parallel verification approach for new technology application in complex radar system.

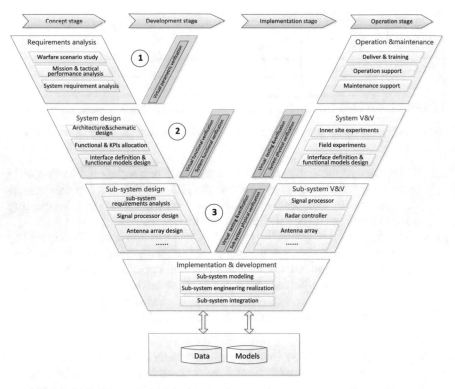

Fig. 3 Typical MBSE process of radar system

3 The Parallel Verification Approach

3.1 The Parallel Verification Process

The proposed verification approach mentioned a parallel verification system. The parallel verification system is developed and assembled in component-based development manner, which means the verified algorithm components have the high-enough TRL and can be directly utilized in the new designed radar. The parallel verification process is shown in Fig. 4.

Firstly, when some newest technology is needed to fulfill the difficult and challenging system requirements in the early concept stage of a new radar, the technology verification requirements should be analyzed and transferred to the algorithm expert as soon as possible. Then the parallel verification system can be developed and integrated according to a whole technical scheme and baseline. Instead of waiting for the V&V process in the new radar or prototype, the parallel verification system can verify and iterative improve the algorithm models and technical baseline as a parallel bypass road in the real operation environment of other delivered radar equipment, after the virtual verification with the simulated data in the radar database. Finally, when the

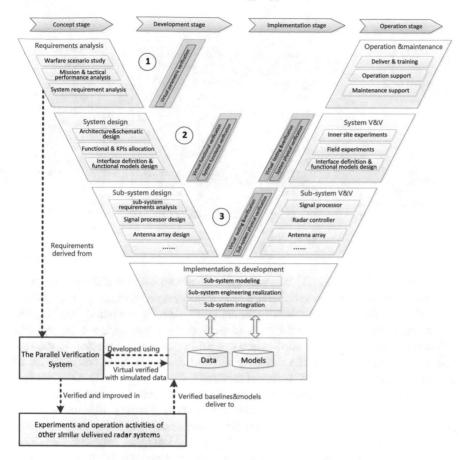

Fig. 4 The parallel verification process

models and technical baseline are well verified and the statement is solidified under the review of the system engineers and algorithm experts, the TRL may reach 6–8 level, and the models and baseline can be deliver to the knowledge database. The components and models can be used directly in the new radar V&V process to fulfill the new technology application requirements.

What's more, the verified signal-level models can also be abstracted to the parametric or functional models to support the requirement analysis and system design activities in regular MBSE process.

3.2 Inputs and Outputs of Parallel Verification System

The inputs and outputs of parallel verification system are shown in Fig. 5. The

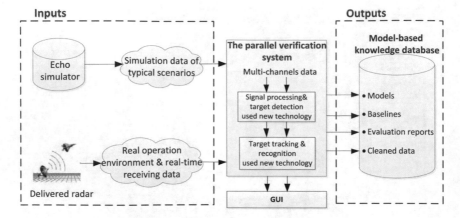

Fig. 5 The inputs and outputs of parallel verification system

technical performance should be virtually verified and evaluated firstly in typical scenarios utilizing the echo simulator, when the parallel verification system is developed. The interface is compatible with the delivered radars, so the parallel verification system could be connected to the radar in service. As a side road, the new applied technology integrated in the verification system-bypass can be verified in the real operation environment with the real-time received data from the operating radar, in the condition of no impact on delivered service.

The results and achievements can export to the model-based knowledge database, which support the implementation and integration of new radars, after the technical performance and baseline are well improved and verified in real operation environment. The achievements mainly including verified models, solidified baselines, technical performance evaluation reports and cleaned radar data covering clutter, jamming, kinds of targets etc.

3.3 The Experience of Applying the Parallel Verification Approach

We develop several parallel verification systems such as target classification, clutter suppression and crowed target tracking etc. Take the target classification bypass-system for example, the application and verification process is shown in Fig. 6.

When we develop an important radar, some new target recognition algorithm is needed. We build a bypass-system according to the new requirements firstly. Then the classification performance of bypass-system is trained and tested offline based the original radar data of several kinds of targets, accumulated in past flight experiments. Continuously, the online test and verification are executed in complete work flow in real operating environment of similar delivered radar. After the offline and online

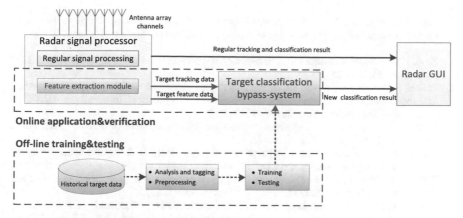

Fig. 6 The parallel application and verification process of target classification

test and verification, the technical status of the synthetic classification baseline could be solidified, and the TRL of target classification algorithm could be high enough to be applied in the new radar.

Benefit from the parallel verification, the key technologies breakthrough activities are executed parallel with the hardware and software develop process of new radar. The research and development cycle could be shortened and the potential technical risk could be handled and released. What's more, the bypass-system also provides a public platform for the domain expert to do technical research and pre-research.

4 Conclusion

According to the trend of rapid increase requirements on new technology application to new radar, we proposed a novel parallel verification approach to release the high technical and schedule risks in this paper. This approach is also a helpful supplement for MBSE mode in use today, making the development process more efficient. The bypass-system used in this approach is beneficial to domain experts for building a uniform algorithm baseline and public research platform. This can lead the radar technology evolution to a virtuous cycle of application-verification-exploration.

References

1. Friedenthal, S., Moore, A., Steiner, R.A.: A Practical Guide to SysML the System Modeling Language. Elsevier, Waltham, MA (2012)
2. David, D.W., Garry, J.R., Kevin, J.F., Douglas Hamelin, R., Thomas, M.S.: System Engineering Handbook: A Guide for System Life Cycle Processes and Activities. Wiley, New Jersey (2015)

3. Bayer, T., Chung, S., Cole, B.B.: Update on the model based systems engineering on the Europe mission concept study. In: Proceedings of Aerospace Conference, BigSky, Montana, 2013, pp. 1–13
4. Yang, F.Y.: Toward a revolution in transportation operations: AI for complex systems. IEEE Intell. Syst. **23**(6), 8–13 (2008)
5. Bonnet, S., Voirin, J.L., Navas, J.: Augmenting requirements with models to improve the articulation between system engineering levels and optimize V&V practices. In: 29th Annual INCOSE International Symposium, Orlando, FL, 2019: 1:15
6. Piaszczyk, C.: Model based systems engineering with department of defense architectural framework. Syst. Eng. **14**(3), 305–326 (2011)
7. Carty, A.: An Approach to Multidisciplinary Design, Analysis & Optimization for Rapid Conceptual Design. AIAA, Reston (2002)
8. Tepper, J.S., Horner, N.C.: Model-based systems engineering in support of complex systems development. J. Hopkins APL Tech. Dig. **32**(1), 419–432 (2013)
9. Andersson, H., Hezog, E., Johansson, G., et al.: Experience from introducing unified modeling language/systems modeling language at Saab aero systems. Syst. Eng. 13(4), 369–380 (2010)

A Preliminary Research on Zonal Safety Analysis Method for Aircraft Complex Systems by Using Virtual Reality and Augmented Reality

Tao Li, Bo Ye, Dawei Wang, and Hu Cao

1 Introduction

Aircraft systems consist of large number of installed components including pipes, lines, cable harnesses, supports and equipment from different sub-systems. These components are normally connected, attached and fixed following by physical layout principles in aircraft bays. Unlike aircraft structural parts, system components have complicated interactions and interdependencies within [1]. Such complicities are shown as the functional aspect of aircraft. Various aircraft systems work together to achieve one function [2]. For example, the flight control function requires hydraulic system, electrical power supply system and navigation system working together to satisfy the aircraft vehicle control requirements in different flighting conditions. Modern aircraft are equipped with advanced functionalities and those functionalities are actually based on the physical installations of individual systems. As many of system components are arranged in a limited space, a good assembly design and reliable installations in production ensure the system components perform their functions correctly. In other words, the physical installations contribute much to the final introduction of functions. A typical example of system installations in bays is shown in Fig. 1.

In the industry practice, the Zonal Safety Analysis (ZSA) tool from SAE Aerospace Recommendation Practice 4761 (ARP4761) Guidelines for Development of Civil Aircraft and Systems, help to identify and analyse how aircraft system installation of individual systems or components could mutually influence between other

T. Li (✉) · B. Ye · D. Wang · H. Cao
AVIC Chengdu Aircraft Industrial (Group) Co., Ltd., No. 105, Chengfei Street, Qingyang District, Chengdu, Sichuan, China
e-mail: chrislt@foxmail.com

B. Ye
e-mail: bo.ye@foxmail.com

© The Author(s), under exclusive license to Springer Nature Switzerland AG 2021
D. Krob et al. (eds.), *Complex Systems Design & Management*,
https://doi.org/10.1007/978-3-030-73539-5_7

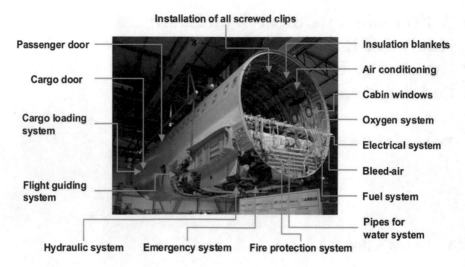

Fig. 1 Aircraft system components installed in bays [3]

systems and components installed in close proximity on the aircraft [4]. Those influences not only include the poor installations like wire bundle riding on structure, hole edge, pipelines, or poor fastened screws, but also indicate the potential risks of functionalities introduction caused by the improper installations. Federal Aviation Administration (FAA) reported that many aircraft accidents and incidents result from adjacent cable harness wirings, pipelines and equipment installations [5].

SAE ARP4761 suggested that ZSA to be applied at all the development stages [6]. However, current industry practice focuses more on the application at design stages, ignoring the use of ZSA in assembly planning at manufacturing stage. ZSA is not well involved in the widely used digital engineering principles in aircraft industry, which leads to the lack of information connection of product, process, aircraft failure modes from different development stages. It is acknowledged that digital engineering tools are crucial for aircraft manufacturers to quick introduce and deliver new aircraft to the airline market. It is therefore necessary to find how these tools like Digital Mock-Up (DMU), Virtual Reality (VR) and Augmented Reality (AR) would be used to support the ZSA implementations.

2   Literature Review

This section investigates the concepts of ZSA, current research and practices in aircraft development process. Aircraft system integration characteristics are also investigated to allow a comprehensive understanding of the needs to apply ZSA. Gaps are the then concluded to help integrating digital engineering tools with traditional ZSA which covers the full product development lifecycle.

2.1 Zonal Analysis-Based Method

There are several zonal based methodologies used in industry to analyse aircraft design. FAA suggested a zonal analysis-based approach in 2006 called Enhanced Zonal Analysis Procedure (EZAP) from Aircraft Certification chapter 25.27A (AC 25.27A) [5]. It provides guidance for developing maintenance and inspection instructions for the Electrical Wiring Interconnect System (EWIS). It also supports the improvement of aircraft at manufacturing stage. A research sponsored by U.S. Department of Transportation in 2006, reported the use of zonal analysis for EWIS and develop risk assessment software tool to help combine four aspect analyses together for the whole aircraft EWIS [7], which are bundle damage potential, system damage potential, subsystem damage potential and fault damage potential. This approach examines the power wires in a bundle and uses experimental data along with parameters, such as wire gauge and insulation type, power source and voltage, and circuit protection, and then it assesses the potential damage by zones. This approach is effective only when containing a EWIS database where the relevant parameters of wires, the systems they support, bundles and zones are brought together. By using this approach, it must be able to integrate the results of the EWIS analysis with the overall aircraft safety analysis.

Another approach mostly used in aircraft industry is ZSA from SAE ARP4761. The analysis begins with the definition of aircraft zones, and then uses design and installation guidelines and criteria to inspect the system component installations and inference in each zone. Inspection records are documented and submitted to relevant departments of the project for problem resolution. Analysis records are also used for resolution process tracking in regular inspections. Figure 2 shows the working process of ZSA.

In the initial ZSA process, the inputs include product design data, system descriptions like schematics, traditional physical mock-up, regulations and other reliable engineering results from FMEA (Failure Modes and Effects Analysis) and PSSA (Preliminary System Safety Assessment). It is also interesting to find that experience, maintenance and operational hazards, and aircraft level requirements are important inputs in preparing of design and installation rules. This partly reflects the fact that current ZSA results are depends on various engineering data sources and the results are scored more from personal experience. According to SAE ARP4761, the general design and installation guidelines for system installations normally include ensuring of no unacceptable stress, minimizing stress of the attachments to moving parts in positioning and mounting, and minimizing water accumulation of the pneumatic pipes and hoses in installations [8]. All these guidelines are used to ensure the systems work correctly regard to the failure of one system affecting the other ones, avoiding both systems failure by separating one system. These detailed check activities are actual done by engineers or mechanics manually with low efficiency. At manufacturing stage, ZSA is applied for early production aircraft units [9]. A sample check list is shown in Table 1 to have a brief view of how the inspection

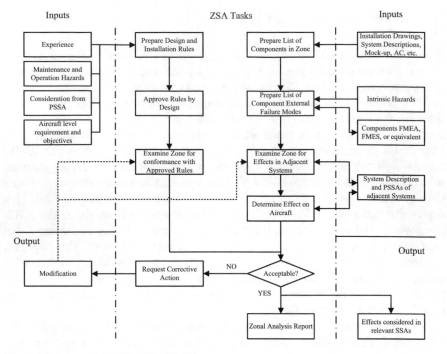

Fig. 2 ZSA process in SAE ARP4761 [6]

Table 1 An example of ZSA worksheet [6]

Zonal safety analysis summary sheet			
Aircraft type/standard	Zone: main landing gear bay	Complied by	Sheet: 01 of: 01
		Date	Issue: 01
Query Sheet No.	Reason for non-compliance	Action taken	Remarks
27.01	Flight control bolt installed head down		Query sheet OPEN
92.01	Long bonding strips		Query sheet OPEN
92.02	Chafing of conduits	Installation drawing modified	Query sheet CLOSED
92.04	Missing drain holes in several conduits		Query sheet OPEN

information are recorded and documented. Traditionally, the information used in the zonal analysis process are mostly in the form of textual-based paperwork.

2.2 Aircraft Digital Engineering Tool

A geometric model or spatial model represents geometric and/or spatial relationships. 3D CAD models or Digital Mock-Up (DMU) are applied in aircraft design for quite a long time that include dimensions, positions and other descriptive non-geometric attributes data or so called as Product and Manufacturing Information (PMI) [10]. Traditionally, PMI contains tolerance, welds types, surface finish, datum target, annotations, etc. In 3D CAD systems, PMI may also contain geometric dimensions, Bill of Materials (BOM) and 3D functional textual instructions. Installation and test notations are also a part of PMI for in aircraft systems 3D models. Dassault Systèmes CATIA V5 is one of the most popular and widely used computer aided engineering tool, which incorporates CAD, CAM and other applications. The typical modelling environment is illustrated in Fig. 3.

To make the most use of DMUs and improve the working experience in design and manufacturing, Augmented Reality (AR) and Virtual Reality (VR) are introduced. AR is a kind of innovative and effective human computer interaction (HCI) technology, used to enhance the users' perception of the real world by superimposing virtual digital information to the relevant real objects [11]. AR has been applied in some area of industry, e.g., inspection [12], maintenance training and operation support [13], assembly [14], design and manufacturing [15]. It shows that the AR can present virtual model and necessary information on the corresponded real objects in time. Thus, it improves the ability of information acquisition for the operator during the manual process, and makes it easy to recognize the real objects. The typical application scenario is showed in Fig. 4.

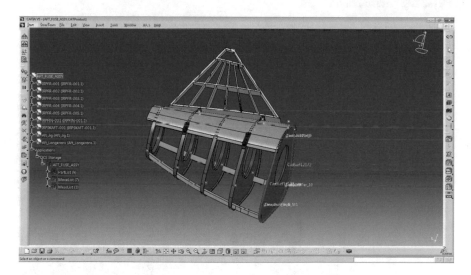

Fig. 3 An example of 3D CAD assembly model with PMI in CATIA V5

Fig. 4 The AR system used in aircraft assembly [16]

VR includes computers, sensors and simulation technologies. Its basic implementation is to simulate the virtual environment with 3D CAD model to give the user a sense of immersion [17]. VR has been used for training and analysing in some fields of industry, for example, assembly training [18], accessibility and ergonomic analysis [19], verification of assembly and maintenance processes [20]. It is proved that VR is effective in improving human perception by establishing an immersive and interactive virtual engineering environment. Figure 5 illustrates the typical virtual interacting environment.

Fig. 5 The virtual engineering scenario for door assembly [20]

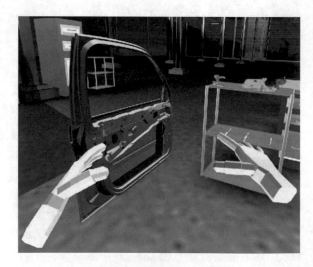

2.3　Aircraft System Integration and Development Process

An aircraft is a system of systems (SoS) [21–23]. The top-level aircraft system structure can be defined as vehicle system, avionic system and mission system [2]. In the design and development process, multidisciplinary are applied on these systems including aerodynamics, materials, mechanical, electrical, information and computer technologies [24]. There are a number of forms of system integration. System like fuel tanks are physically integrated with structures [25]. Some systems receive, send and exchange internal and external information for system control and display purpose [26, 27]. Typical examples include the need of information about valves open or closed state, the display of fuel mass and engine speed, and usage of the aircrew commands [28]. The system interdependencies can be generally concluded as two aspects which are physical and functional interdependence. The physical aspect has strong links with weight, installation and loads, while the functional aspect is specified as information based integration [29]. From the system integration point of view, the principles of ZSA introduced in Sect. 2.1 actually guarantee the aircraft system functionalities by ensure the right physical integrations.

Modern large-scaled aircrafts are long life complex systems following by system integration and validation [30]. A widely accepted simplified aircraft development lifecycle model is shown in Fig. 6, which covers from definition, design, build, test, operate to refurbish or retire.

To meet the rapid growing market requirement and reduce the high product change cost, concurrent engineering was applied in many companies since 1990s'. In 1991 the Concurrent Engineering Forum that took place at Cranfield University defined concurrent engineering as "the delivery of better, cheaper, faster products to market by a lean way of working using multi-discipline teams, right first-time methods and parallel processing activities to continuously consider all constraints" [32].

It is clear from Fig. 7 that a concurrent engineering lifecycle allows the design and manufacturing activities to be taken in parallel. The advantage of this model can be concluded as more efficient information exchange between different stages, thus the growing information certainties with development process going create opportunities for more system verifications be taken at earlier stages.

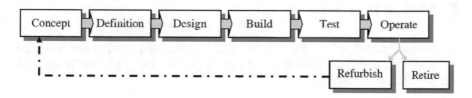

Fig. 6 The aircraft full lifecycle [31]

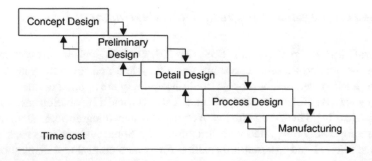

Fig. 7 A concurrent engineering lifecycle model for design and manufacturing

2.4 Gap Summary

Further investigation found that previous ZSA research concentrate more on aircraft design stages, especially the conceptual and preliminarily stage [4, 33]. ZSA is a methodology that ensures safety analysis towards physical installations, early design stages may not have adequate physical integration information to support a robust analysis result. This requires the product design and manufacturing engineering to be taken in concurrent engineering mode, thus information can be transmitted and iterated through the development processes. Another gap is traditional ZSA relies much to personal experience, which leads to different analysis results. The same installations in a zone may have different hazard analysis results due to different understanding and implementation of criteria. Immersive digital tools like VR and AR are in the advantage of real-time and real-world experience. This would make engineers and mechanics make more reliable decision in a ZSA process. Besides, the guidelines in SAE ARP4761 are published in 1990s and digital engineering principles are getting profound influence on the aircraft design and manufacturing. The initial ZSA should be expanded to support DMU as a main master engineering data source than previous 2D drawings.

3 Method Development

This research aims to propose an improved ZSA methodology by integration digital engineering tools into the inspection and analysis process. The gaps found in Sect. 2.4 indicate the new method should:

- Supports DMU as a master model covering the lifecycle of aircraft system development stages including assembly process planning and product serial manufacturing.
- Has the advantage of fast identifying and analysing the physical installation objects in ZSA process.

- Integrates system both DMUs and installations associated information to allow further zonal analysis.

3.1   Integrate VR and AR into ZSA

Previous literature indicates that VR and AR have the potential to bring the information in an immersive way to people, which improves the accuracy of inspection and decision making.

VR provides a real-world sense to allow people verify the adjacent system components in assembly design and installation plans regardless it is at early or late development stage. As the DMUs are the master assembly model in aircraft design and manufacturing, physical installations are represented not only in static fixed positions, but also can be simulated the all the installation activities such as fastening, attaching, holding and fixing. In other words, the hazard factors caused by human activities, process or tool selection can be also included in the ZSA process. Previous investigation reveals the system complexity comes from the large number of installed components in a limited space.

When applying ZSA inspections, it is not easy to identify the specified system component or equipment like a hydraulic line, sub-segment cable bundle or a small connector. AR pushes notifications to people to easier identify individual component and access at the right position in the zone. Additional information that required by ZSA can be also embedded into the DMUs including process plans, tolerance, system schematics, material information and other kinds of PMI. According to the initial ZSA process, much of the inputs are non-geometric information. The AR-based information delivering would bring more convenient in ZSA inspections and analysis to engineers and mechanics working.

3.2   Engineering Data Source that Support ZSA in Different Stages

3D DMU is considered as a master engineering data source in the developing method, which is the major difference compared to the traditional 2D drawings input. A DMU integrates detailed geometric information and the embedded PMI expands models with more information needed for applications. SAE ARP4761 requires ZSA to be undertaken in all the development stages. If follow the general principles of the concurrent engineering model mentioned in Sect. 2.3, the ZSA processes involved in each stage can be presented in Fig. 8.

Generally, there are three main stages in the development process if classifying by the generating and use of engineering data source: data in aircraft design, data in process design or known as industrial design, and the data used in manufacturing. Figure 8 explains the data flows transmitting between these major stages. Firstly, ZSA

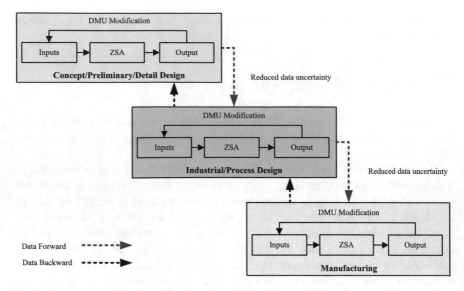

Fig. 8 The inputs and outputs of ZSA and their relationships in different stages

is applied in each stage following the development process. Once the ZSA outputs indicate modifications should be done in the 3D DMUs, DMUs are then improved and allow a further inspection until it is completely satisfied. The analysis result in each stage will be used as basis for the next stage work. Besides, the information uncertainties are reducing with the aircraft system development process going. As the stages progresses and the engineering data become fixed or with lower-level uncertainties, more of useful information can be extracted from the DMU to support more reliable and robust decision making. It should be pointed out that the DMU modifications include both the geometric and non-geometric information. However, there are more geometric or assembly design changes at design stages, and more non-geometric PMI changes like processes, assembly sequences and tooling information at the latter two major stages. This framework also allows communication chances for every two stages in parallel.

3.3   Expand ZSA Process with Digital Engineering

The ZSA process requires design and installation guidelines, product data, FMEA and system descriptions, which both contains physical and functional integration information. For instance, the installation criteria "hydraulic lines to structure distance should be minimum to 5 mm" requires people to identify the specified hydraulic line installations first. The description of current zone information, and other functional information from system schematics should also be prepared to support a dedicated inspection. In this research, the information that support ZSA process are defined as

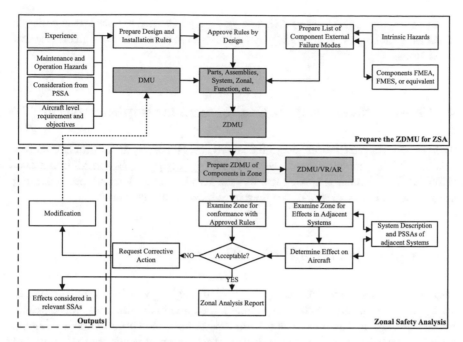

Fig. 9 Proposed ZSA methodology with digital engineering tools

ZDMU, which aims connecting required aircraft zones, system components, human factors, and processes to be a unified engineering data source. It helps the information transfer between development stages by model improving and reusing. The proposed new ZSA methodology with digital engineering tools is shown in Fig. 9.

The expanded ZSA methodology consists of three parts, the preparation of ZDMU, detailed ZSA, and results outputs. The ZDMU are defined by adding product Bill of Materials (BOM), specified assembly process, zone partitioning breakdowns, and system function descriptions. The list of component external failure modes can be also included in a ZDMU. When a ZDMU goes into the dedicated ZSA process, VR and AR are introduced to help examining hazard effects in adjacent system components and structure by zones with approved rules. In digital engineering, modifications are made to improve the DMU if required by the analysis results. Design and manufacturing activities in different stages would all benefit from this methodology. At early design and development stages especially the conceptual design stage, much information is connected to system requirements and functions. The early stage ZSA inspections actually refer to general system layouts and verifications of layouts towards aircraft functionalities partitioning. VR will help fast iteration of such zonal-based analysis. When it comes to later process planning and manufacturing stages, AR helps to have easier and more accuracy information access, as well as more efficiency ZSA practice.

By following the AR-based ZSA analysis, people would require less experience in decision making of whether the physical installations meet the design criteria and regulations.

4 Construction of VR/AR Environment for Implementation

The proposed approach uses integrated ZDMU that links physical installation information and non-geometric inputs required by ZSA. To support VR and AR implementation, an environment is created towards the activities in design and manufacturing stages. Two initial case studies are then introduced in this section.

4.1 Implementation of VR in ZSA

Figure 10 shows a detailed assembly model of the engine pylon bay zone. The system components in this area include equipment, tube and hydraulic lines, valves, cable harness wirings, and pylon skin. Real-time operation related information like human postures, system descriptions, assembly processes, ground facilities are also created as associated data as part of the ZDMU.

In the virtual world, engineers analyse the adjacent effect of one component to the others. For example, the cable harness wirings on the left-hand side in Fig. 10 are closed to the big tubes. Further inspection found the wirings are electrical power supply wirings and thus following the safety distance for crucial systems, it does not meet the installation criteria in current assembly design. Similarly, attachments to moving parts in this bay are also checked in VR to ensure these components have the minimized functional hazard caused by poor installation positioning. Figure 11

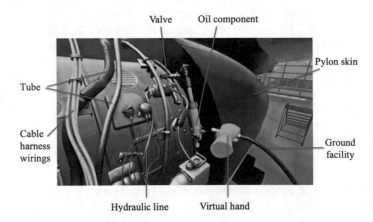

Fig. 10 A ZDMU created for ZSA

Pylon bay   Hatch lock

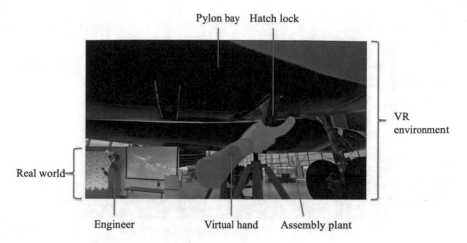

VR environment

Real world

Engineer   Virtual hand   Assembly plant

Fig. 11 ZSA inspection in VR environment

illustrates the implementation sense of ZSA using VR. The virtual hand activities in Figs. 10 and 11 allows the hazard checking in assembly and even maintenance operations. The accessibly checking of a ground facility to oil system, and hatch locking operations in a virtual assembly plant is available in this ZSA implementation environment.

4.2 Implementation of AR in ZSA

The AR is used to superimpose virtual information on the specified real objects during the ZSA process, such as the 3D model, texture and image data. As shown in Fig. 12, in checking the hydraulic lines stand-offs between the floor, BOM information and other working instructions from the ZDMU are pushed in the implementation environment.

In manufacturing phase, engineers or mechanics will carry out the zonal based analysis task in the ZDMU or a real-world aircraft. For instance, the ZSA process needs to be applied for the hydraulic lines and floor in the same zonal showed in Fig. 12. Thanks to AR, the hydraulic line ID numbers like "YT001", "YT002", "YT003", and the inspection instructions are pushed in the AR terminations, helping people to take the ZSA tasks directly without preparing much more texture-based paperwork and acquire the ZSA guild information in time.

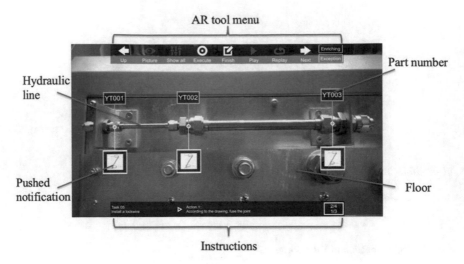

Fig. 12 AR environment support ZSA process

5 Conclusion and Future Work

This research investigates the problems of traditional ZSA process in SAE APR4761 and provides a novel method to implement ZSA that would be used towards digital engineering. With the help of VR and AR tool, it is possible to support ZSA implementations both in design and manufacturing. Another novelty in this research is to build a new ZDMU, which integrates geometric and non-geometric information required for ZSA inputs. By applying the new ZSA methodology, engineers and mechanics will obtain sufficient physical and functional integration information allowing less experience needed in the ZSA process. It changes the initial ZSA process from paperwork based to ZDMU based. The concurrent engineering framework allows information exchange and iterations to be taken efficiently. This ensures the analysis results better meets the requirements of zonal hazard analysis in different stages, avoiding the risks of poor decision making based on inadequate product and assembly information. Implementation environment are also introduced to examine VR and AR in the expanded ZSA methodology. Two initial case studies are created for implementations for design and assembly respectively. The results show the feasible of the methodology and VR/AR tools improve the accuracy in hazard identification in the immersive and augmented world, which brings better experience in ZSA process.

Future work would be required to develop the more detailed modelling regulations for ZDMUs. Currently, the new ZSA methodology is only tested in one aircraft bay with simplified and limited numbers of system components. It is suggested to use more bays with high-level complicities installations to examine the workflow and efficiency. The effects between adjacent bays in the digital engineering are also very interesting to be analysed by using VR/AR tools.

Acknowledgements This research is a fundamental research funded by SASTIND China under Grant JCKY2018205B021 and JCKY2019205A004.

References

1. Li, T., Lockett, H.: An investigation into the interrelationship between aircraft systems and final assembly process design. Procedia CIRP **60**, 62–67 (2017)
2. Moir, I., Seabridge, A.: Aircraft Systems: Mechanical, Electrical, and Avionics Subsystems Integration, 3rd edn. Wiley, Chichester (2008)
3. Frankenberger, E.: Concurrent design and realization of aircraft production flow lines—process challenges and successful. In: International Conference on Engineering Design ICED' 07 Paris, 2007, August, pp. 1–11.
4. Chen, Z., Fielding, J.P.: A zonal safety analysis methodology for preliminary aircraft systems and structural design. Aeronaut. J. **122**(1255), 1330–1351 (2018)
5. Federal Aviation Administration: Aircraft Electrical Wiring Interconnect System (EWIS) Best Practices—Job Aid 2.0 (2010)
6. Society of Automotive Engineers: ARP4761: Guidelines and Methods for Conducting the Safety Assessment Process on Civil Airborne Systems and Equipment. SAE International (1996)
7. Linzey, W.G.: Development of an Electrical Wire Interconnect System Risk Assessment Tool (2006)
8. Wilson, M., Jones, J.: Guidance on the Conduct of Aircraft Zonal Hazard Analysis/Assessment (2016)
9. Caldwell, R.E., Merdgen, D.B.: Zonal analysis: the final step in system safety assessment (of aircraft). In: Proceedings of Annual Reliability and Maintainability Symposium, pp. 277–279, 1991.
10. Andre, P., Sorito, R.: Product manufacturing information (PMI) in 3D models: a basis for collaborative engineering in product creation process (PCP). In: Proceedings 14th European Simulation Symposium, 2002. c
11. Wang, X., Ong, S.K., Nee, A.Y.C.: A comprehensive survey of augmented reality assembly research. Adv. Manuf. **4**(1), 1–22 (2016)
12. Eschen, H., Kötter, T., Rodeck, R., Harnisch, M., Schüppstuhl, T.: Augmented and virtual reality for inspection and maintenance processes in the aviation industry. Procedia Manuf. **19**(2017), 156–163 (2018)
13. De Crescenzio, F., Fantini, M., Persiani, F., Di Stefano, L., Azzari, P., Salti, S.: Augmented reality for aircraft maintenance training and operations support. IEEE Comput. Graph. Appl. **31**(1), 96–101 (2011)
14. Mas, F., Menéndez, J.L., Serván, J., Gomez, A., Ríos, J.: Aerospace industrial digital mock-up exploitation to generate assembly shopfloor documentation. In: 29th Intl. Manufacturing Conference (IMC29), 2012.
15. Bottani, E., Vignali, G.: Augmented reality technology in the manufacturing industry: a review of the last decade. IISE Trans. **51**(3), 284–310 (2019)
16. Mas, F., Gomez, A., Menéndez, J.L., Serván, J.: Aerospace industrial digital mock-up exploitation to generate assembly shopfloor documentation
17. Jayaram, S., Connacher, H.I., Lyons, K.W.: Virtual assembly using virtual reality techniques. CAD Comput. Aided Des. **29**(8), 575–584 (1997)
18. Adams, R.J., et al.: Virtual training for a manual assembly task. Haptics-E Electron. J. Haptics Res. **2**(2), 1–7 (2001)
19. Rajan, V.N., Sivasubramanian, K., Fernandez, J.E.: Accessibility and ergonomic analysis of assembly product and jig designs. Int. J. Ind. Ergon. **23**(5–6), 473–487 (1999)

20. De Sá, A.G., Zachmann, G.: Virtual reality as a tool for verification of assembly and maintenance processes. Comput. Graph. **23**(3), 389–403 (1999)
21. Kossiakoff, A., Sweet, N.W., Samuel, J.S., Biemer, M.S.: Systems Engineering Principles and Practice, 2nd edn. Wiley, New Jersey (2011)
22. Liu, H., Tian, Y., Gao, Y., Bai, J., Zheng, J.: System of systems oriented flight vehicle conceptual design: perspectives and progresses. Chinese J. Aeronaut. **28**(3), 617–635 (2015)
23. INCOSE: Systems Engineering Handbook: A Guide for System Life Cycle Processes and Activities, 4th edn. Wiley, Hoboken (2015)
24. Altfeld, H.-H.: Commercial Aircraft Projects: Managing the Development of Highly Complex Products, 1st edn. Routledge, London (2010)
25. Langton, R., Clark, C., Hewitt, M., Richards, L.: Aircraft Fuel Systems. Wiley, Chichester, UK (2009)
26. Seabridge, A.: Aircraft sub-systems introduction and overview. In: Encyclopedia of Aerospace Engineering. Wiley, Chichester (2010)
27. Scott, W.: A330/A340 Cabin Avionics. In Advanced Avionics on the Airbus A330/A340 and the Boeing-777 Aircraft Conference Proceedings, 1993.
28. Wainwright, W.: A330/A340 Flight Deck In: Advanced Avionics on the Airbus A330/A340 and the Boeing-777 Aircraft Conference Proceedings, 1993.
29. Moir, I., Seabridge, A.: Vehicle systems management. Encycl. Aerosp. Eng., pp. 1–16 (2010)
30. Ulrich, K.T.: Product Design and Development, 4th edn. McGraw-Hill, London (2008)
31. Moir, I., Seabridge, A.: Design and Development of Aircraft Systems, 2nd edn. Wiley, Chichester (2013)
32. Delchambre, A. (ed.): CAD Method for Industrial Assembly: Concurrent Design of Products, Equipment and Control Systems. Wiley, Chichester (1996)
33. Chiesa, S., Corpino, S., Fioriti, M., Rougier, A., Viola, N.: Zonal safety analysis in aircraft conceptual design: Application to SAvE aircraft. Proc. Inst. Mech. Eng. Part G J. Aerosp. Eng. **227**(4), 714–733 (2013)

A Quantitative Analysis Method of Itemized Requirement Traceability for Aircraft Development

Dong Liang, Liu Kanwang, Zhao Pan, and Dai Jingli

1 Introduction

With the application and promotion of itemized requirement management technology in aircraft development, a large number of itemized requirements are defined, refined, and decomposed as the further decomposition of aircraft development hierarchy, forming a rich traceability relationship [1, 2]. A more complex and richer relationship between top and bottom requirements, different types of requirements, requirements and different design features is shown, forming a complex evolutionary process of "top–bottom decomposition and transmission" and "bottom-top derivative trade-offs" [3]. This complex traceability relationship and its changes reflect not only the requirement transmission process, but also the decomposition and refinement quality of the requirement along with the aircraft development process. Therefore, it is necessary to conduct quantitative analysis and evaluation research on the requirement traceability relationship by building a quantitative traceability analysis model for requirement items to better identify and analyze requirement decomposition granularity, requirement transmission route status, and requirement definition quality status, etc.

D. Liang (✉) · L. Kanwang · Z. Pan · D. Jingli
AVIC The First Aircraft Institute, Xi'an 710089, China
e-mail: coury2002@163.com

L. Kanwang
e-mail: liukan_wang@sina.com

Z. Pan
Northwestern Polytechnical University, Xi'an 710072, China

2 Analysis of Itemized Requirement Decomposition and Allocation Process

2.1 Requirement Decomposition and Allocation

The requirements are decomposed layer by layer along with the architecture design process, that is, the requirements of the upper product structure are detailed and decomposed due to the decomposition of the product structure to its next layer. After the upper-level requirement is decomposed, the requirement can be allocated. The type of requirement traceability relationship formed by decomposition is a satisfaction relationship [4], that is, the set of the lower-level requirements should be able to meet all the upper-level requirements, see Formula (1), where R_i is an upper-level requirement, and R_{ij} is one of the set of the lower-level requirements, and the maximum value of n is the number of components of the lower product structure.

$$R_i = \sum_{j=1}^{n} R_{ij} \tag{1}$$

The mapping relationship established by requirement allocation is the basis for traceability. The allocation of requirements in aircraft development can be categorized into vertical and horizontal aspects. Vertical allocation is the decomposition and allocation of requirements from the aircraft level, to the system level, to the subsystem level, and then to the equipment level, and so on. Horizontal allocation is the distribution relationship between requirements and architectural design elements, which can be divided into 2 categories (Fig. 1).

(a)   Allocation of non-functional requirements such as performance and safety to functional requirements

After the functional requirements of each product object is determined, its corresponding non-functional requirements like performance and safety have to be defined and fulfilled, and "Trade Studies" is used to quantify these requirements. All non-functional requirements should be traced to its corresponding function, and all functions have performance requirements. At the same time, a traceability relationship between performance requirements and its upper-level requirements should be established.

(b)   Decomposition and allocation of requirements to architectural design elements

For a complex system, the architecture design of the product objects at each layer involves the evolutionary process of functional architecture, logical architecture, and physical architecture. A specific architecture scheme is finally formed through the trade-off analysis with system architecture design. The evolution process of architecture design is closely related to the decomposition and allocation process of requirements. Requirements are the input of architecture design. At the same time, the architecture trade-off results will

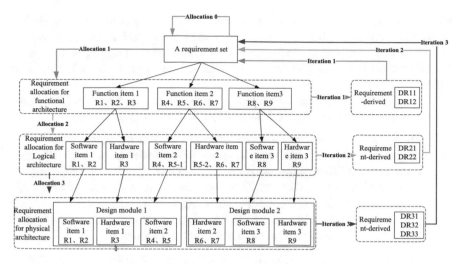

Fig. 1 Four requirement allocation processes

derive corresponding requirements to achieve constraints on the architecture. The process can be further divided into the following three aspects:

(1) Decomposition and allocation of requirements to functional architecture. Requirements are allocated to functional architecture elements. After architectural trade-offs, derivative requirements and interface requirements based on a specific functional architecture will be generated, which is helpful to determine the functional baseline.

(2) Decomposition and allocation of requirements to logical architecture. Based on a specific functional architecture, the requirements are allocated to logical architecture elements. After architectural trade-offs, derivative requirements and interface requirements based on the specific logical architecture will be generated, which contributes to determine the allocation baseline.

(3) Decomposition and allocation of requirements to physical architecture. Based on a specific logical architecture, the requirements are allocated to the physical architecture elements. After architectural trade-offs, derivative requirements and interface requirements based on the specific physical architecture will be generated, which helps to determine the product design baseline.

During the allocation process above, derivative requirements will be correspondingly generated and they are of uncertainty, for they are usually determined according to the architecture scheme of its sub-layer product. For example, when the flight control hydraulic system goes to actuating actuators, it is necessary to define the travel requirement of the actuators. Therefore, it is generally considered that derivative requirement is a requirement that depends on the solution. The derivative requirement needs to be evaluated and analyzed by taking the decomposition process of the upper layer into consideration to determine the rationality and influence of its existence.

To sum up, the requirements decomposition process in the horizontal dimension can be simply described as the following four allocation steps:

(1)   The parent non-functional requirements to functional requirements;
(2)   The functional requirement sets to the child functional architecture, and then some new requirements are derived;
(3)   New requirement sets derived from the functional architecture to the child logical architecture, and then some other requirements are derived;
(4)   The derived requirement sets from logical architecture to the child physical structure, which will also generate some new requirements.

In this process, it is also necessary to combine the requirements derived from different architecture to eliminate conflicts and redundancy. Therefore, the requirements traceability analysis is an effective technique to discover requirements contradictions and overlaps.

2.2   Top-Down and Bottom-Up Tracing Relationships

The requirement evolution process is a complex one of "top-down decomposition and transmission" and "bottom-up derivative trade-offs". It is not always feasible to establish a complete traceability of top-down requirements, but it can ensure that the source of each requirement is clear and can enable high-level requirements to be transmitted along with the decomposition of aircraft architecture. The derivative requirements arising from architectural trade-offs (for example, that an off-the-shelf product is used to realize a certain function of the upper product is a bottom-up transmission process of requirement) cannot be directly traced to its upper layer requirements, but corresponding design decisions can be found. These design decisions themselves can be traced back to upper-level requirements. Therefore, to a certain extent, derivative requirements can be traced to upper-level requirements indirectly. In this process, it is necessary to confirm that there are no conflicts between the lower-level and upper-level requirements.

3 Itemized Requirement Traceability Analysis Method

3.1 Perspective of Requirement Traceability

In the aircraft development process, as the product life cycle advances, traceability relationship will be established between requirements and other design results, mainly including:

(a) Requirement-Source (RS): the traceability relationship between requirements and the stakeholders or documents that put forward the requirements.
(b) Requirement-Basis (RB): The traceability relationship between requirements and the reasons to propose those.
(c) Requirement–Requirement (RR): the traceability relationship between requirements and other requirements, and this is a two-way relationship.
(d) Requirement-Architecture (RA): the traceability relationship between requirements and the next-level system architecture elements. Different subsystems are developed by different suppliers, so such traceability relationship is very important.
(e) Requirement-Design (RD): The traceability relationship between requirements and specific software and hardware components, which are used to achieve requirements.
(f) Requirement-Interface (RI): The traceability relationship between requirements and external system interfaces which are also the source of requirements.
(g) Requirement-Verification (RV): The traceability relationship between requirements and bottom-top integrated verification of a system.

These traceability relationships reflect different requirement traceability views and can be used for requirement traceability analysis from different perspectives.

3.2 Types of Requirement Traceability

The traceability of requirements down to the lowest level can assist in analyzing whether each requirement comes from or is related to the expectations of stakeholders. If the requirements are not allocated to a lower level or are not achieved at a lower level, the design goal will not be met, which leads to invalid design. Conversely, if the lower-level requirements cannot be traced to the upper-level ones, it will lead to over-standard design. In aircraft development, the traceability relationship of requirement-related elements covers satisfaction, interface, compliance, derivation, verification, etc. It will be shown in detail as follows:

(a) Requirement satisfaction relationship: It means that the parent requirement is decomposed into several child requirements, and the execution of child requirements will trigger the achievement of the parent requirement. This is the widely used type of relationship.

(b)   Requirement interface relationship: It indicates the interface constraint rela-
      tionship between the requirement items composed of different products.
(c)   Requirement compliance relationship: It indicates the compliance relation-
      ship of the requirement item to the terms of standards, regulations and other
      documents.
(d)   Requirement derivation relationship: It is used to identify derivative require-
      ments (i.e., a solution-based requirement). For establishment of a derivative
      relationship, it is necessary to carry out an impact analysis of the derivative
      requirement and define its influence on other requirements.
(e)   Requirement verification relationship: It indicates the relationship between a
      requirement item and its requirement verification.

3.3   Representation of Requirement Traceability

Generally, there are four ways to express the requirement traceability relationship
[5, 6].

a.   Matrix representation. The rows and columns of the matrix represent the trace-
      ability relationship between different requirements. This method is found to
      have the following problems:

      (1)   The requirement description in both directions of the matrix is too long
            and the display effect is poor.
      (2)   The traceability relationship in the matrix is usually sparse, and space
            waste will be made.
      (3)   Multi-layer traceability needs to be represented by multiple separate
            matrices, which is poor in efficiency and overall display.

b.   Link notation. Based on the Web environment, the requirement traceability
      relationship can be expressed in the form of hyperlinks. But the problem is as
      follows:

      (1)   The link object at the other end must be made visible to ensure that the
            link is valid.
      (2)   When the hyperlink is deleted at one end, it is difficult to find out in time,
            and the link is prone to be dangled.
      (3)   It is difficult to conduct global requirement tracking analysis.

c.   Logical operation notation. Use logical AND, OR, and NOT to express require-
      ment link relations, also known as requirement And/or tree. For example, R1
      can be decomposed into R11 and R12, or decomposed into R13, R14, and R15.
      The logical representation has the following defects:

      (1)   The expression method suits for the computer processing mechanism, but
            it is difficult to express intuitively.

(2) For a simple traceability relationship, it is inevitably too complicated to use, and the operation is not convenient enough.

d. Graphical representation. Graphical representation is a relatively ideal way to represent the traceability of requirements, and the traceability relationship is represented by visual graphical links (Fig. 2). Its characteristics are as follows:

(1) Create links between requirements to describe the corresponding traceability relationship.
(2) Delete the links between requirements in a controlled manner.
(3) Simultaneously observe the text (or other attributes) of the descriptions at both ends of the selected relationship.
(4) Do Coverage analysis to show requirements that are covered and not covered by the selected relationship.
(5) Perform single-layer or multi-layer impact analysis to show the set of affected requirements.
(6) Perform single-layer or multi-layer export analysis to show the original set of requirements.
(7) Perform upward and downward retrospective relationship analysis to show the objects covered and not covered by the selected relationship.

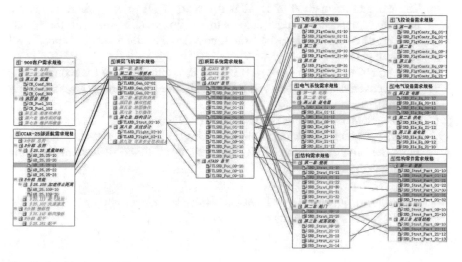

Fig. 2 Graphical traceability

4 Quantification and Evaluation Method of Itemized Requirement Traceability

4.1 A Quantitative Models of Requirements Traceability (QTMR)

The complex traceability relationship between requirements can be analyzed and evaluated using quantitative indicators, and the traceability model formed is as follows.

$$QMRT(RN) = \begin{cases} RCR(RN), \\ RTDD(R_i), R_i \in RN \\ RTGD(R_i), R_i \in RN \end{cases} \tag{2}$$

(a) Requirement Cover Rate (RCR)

RCR is a partial quantitative index of a collection of requirements items, which represents the degree of coverage of the traceability relationship to the next level, and is used to measure the completion of the traceability relationship between the requirements of this level and the next level.

The calculation method of RCR is the ratio of NI to M1, where N1 is the number of upper-level requirements that a link relationship with the lower-level requirements can be established and M1 is the total number of upper-level requirements.

That is, $RCR = \frac{N_1}{M_1}, 0 \leq R \leq 1$. If $R = 0$, it indicates that the requirements are not covered at all; if $R = 1$, it indicates that the requirements are completely covered.

(b) Requirement Traceability Depth Degree (RTDD)

RTDD is a global indicator oriented to a single requirement item, which represents the number of layers that it extends upward and downward from the requirement item [6].

$$RTDD(R_i) \begin{cases} D_{down}, D_{up} \\ D_{down} = 0, 1, 2, 3, \ldots, N \\ D_{up} = 0, -1, -2, -3, \ldots, -N \end{cases} \tag{3}$$

whereas,

(1) D_{down} indicates the number of layers that R_i extends downward.
(2) D_{up} indicates the number of layers that R_i extends upward.
     In aircraft development, the levels of requirements decomposition will not be so many, and usually $N \leq 10$, where N represents the total number of levels of requirements.

(c) Requirement Traceability Growth Degree (RTGD)

RTGD is also a global indicator for a single requirement item, which indicates the number of requirement items that can be traced upward or downward, so it is also called the traceability width [6]. When traced downward, it is called the out-degree of requirement and takes a positive value; when traced upward, it is called the in-degree of requirement and takes a negative value.

$$RTGD(R_i) = \begin{cases} (G_{out}, G_{in}) \\ G_{out} = 0, 1, 2, 3, \ldots, N \\ G_{in} = 0, -1, -2, -3, \ldots, -N \end{cases} \tag{4}$$

whereas:

(1) $G_{out} > 0$ indicates the growth of Ri to the next layer. $G_{out} = 0$ indicates that Ri is the bottom requirement, or that Ri is not decomposed and allocated;

(2) $G_{in} < 0$ indicates the retrospective relationship between Ri and the requirements of the upper layer, that is, reverse growth. $G_{in} = 0$ indicates that Ri is the bottom requirement, or that Ri is not decomposed.

In Fig. 3, Ra is met by a requirement of its next layer, $RTGD(R_a) = (1, 0)$. In Fig. 4, the requirement R_b is satisfied by the next five requirements, $RTGD(R_b) = (5, 0)$. Then the following conclusions can be got through relative analysis:

(a)   The requirement R_a is decomposed into one child requirement. It is necessary to consider whether Ra1 is the requirement of this layer.

(b)   The requirement R_b is broken down into multiple child requirements. The quality of its definition might not be high.

Fig. 3 G_{out} is 1

Fig. 4 G_{out} is 5

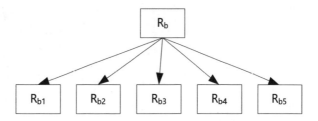

(c)   The requirement R_b is more complex in nature than Ra and requires special attention.

(d)   The impact of changing R_b is greater than changing Ra, which requires attention.

4.2   Evaluation of Requirement Traceability

The quantitative model of requirement traceability can assist designers in quantitative analysis and evaluation of requirement decomposition and distribution, and it can help evaluate and predict the impact of requirement changes.

The rules applied are as follows:

(a)   Apply RCR to assess the completion of a set of requirements to establish a traceability relationship. The closer the RCR is to 1, the more complete the traceability relationship is.

(b)   Use RTDD to assess the importance of requirements. Generally, the smaller D_{down} is, the lower the level of requirement and the smaller the scope of influence during changes. On the contrary, the larger the D_{down} is, the higher the level of needs and the larger the scope of influence during changes. The analysis should be focused.

(c)   Use RTGD to evaluate the granularity of requirement decomposition and distribution. Usually, if G_{out} is too large or too small, it means that the quality of requirement decomposition and distribution is not good. If G_{in} is too large, it means that the granularity of requirement is too large and should be further subdivided.

Figures 5, 6, 7, and 8 show four typical requirement decomposition and allocation examples

In example 1, $RTGD(R_{11}) = (5, -1)$ indicates that the granularity of R_{11} is too large and is decomposed into multiple requirements.

Fig. 5 Example 1 of requirement decomposition and allocation

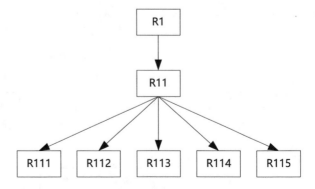

Fig. 6 Example 2 of requirement decomposition and allocation

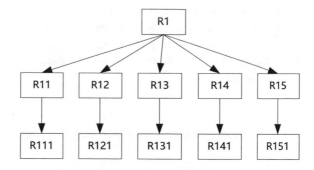

Fig. 7 Example 3 of requirement decomposition and allocation

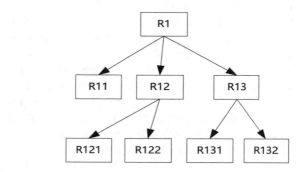

Fig. 8 Example 4 of requirement decomposition and allocation

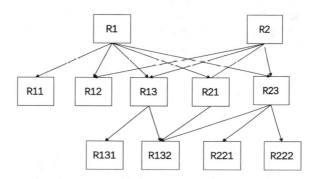

In example 2, RTGD(R_{11}) = $(1, -1)$, RTGD(R_{12}) = $(1, -1)$, RTGD(R_{13}) = $(1, -1)$, RTGD(R_{14}) = $(1, -1)$ and RTGD(R_{15}) = $(1, -1)$ indicate that the granularity of R_{11}, R_{12}, R_{13}, R_{14} and R_{15} is too small and they should be processed together.

In example 3, $RTGD(R_{11}) = 0, -1$ means that R_{11} has not been decomposed. $RTGD(R_{12}) = 2, -1$ and $RTGD(R_{13}) = (2, -1)$ mean that the decomposition and distribution of R_{12} and R_{13} are more reasonable.

In Example 4, $RTGD(R_{132}) = (0, -3)$ indicates that B comes from multiple parent requirements and requires special attention.

In addition, for example 1, R11 may be positioned as too high, and it should be grouped to its next layer. For example 2, R11~R15 may be located too low and it should be affiliated to its upper layer. Of course, too high or too low here is a comparative statement, and such positioning imbalance can be digested and coordinated through its next level.

4.3 Balance Analysis of Requirement Decomposition

The requirement decomposition and distribution balance ensure the stability of the requirement at each level in the evolution process, and the reasonability of granularity of the requirement in the traceability control process, and satisfaction of requirements for plan design and control requirement of subsequent verification process. The granularity of requirements will be an important factor in the control of requirements. The appropriate granularity of requirements will provide an extremely important consistency guarantee for module design. Therefore, the modularization of aircraft development is the modularization of the entire process of aircraft development, covering modular design, modular management, modular production task decomposition, modular verification and analysis, modular maintenance support, and modular Life cycle processes such as scrapping and reuse (Fig. 9).

At the same time, RTGD can be represented with histogram. The lower part of the horizontal axis represents the upper-level input of the requirement, and the upper part of the horizontal axis represents the lower-level decomposition of the requirement. This can make evaluation of analysis of the granularity of requirement decomposition in a more intuitive way.

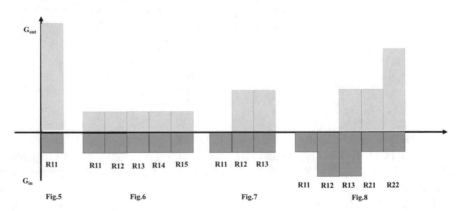

Fig. 9 Example of RTGD histogram

5 Conclusion

With the application and promotion of itemized requirement management for aircraft development, a large number of itemized requirements are continuously decomposed with the gradual decomposition of aircraft development levels, and a rich and complex traceability relationship has been established. The quantitative analysis and evaluation method for itemized requirement traceability proposed in this paper can assist designers to better identify and analyze requirement decomposition granularity, requirement transmission route, and requirement definition quality status, and this has a significant role in improving the refined level of itemized requirement management.

References

1. Liang, D., Kanwang, L.: Overview of requirement management in Airbus Industries. Aeronautical Sci. Technol. **10**, 8–13 (2017)
2. Liang, D., Juntang, L.: Discussion on itemized requirement management in aviation product development. Aeronautical Sci. Technol. **11**, 5–9 (2017)
3. INCOSE: System Engineering Handbook, vol. 4, pp. 465–470. China Machine Press, Beijing (2017)
4. Voros, R.E.: Planning for the application of ARP4754A for new and modified aircraft projects with new, simple, and reused systems. SAE Int. J. Aerosp. **8**(1) (2015)
5. Alexander, I.: Requirements Management with DOORS A Success Story [EB/OL]. https://easyweb.easynet.co.uk/~iany/index.htm
6. Elizabeth, M., Hull, C., Jackson, K., Jeremy, A., Dick, J.: Requirement Engineering, pp. 110–115. Springer Press (2004)

Acknowledged SoS Architecture Design Based on Agent-Based Modelling

Qing Shen and Lefei Li

1 Introduction

In today's interconnected word, System of Systems (SoS) is widespread. A single system can hardly meet all needs in different scenarios without the collaboration of other systems. Therefore, at the SoS Architecture design phase, it's important to consider the interaction of SoS and Constituent Systems (CS), in addition to the design requirements of a single constituent system itself. SoS can be classified into four types by the degree of managerial control of SoS [1]. The Acknowledged SoS is one of them and it commonly exists in reality, like Transportation SoS, Military SoS. Acknowledged SoS Architecture has recognized goal and focus on the interaction and collaboration of SoS and CSs to achieve these SoS-level common goals. Therefore, the research on Acknowledged SoS Architecture design is conductive to explicate the interaction impact between CSs and SoS.

There are already several architecture frameworks dedicated to solving SoS architecture design problems, e.g. Zachman Framework, DoDAF, TOGAF, C4ISR and many more [2]. However, these models are mainly used to display static information at top-level design phase, which is hard to synchronize performance on different operational scenarios [3]. Based on this limitation, it's difficult to involve real operational data to evaluate the effectiveness of different Architecture Design. Some scholars propose executable model to study interaction and evolution of SoS [4, 5]. Darabi comes up with the SoS architecture conceptual model based on Agent-based Modelling (ABM) [6]. In general, SoS is composed of many systems in different domains and operational scenarios are complex and changeable. Therefore, considering the implementation difficulty and practical application value, there are larger advantages on mixing Agent-Based Modeling into traditional architecture structure modeling. The method of modeling and simulation is more suitable to capture the

Q. Shen · L. Li (✉)
Tsinghua University, Beijing, China
e-mail: lilefei@tsinghua.edu.cn

© The Author(s), under exclusive license to Springer Nature Switzerland AG 2021
D. Krob et al. (eds.), *Complex Systems Design & Management*,
https://doi.org/10.1007/978-3-030-73539-5_9

emergence of SoS. However, there are few research to provide a comprehensive perspective on SoS architecture framework and SoS operational scenario modeling. In this paper, we focus on Acknowledged SoS architecture design problem, seeking to provide a set of process to model real Acknowledged SoS cases, abstract SoS architecture model to explicate interaction between CSs and SoS, build exactable model to analyze the impact of constituent system motivations and evaluate effectiveness of alternative architectures.

The rest of this paper is as follows: Sect. 2 introduces related work on SoS architecture framework and executable modelling. Section 3 describes the set of process to model Acknowledged SoS with the case of Epidemic Prevention and Control SOS. Section 4 provides the simulation results to evaluate alternative architectures. Section 5 concludes our work and summarize the future research.

2   Related Works

2.1   SoS Architecture Framework

The architecture in a SoS is used to designate how the CSs coordinate in different operational scenarios [7]. It includes the definition of cooperation mechanism among CSs, their functions, date flow, communications between CSs and SoS. Architecture modeling is a process to describe the architecture development through a series of architecture models in different perspectives [8]. Architecture framework provides unified terms and rules to guarantee consistent understanding of architecture integration.

Over the past few decades, there has been a number of architecture works and the earliest and the widely accepted one is Zachman framework [9]. It uses a 2-dimension matrix to describe six perspectives (what, how, where, who, when, why) from viewpoints of five stakeholders (planners, owner, designers, implementers, subconstructors). It emphasizes comprehensive presentation of systems regardless of the guidance on implementation process, which has no compliance rules to guide development process [10].

Besides, the research on architecture framework in the last two decades is driven by the U.S. department of defense, which has released a series of C4ISRAF (Command, Control, Communications, Intelligence, Surveillance, and Reconnaissance Architecture Framework) and DoDAF (Department of Defense Architecture Framework) [11]. C4ISRAF and DoDAF are also regarded as benchmark of other architecture in the military field, such as MODAF (the Ministry of Defense Architecture Framework) and NAF (the NATO Architecture Framework). Take DoDAF as an example, it includes three views in operational, system and technical standards to show a detail explanation. However, it lacks description of development and maintenance. Therefore, DoDAF is more like the statement on the final product.

TOGAF (The Open Group Architectural Framework) was developed in 1995 [12–14]. It focuses more on the architecture developing principles. The key element of TOGAF is Architecture Development Method (ADM) which provides guidance on step-by-step process for development. Therefore, it's more applicable at the architecture design phase. In addition, TOGAF emphasizes information management, which supports more on the study of the communication mechanism of SoS.

2.2 Executable SoS Modeling and Simulation

Our final goal is not only to establish different perspective representation by the architecture framework mentioned before, but to improve interoperability mechanism among CSs by the analysis of standardized architecture modeling and guarantee performance of desired functions [15]. Although these architecture frameworks are able to describe SoS structure and behaviors, the information is static and hard to help the verification of the dynamic interactions [3]. Therefore, an executable model is needed to simulate the behavior change in SoS evolution process, creating the bridge of architecture design level and operational level to evaluate whether effectiveness of SoS is in line with expectations [16]. Acheson et al. proposes to use wave model to simulate the evolution of the SoS development process [4]. Based on this, Acheson proposes that the Acknowledged SoS can be modeled by the Object-Oriented Systems Approach (OOSA). She points that the objects here can be regarded as the agents in ABM [5]. Since UML is the earliest tool used for software development, there are many UML-based extension tools to support the transformation of static model into executable model. However, these modeling applications are limited because there's no standardized semantics [17]. Levis established a three-process (architecture design, executable model modeling, and architecture evaluation) framework, which transform DoDAF model into a colored Petri-net executable model for architecture design and evaluation [18]. Huynh comes up with a method of using SysML modeling to achieve system analysis based on the idea of system engineering. The main process includes SysML modeling, using extension tools to convert SysML models into executable simulation models, and analyzing simulation results by design of experiments [19]. Dagli further proposes a case of an executable architecture model based on SysML modeling [17]. In addition, Zinn tries to transform DoDAF model into agent-based models. After that, many scholars tries to deal with complex adaptive system problems by ABM [20].

ABM is a modeling and simulation method which uses autonomous agents with complex interaction behaviors to describe complex systems [21]. ABM emphasizes the autonomy of agents, the impact of interaction among agents on system emergence.

An agent-based model is composed of individual agents which have their own operational state and behavior rules. We can monitor the status and interactions when running the model. Compared with other modeling technologies, ABM has the following characteristics which is appropriate for SoS modeling [22]: (1) ABM can capture the emergence of complex system. Emergence refers to the macroscopic behavior of the system resulting from individual interactions, which doesn't have when the system are broken down into components. It's difficult to predict this phenomenon. ABM is a standardized method for modeling emergent phenomena. It describes the macro behavior of SoS by modeling and simulation of behavior of components and interaction between them. (2) ABM supports modeling and simulation of initiative behavior. ABM provides an effective description of individual initiative. Agents can receive the stimulus from the environment and other agents, then they modify their own rules and states according to the internal processing mechanism. Based on advantages above, ABM is regarded as an effective way to solve complex adaptive system problems. It's also one of the hotspots in research field. Darabi proposes a conceptual model of SoS architecture based on ABM, focusing on the problem of collaboration and interaction between CSs. He also comes up with a CS negotiation model based on the effectiveness theory [6]. Agarwal uses ABM to verify the influence of the lack of one single constituent system on the whole SoS [23]. Nigam et al. evaluate the robustness and flexibility of SoS by ABM [24]. In addition, ABM is applicable to system modeling in different field. Kewley simulates the military war scenarios and analyze the effects of non-centralized military command and control by ABM. It can also predict the impact of different tactics in a given scenario [25]. Xu evaluates the effectiveness of cooperation policy that shares doctors and beds in urban health management by ABM [26]. Dekker studied on the organizations composed of members from different cultural backgrounds, such as the NATO organization [27]. He uses ABM to analyze the military strength difference among different military configurations. This model designs a set of schemes to evaluate the performance of different organizational structures. These papers focus more on the influence of shared-information accuracy and timeliness in the same discipline fields and the experiment often specializes in a particular area. We try to involve more shared information by the changes of SoS architecture and evaluate effectiveness of improving interoperability among CSs.

3 Modelling Acknowledged SoS Architecture

We define the modelling process of Acknowledged SoS architecture design as five steps, as shown in Fig. 1. First, we need to define and abstract typical Acknowledged SoS characteristics in Real Case. Second, we try to build a conceptual model to explicate relationship between components of Acknowledged SoS, especially interaction interface to help subsequent analysis. Third, the Acknowledged SoS Architecture and operational scenario are specified by The Open Group Architecture Framework (TOGAF). Fourth, we try to transform it into executable model by Agent-based Modelling. Then, in the end, alternative architectures can be evaluated by simulation. To make our point, we use Epidemic Prevention and Control SOS (EPC-SoS) as the real case to explain each of these steps in detail.

Fig. 1 Modelling process of acknowledged SoS architecture design

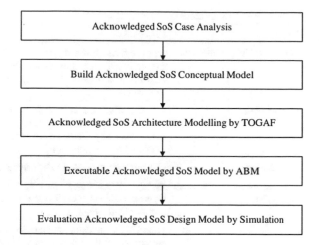

3.1 Acknowledged SoS Case Analysis

At first, we need to find the eligible real case according to Acknowledged SoS characteristics. Even though there's no absolutely precise criterion to define SoS, lots of literature provide a rich set of SoS prosperities' description. *Nielsen* et al. proposes eight dimensions from these literature to help analyze SoS modelling patterns [1]. In this paper, we will adopt them as the SoS criterion. They are Autonomy of Constituents, Independence, Distribution, Evolution, Dynamic Reconfiguration, Emergence of Behavior, Interdependence and Interoperability. Besides, we should guarantee the collaborative management in SoS-level, while CSs still keep managerial and operational independence. In this paper, we regard the SoS for Epidemic Prevention and Control as EPC-SoS real case. The following will show why we use it and how it meets the standard of Acknowledged SoS.

The epidemic has the characteristics of high harmfulness, fast propagation and wide distribution, which is difficult to prevent and control. Unlike other diseases, once the epidemic has been declared, there must be shortage of regional medical resources, including specialist and medical supplies. Health system, transportation system, community system should be all involved to execute infection control, measures such as reducing people flow, isolating clos contacts, etc. It follows that the single regional health system is hard to control the spread of epidemic by itself. Therefore, the prevention and control of epidemic cannot be solved by an individual system. Table 1 shows our 8 SoS dimensions analysis of EPC-SoS. In addition, we regard Government as SoS Manager who is responsible for resources and authority at SoS level, while the constituent systems keeps their own independent control. EPC-SoS can be regarded as Acknowledged SoS. We clarify concrete SoS components and their interactions in the following step.

Table 1 8 SoS dimensions analysis of EPC-SoS

Dimension	EPC-SoS
Autonomy of Constituents	SoS only care about services contribute to recognized goals. For example, Health System would organize clinical training, periodic inspection to improve service quality, which have no need to be reported
Independence	CSs needn't to report how they do and only show result and changeable status to others. For example, Health System notices the number of patients and available beds, and they have no need to show how to allocate it
Distribution	CSs are dispersed, so the information needed to be shared are put on the infrastructure provided by government
Evolution	Operational logs at each step for each CS will be recorded to monitor state changes
Dynamic Reconfiguration	Each CS has decision mechanism and their decision will chage the structure of SoSwithout planned intervention
Emergence Behavior	Emergent behavior can be evaluated by SoS-level goal. For example, the goal for decreasing the number of infected can only be realized by allocation of Health System and Community System
Interdependence	For some SoS-level service, more than 2 CSs will be involved to complete the task. Operational logs help to trace dependency and links between CSs
Interoperability	In order to better coordinate with each other and conform to SoS manager. CSs have consistent semantics to define communication interface

3.2 Build Acknowledged SoS Conceptual Model

In order to provide a holistic view of SoS and facilitate analysis, we need to build a conceptual model to represent Acknowledged SoS. Note that the model is mainly used for analysis at design phase, not for development and implementation. It should emphasize more on function and information architecture, also highlight interactions between CSs and SoS which will be our research focus. Baek et al. proposed M2SoS metamodel for SoS. The key ontology in our conceptual model is a subset of M2SoS [28]. However, in order to build the model connection with the architecture model by TOGAF and executable model by ABM, we modify it as needed, which is shown in Fig. 2.

In this conceptual model, there are four major component units, SoS-level Target Problem, SoS-level Organization, SoS-level Environment and Constituent Systems. Then entities in SoS-level and CS-level will be introduced. Table 2 shows the components using EPC-SoS case.

In addition, we pay more attention to the relationships between SoS-level and CS-level entities. The Acknowledged SoS manager coordinates common SoS-level goals with CSs, which satisfy requirements required by SoS-level stakeholders. In order to achieve SoS goals, SoS need a set of capabilities which can be demonstrated

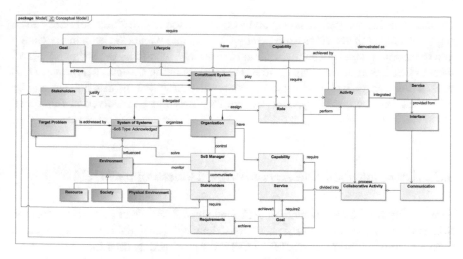

Fig. 2 Conceptual model to analyze acknowledged SoS

Table 2 Elements of conceptual model in EPC-SoS case

Component units of Acknowledged SoS	
SoS-level Target Problem	Prevent and control epidemic
SoS-level Organization	Epidemic Prevention and Control SoS
SoS-level Environment	Epidemic situation, investment in medical resource, people, policy
Constituent Systems	Health System Transportation System Community System
SoS-level Entity	
SoS-level Goal	Protect public health and safety through the control and prevention of epidemic
SoS-level Service	Prevention; quarantine; medical care
SoS-level Role	Health System: researcher, therapist Transportation System: freight dispatcher, passenger dispatcher Community System: administrator
SoS-level Manager	Government
CS-level Entity (take Health System as example)	
CS-level Goal	Decrease fatality, allocate more medical resource in time with the help of government
CS-level Service	Patient information service, clinical service, medical information service, research service, surveillance service
CS-level Activity	Surveillance, diagnosis, triage, treatment
CS-level Environment	Medical supplies, medical staff, hospital

by SoS-level services which achieved by collaboration of CSs. Each services can be regarded as a set of activities executed by CSs. We know that CSs keep management and operation independence. In order to clarify CSs' functions, we assign SoS-level roles to CSs based on their capabilities, the corresponding CS-level services are also broken down from SoS-level services. It's worth noting that information sharing by other CSs may be necessary, even though they may not be involved in functional level.

3.3 Acknowledged SoS Architecture Modelling by TOGAF

After clarifying the components of Acknowledged SoS. We need to know specific SoS architecture and operational scenarios to help build executable model. The Open Group Architecture Framework (TOGAF) is the most widely used Enterprise Architecture today and contains all needed pieces to guide any step of business. In addition, it specifies information management which is in favor of system interaction study. For these reasons, we choose TOGAF to support the following research. One key element of TOGAF is the Architecture Development Method (ADM), it shows step-by-step process to develop an enterprise architecture, Since we focus on the part of SoS architecture design, we only go through first three phases: Phase A: Architecture Vision identifies SoS-level goals. Phase B: Business Architecture describes specific services and roles of CSs to support SoS-level services. Phase C: Information Systems Architectures describes information interaction mechanism and shared date between CSs and SoS. We still use the case of EPC-SoS to elaborate each phase concretely and the related TOGAF diagrams are drawn in Cameo Systems Modeler. Figure 3 shows the specific phases of ADM, we go through the first three phases and represent dependencies between them.

At Phase A, we clarify the goal of EPC-SoS, which is to protect public health and safety through the control and prevention of epidemic. It includes two sub-goals, control the number of people with confirmed epidemic at first and then reduce fatality. Several SoS-level services are needed to meet each sub-goals, so we further break them down. The final goal tree diagram is represented in Fig. 4.

At Phase B, we should identify SoS organization by selecting CSs which have capabilities required by goal tree and clarify services they provide. Therefore, The business architecture diagram showed in Fig. 5 includes three parts: SoS organization diagram, SoS-level services diagram and CS-level services assignment diagram. We introduce them one by one. The SoS organization shows specific CSs and their roles in SoS. Take EPC-SoS as example, the SoS manager is Government, which coordinate CSs in SoS-level. There are three CSs, Health System are assigned roles of researcher and therapist, its environment objects are also shown to help with subsequent analysis. Transportation System take the roles of freight dispatcher and passenger dispatcher. Community System plays the role of administrator to do daily management. Then we abstract three SoS-level services performed by CSs: prevention service, quarantine service and medical care service. In the SoS-level services

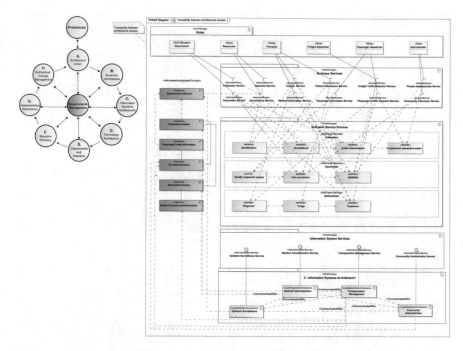

Fig. 3 Architecture development method overview

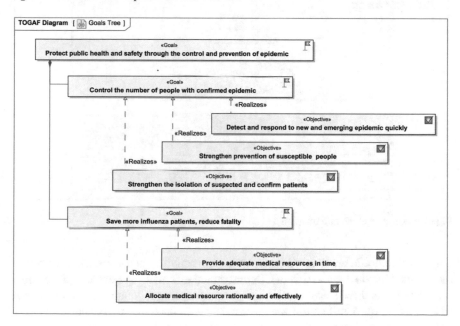

Fig. 4 Goal tree diagram

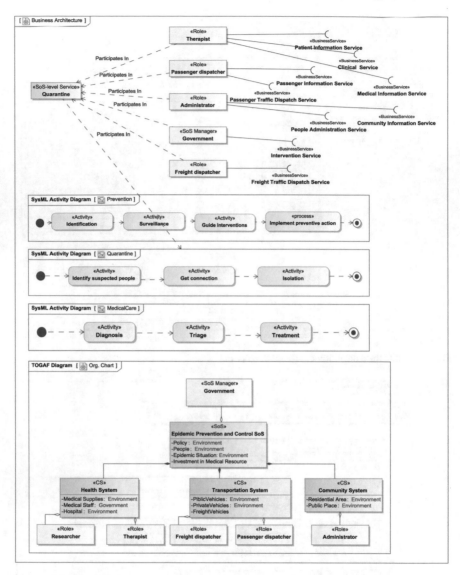

Fig. 5 Business architecture diagram

diagram, we should break down each service into activities because each activity may by executed by the cooperation of CSs and shared information. For example, quarantine service has three steps: identify suspected people, get connection and isolation. In the process of identification, we are not only with the help of virus detection technology, available resource for detection, but also shared travel information to screen high-risk populations. Therefore, we dig into each process in quarantines, clarify CS-level service needed.

At Phase C, we complete key element of the communication to connect SoS and CSs. We regard it as Information Architecture in Fig. 6. CSs provide not only business service but also integrate information service to communicate with others. The key point is to define what data needed and who can use it. Correspondingly, we set a series of data tables and use connection between CSs to simulate data flow which forms the CS-level services. We can see the whole frame in Fig. 3.

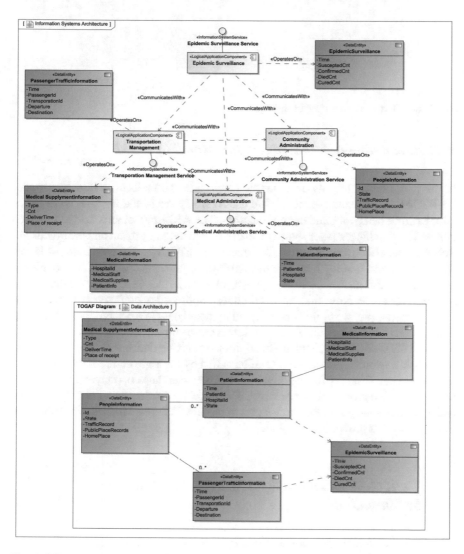

Fig. 6 Information architecture diagram

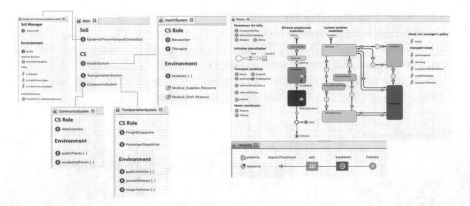

Fig. 7 EPC-SoS model interface in anylogic

3.4 Executable Acknowledged SoS Model by ABM

In this part, we aim to provide a mapping approach to transform architecture model by TOGAF into executable model. We use Agent-Based Modeling (ABM) to simulate dynamic behaviors and communication. The ABM model are divided into two layers. The first higher layer shows the basic organization by agent and the quantified SoS goals as effectiveness evaluation indexes. In EPC-SoS, the core index is infection number and fatality rate. The second layer represents specific CS operational environment. For each constituent system, we assign roles to agent and define environment variables. Note that we need choose appropriate component to represent them. For example, we build Person Agent to model people's behavior and operating status in SoS environment. Decision variables can be represented by executable event to model CS decision mechanism. In addition, we create data based on data tables in Information Architecture and they can be called by message communication mechanism between CSs. As for the collaboration between these two layers, we transfer business service process diagram into ABM model by discrete event library modules and agent message ports. We show a part of basic ABM model interface of EPC-SoS in Fig. 7 and it's drawn in Anylogic.

3.5 Evaluation Acknowledged SoS Design Model by Simulation

Considering different operational scenario, each SoS-level service has a different set of CS-level services enabled in functional level, each CS-level service has a different set of approaches to complete by different cooperation ways in business-realization level. Therefore, alternative architecture solution can be extended at the design phase. With the help of ABM simulation, we can evaluation SoS effectiveness of alternative architectures, which can not only facilitate evaluation of CS contribution in global but also optimization of architecture design with cost-benefit analysis.

In the next section, we will set up a set of alternative architectures of EPC-SoS and evaluate their effectiveness through simulation.

4 Modeling and Simulation Results of EPC-SoS Architecture

4.1 EPC-SoS Architecture Model Description

In this illustrative example, we are working on anti-infection measures of EPC-SoS in epidemics. At fist, the basic epidemic spread model in a small city should be built. We apply SEIR model to capture the progression of epidemic. Suppose there is a small city, a population of 100,000 people. 2–5 people form a family. Everyday, they go to company or other public places in the morning and go back to home in the evening. They have 70% probability to take public transit. Besides, there is initially totally 500 beds in hospital. There's no need to study geographical distribution of medical services and the influence of traffic resource. So we simplified model in the following ways: use the number of beds to represent the available medical resource and public traffic resource is ample all the time.

As for epidemic spread simulation, each agent is assigned to a disease state. (1) susceptible agents which can contract epidemic. (2) exposed agents which are in latency stage. (3) infectious agents which can transmit virus but show no symptoms. (4) symptoms agents which can also transmit virus and have symptoms. (5) died agents which are not able to recover after illness duration, (6) immune agent which have recovered and will not be infected again. Figure 8 shows the ABM model of a person in EPC-SoS. We use two state charts to represent disease progression and current location separately.

Next, we try to compare the effectiveness of different epidemic prevention measures. Suppose at first there are 5 people has been in infection status.

Scheme 1: no extra intervention measures
People go to work or other public places as normal. When they go into symptoms status, they will go to hospital if their disease concern is more than they can bear. We suppose there are 30% probability to be severe and others are mild. Once the value of it is more than the people can stand, they will go to hospital. If there are no enough beds for treatment, they will go back home. Note that the illness duration and fatality rate is different according to the severity and whether in hospital.

Scheme 2: isolate connected people of confirmed patients
One of the key element is to isolate infectious in time. With the help of big data systems, we assume that once a person has been diagnosed, the itinerary information

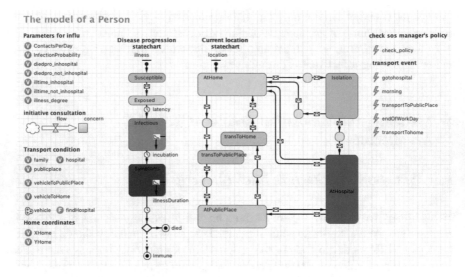

Fig. 8 The ABM model of a person in EPC-SoS

can be used to assist in finding clos contacts in his infection duration. Then these close contacts will be isolated and have no infectivity any more. We suppose the room for isolation is sufficient, when people are found in infectious state, they will be treated, and if they are not infected after quarantine period, they will go home.

Scheme 3: all people are segregated at home after a state of emergency is declared

The measure of forced quarantine is an effective way to decrease turnover rate. After clarify epidemic property, we suppose the government, as the SoS manager, has the right to require people segregate at home for a period of time. The policy can be executed with the help of administrator in community.

Scheme 4: gradually increase medical resources

In addition to the effectiveness index of the numbers of infections, we still want to decrease fatality rate. The core element is to allocate adequate medical resource to save more people. Therefore, we suppose government have the ability to mobilize certain resource to the healthcare system.

4.2 Experiments and Results

In the following part, we will introduce the operational scenarios' data in 4 schemes and show the simulation results. Some of the parameters are shown in Table 3. We suppose that if the cumulative number of people who go to hospital is more than 1000

Table 3 Parameter settings in EPC-SoS

Parameter	Value
Total population	100,000
Initially infected	5
Contact rate	10
Infectivity	0.05
Latency duration	Triangular (1, 7, 4)
Incubation duration	Triangular (1, 7, 4)
Severe illness duration in hospital	Triangular (10, 25, 20)
Severe illness duration not in hospital	Triangular (20, 30, 25)
Mild illness duration in hospital	Triangular(7,14,10)
Mild illness duration not in hospital	Triangular (10, 20, 13)
Severe illness fatality rate in hospital	10%
Severe illness fatality rate not in hospital	20%
Mild illness fatality rate in hospital	1%
Mild illness fatality rate not in hospital	2%

because of epidemic, some measures would be taken. In Scheme 2, we would isolate connected people. once the person go to hospital, the people who has connection with him in latency duration will be isolated immediately and have no infectivity later on. In Scheme 3, government would commend people to stay at home until epidemic end. In Scheme 4, government allocate more medical resource, 100,000 new hospital beds will be gradually ready after 10 days later. 7000 for mild patients and 3000 for severe patients. If there are more confirmed patient than what we can receive, they are still not being treated.

We run the model and the simulation results is shown in Figs. 9, 10, 11 and

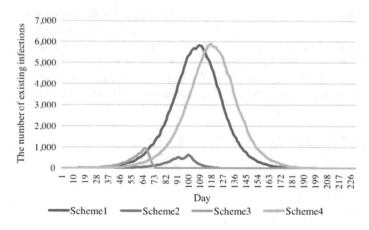

Fig. 9 The number of existing infections per day

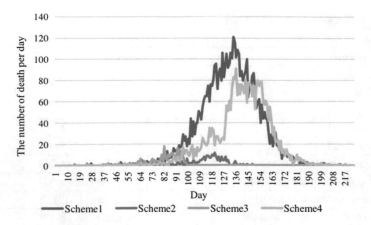

Fig. 10 The number of death per day

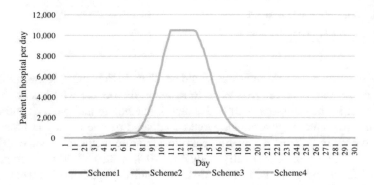

Fig. 11 The number of patients in hospital per day

Table 4. We can observe that universal compulsory segregation is a strong action to break epidemic directly, the infection rate goes down to 4.04%. In addition, if we can detect virus timely and find all of the connected people to isolate them immediately, the effectiveness of this measure is nearly the same as universal segregation and connected people can be detected earlier and be treated earlier to decrease fatality rate. However, it's hard to do such comprehensive inspection and guarantee precise

Table 4 Simulation results of EPC-SoS in different schemes

Parameter	Scheme 1	Scheme 2	Scheme 3	Scheme 4
Cumulative infections	68,023	4402	4035	67,795
Total infection rate (%)	68.02	4.40	4.04	67.80
Cumulative death	4493	220	250	2857
Fatality rate (%)	6.6	4.99	6.2	4.2

tracking of connected people. Therefore, Scheme 2 is ideal situation. In real life, the effectiveness is depended on accuracy of detection and information traceability, which can be optimized by improving collaboration of CSs or introducing new CSs with high performance. In Scheme 4, we can see that even though we have increase 10,000 available beds to treat more patients, the medical resource is still not sufficient when epidemic is exploded. The hospital is full load from Day 63 when we start to add beds gradually and it continues to Day 135. Therefore, effective preventive control measures must be combined to decrease fatality rate. It also shows that effective information sharing among CSs is crucial to improve the implementation in intervention policy.

5   Conclusions and Future Work

In this paper, we are seeking to provide a method for the analysis of acknowledged SoS, which establishes a link between top-level architecture modeling and operational scenarios simulation. The simulation result is a key element to evaluate effectiveness of alternative architectures. Specifically, we come up with Acknowledged SoS conceptual model to abstract a real case, develop high level and more generic architecture with TOGAF. A mapping approach is applied to transform architecture model into executable model by ABM. In the case study, we show a simple evaluation on the effectiveness of an Acknowledged SoS. Furthermore, as a part of future work, we will try to optimize the acknowledged SoS architecture design based on the proposed modeling method in this paper.

Acknowledgements This study was co-supported by "National Key R&D Program of China" (No. 2017YFF0209400) and "Key Laboratory of Quality Infrastructure Efficacy Research Funding" (No. KF20180401).

References

1. Nielsen, Claus Ballegaard, Larsen, Peter Gorm, Fitzgerald, John, Woodcock, Jim, Peleska, Jan: Systems of systems engineering. ACM Comput. Surv. **48**(2), 1–41 (2015)
2. Martin, R., Robertson, E.L.: A comparison of frameworks for enterprise architecture modeling (2003)
3. Schwartz, M.: Defense acquisitions: how dod acquires weapon systems and recent efforts to reform the process [May 23, 2014]. Congressional Research Service Reports (2014)
4. Acheson, P., Pape, L., Dagli, C., Kilicay-Ergin, N., Columbi, J., Haris, K.: Understanding system of systems development using an agent-based wave model. Procedia Comput. Sci. **12**(Complete), 21–30 (2012)
5. Kilicayergin, N., Acheson, P., Colombi, J., Dagli, C.H.: Modeling system of systems acquisition (2012)
6. Darabi, H.R., Mansouri, M.: The role of competition and collaboration in influencing the level of autonomy and belonging in system of systems. IEEE Syst. J. **7**(4), 520–527 (2013)

7. Maier, M.W., Emery, D., Hilliard, R.: Software architecture: introducing ieee standard 1471. Computer **34**(4), 107–109 (2001)
8. Handley, H.A.H.: Incorporating the nato human view in the dodaf 2.0 meta model. Syst. Eng. **15**(1), 108–117
9. Noran, O.: An analysis of the zachman framework for enterprise architecture from the geram perspective. Ann. Rev. Control **27**(2), 163–183
10. Sowa, J.F., Zachman, J.A.: Extending and formalizing the framework for information systems architecture. IBM Syst. J. **31**(3), 590–616 (1992)
11. Hause, M.: The unified profile for DoDAF/MODAF (UPDM) enabling systems of systems on many levels. In: Systems Conference, 2010 4th Annual IEEE. IEEE. (2010)
12. Desfray, & Philippe.: Modeling enterprise architecture with togaf. Testimonials. 237–247
13. Blevins, T., Dandashi, F., Tolbert, M.: The Open Group Architecture Framework (TOGAF 9) and the US Department of Defense Architecture Framework 2.0 (DoDAF 2.0). The Open Group. 2010. 4
14. Dandashi, F., Siegers, R., Jones, J., Blevins, T.: The Open Group Architecture Framework (TOGAF) and the US Department of Defense Architecture Framework (DoDAF). The Open Group. 2006. 4
15. Maier, & Mark, W.: Disentangling modeling, architectures, and architecture descriptions. INSIGHT **8**(2), 24–25 (2006)
16. Carlomusto, M., Giammarco, K., Lock, J.D.: Development and analysis of integrated C4ISR architectures. In: Military Communications Conference, 2005. MILCOM 2005. IEEE. IEEE (2005)
17. Wang, R., Dagli. C.H.: Executable system architecting using systems modeling language in conjunction with colored petri nets in a model-driven systems development process. Syst. Eng. **14**(4), 383–409
18. Wagenhals, L.W., Levis, A.H.: Service oriented architectures, the dod architecture framework 1.5, and executable architectures. Syst. Eng. **12**(4), 312–343 (2009)
19. Huynh, T.V., Osmundson, J.S.: A Systems Engineering Methodology for Analyzing Systems of Systems Using the Systems Modeling Language (SysML). Department of Systems Engineering, Naval Postgraduate School, Monterey (2006)
20. Zinn, A., DeStefano, Greg, Jacques, David.. 5.4.1 the use of integrated architectures to support agent based simulation: an initial investigation. In: Incose International Symposium, 14(1), 1015–1031
21. Bonabeau, E.: Agent-based modeling: methods and techniques for simulating human systems. Proc. Natl. Acad. Sci. **99**(Supplement 3), 7280–7287. (2002)
22. North M.J., Macal, C.M. Managing Business Complexity: Discovering Strategic Solutions with Agent-Based Modeling and Simulation. Oxford University Press (2007)
23. Agarwal, S., Pape, L.E., Kilicay-Ergin, N., Dagli, C.H.: Multi-agent based architecture for acknowledged system of systems. Procedia Comput. Sci. **28**, 1–10 (2014)
24. Nigam, N. et al.: Sufficient Statistics for Optimal Decentralized Control in System of Systems. In: 2018 First International Conference on Artificial Intelligence for Industries (AI4I), Laguna Hills, CA, USA, 2018, pp. 92–95
25. Kewley, R.H.: Agent-Based Model of Auftragstaktik: Self Organization in Command and Control of Future Combat Forces. wsc. IEEE Computer Society (2004)
26. Xu, X., Li, L., Wu, H.: Cooperation policy simulation in urban health care system. In: IEEE International Conference on Service Operations and Logistics, and Informatics, 2008. IEEE/SOLI 2008. IEEE (2008)
27. Dekker, A.H.: Using agent-based modelling to study organisational performance and cultural differences. In: Proceedings MODSIM 2003 International Congress on Modelling and Simulation. 2003, pp. 1793–1798
28. Baek, Y.M., Song, J., Shin, Y.J., Park, S., Bae, D.H.: A meta-model for representing system-of-systems ontologies. In: The 6th International Workshop. IEEE Computer Society (2018)

Advanced Helicopter Cockpit Ergonomic Design Concepts

Xiaowei Mu, Gongnan Li, and Yang Yu

1 Introduction

As a complex mechanical and electrical product integrating multiple systems, the initial design idea of the aircraft is very simple, which is to complete the flight tasks accurately and efficiently on the basis of safe flight. At this time, the cockpit provides the pilot with airspeed, altitude and other data through the instrument, and the pilot needs to drive the aircraft according to the instrument data and complete the flight mission. At this point, the pilot's flight load is heavy, and multiple crew members are required to complete a flight mission together [1–3].

With the great development of avionics technology, aircraft appeared a variety of electronic instruments and equipment, the relevant system design is also more complex and advanced instrument will collected a large amount of information output and display, give the pilot a variety of flight information in order to ensure flight safety, but also increase the workload of the pilot in the pilot's quality put forward higher requirements.

Since the 1970s, with the development of electronic technology and information technology, aerospace electronics integration degree unceasing enhancement, the original discrete instrumentation cockpit design is gradually replaced by comprehensive display system, aircraft installation is more advanced and more intuitive electronic data display system, digital autopilot and flight management system formed our common glass cockpit design today [4]. At this time, the aircraft only needs two pilots to complete all the routine flight missions. Helicopters have also undergone such a change, and the design of modern helicopter cockpit is developing towards a more advanced, convenient and efficient direction [5].

X. Mu (✉)
Chinese Aeronautical Establishment, Beijing 100129, China
e-mail: muxiaowei316@126.com

G. Li · Y. Yu
Hafei Aviation Industry Co., Ltd., Haerbin 150066, China

© The Author(s), under exclusive license to Springer Nature Switzerland AG 2021
D. Krob et al. (eds.), *Complex Systems Design & Management*,
https://doi.org/10.1007/978-3-030-73539-5_10

2 Cockpit Information Display Design Principles

As a human–computer interaction platform, the cockpit needs to display all effective and necessary information about the flight mission and flight safety to the pilot. From this concept, we need to divide the information into flight mission information, flight safety information and necessary information. Flight mission information is not directly related to flight safety, and such fault information belongs to warning or general information, which is not emphasized in this paper. There are three levels of flight safety information, which can be defined as red warning information, amber warning information and green general information. This definition is consistent with the information display in the helicopter flight manual. For all the useful information, including the necessary information is need to know and take steps to the pilot, or in other normal operation or fault disposal will use the information in the operation, the necessary information is not need to know the pilot, don't need to display to the pilot, this kind of information generally provide maintenance for ground maintenance personnel information [6].

The information system of the helicopter includes the information of each system. The main system of the helicopter includes rotor, tail rotor, transmission system, power system, fuel system, lubricating oil system, fire extinguishing system, environmental control system, hydraulic system, landing gear system, control system and avionics system. All these systems have the necessary information that needs the pilot's constant attention, which is transmitted to the display system through the sensor and collector and displayed to the pilot on the display terminal. The main warning level system information displayed to the pilot by modern helicopters is shown in Table 1. For the system to detect the system state and to the early warning information system [7], mostly with amber information display, because of the difference of each type of system design idea, the amber information differ in large quantities, it is hard to all content with a form, thoughtful, and this part of the information security level below red information, so this article part not the in-depth discussion. Through the simple division of information, we will have a general concept of what

Table 1 List of helicopter red warning information

Ordinal	Information	Name	Type security	Required information
1.	Rotor speed over limit	Flight safety information	Red	Necessary
2.	VNE Insuperable speed	Flight safety information	Red	Necessary
3.	Engine power turbine speed exceeding limit	Flight safety information	Red	Necessary
4.	Engine free turbine speed exceeding limit	Flight safety information	Red	Necessary
5.	Engine torque overshoot	Flight safety information	Red	Necessary
6.	Engine combustion chamber temperature overshoot	Flight safety information	Red	Necessary
7.	Engine oil pressure overshoot	Flight safety information	Red	Necessary
8.	Engine oil temperature overshoot	Flight safety information	Red	Necessary
9.	Total failure of engine control system	Flight safety information	Red	Necessary
10.	Oil shortage	Flight safety information	Red	Necessary
11.	Reducer lubricating oil pressure overshoot	Flight safety information	Red	Necessary
12.	Reducer lubricating oil temperature overshoot	Flight safety information	Red	Necessary
13.	The fire alarm	Flight safety information	Red	Necessary
14.	Generator overheating	Flight safety information	Red	Necessary
15.	Battery overheating	Flight safety information	Red	Necessary
16.	The landing gear is not down	Flight safety information	Red	Necessary
17.	Double generator disconnection	Flight safety information	Red	Necessary

kind of information needs to be displayed to the pilot, and the level of display can also be defined by Table 1.

3 Several Helicopter Cockpit Information Display Analysis

The information display of modern helicopters is displayed through the cockpit pilot display, and the size of the display directly determines the amount of red information it can display. For a display with a small display range, the designer has to cut or combine the display information reasonably, or show whether the display information exceeds the limit through a converted parameter. PI parameters such as the AW139 are calculated by converting engine temperature, torque and speed (Fig. 1).

Fault information, in addition to the above said other amber alert information system on the display system. According to the unified because of its relatively low degree of emergency, requires no additional warning, modern means of information technology makes the system design principle diagram of the system can be loaded

Fig. 1 AW139 helicopter engine parameter information display (PI)

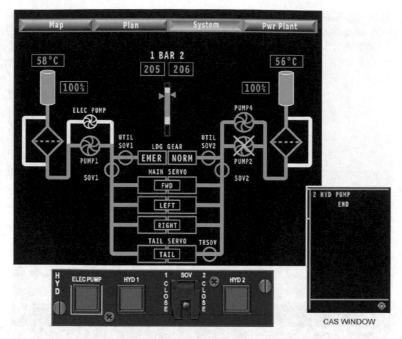

Fig. 2 AW139 helicopter hydraulic system information display

into the display system. The system failure through the display system can query to a specific fault condition, enhance the pilot's understanding of fault, fault properly disposed of. Below is the failure diagram of the display after the failure of the no. 2 hydraulic pump of the AW139 helicopter (Fig. 2).

The EC175 also takes this design concept and applies it to all systems. In the case of system failure, the schematic diagram of system failure can be viewed through the keys on the display, which makes the pilot have a more intuitive understanding and provides further guarantee for enhancing flight safety (Fig. 3).

Despite the modern means of display, display system must also meet the requirements of airworthiness regulations for redundancy. In order to save costs, on the part of the helicopter will still choose to install independent display equipment for some special parameters, including the airspeed indicator and altimeter, rotor speed table, alternate magnetic compass and standby instrument, the equipment is arranged in a position convenient for both drivers, but is closer to the pilot, this is mainly from double to fly a helicopter pilot division of labor.

Although the advanced avionics mentioned above can display all the information that can be detected by the helicopter to the pilots, modern helicopters still do not completely discard the display mode of the fault lamp indicator, as shown in Fig. 4 on the instrument panel of EC175.

As a means of fault indication, the position of fault indicator is very important. According to relevant requirements of GJB 1471A, through analysis, the maximum

Fig. 3 Hydraulic system information display of EC175 helicopter

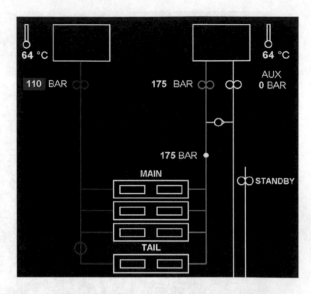

Fig. 4 Failure indication information display of EC175 helicopter

line of sight of human eye activity is ±35° for horizontal line of sight and 25° ~ − 35° for vertical line of sight. Through our analysis of the maximum line of sight of the AC312E pilot at the standard eye site, we can obtain the horizon region as shown in Fig. 5. In normal flight, the scope of pilot's horizon sight scope is concentrated in the upper part near the side of the dashboard, and a helicopter of design is to make the system of fault information is displayed in the right bottom corner of the display, such design concept is bound to affect the normal pilot's manipulation of the helicopter, at least also can increase the burden of the pilot, go against the flight safety, so the pilot's sight on the upper panel increase warning lamp board or very be necessary, it can be when the system fails to inform the pilot, and this kind of warning method is feasible and efficient, so the cockpit design concept has been used today.

However, not all fault information needs to be displayed on the alarm panel. Only the red warning information with high alarm level is displayed in this additional way. After adopting the design concept of glass cockpit, the European helicopter series will still adopt the design method of central alarm panel. Among them, the EC175 helicopter displays its main red warning message on the integrated display, gives

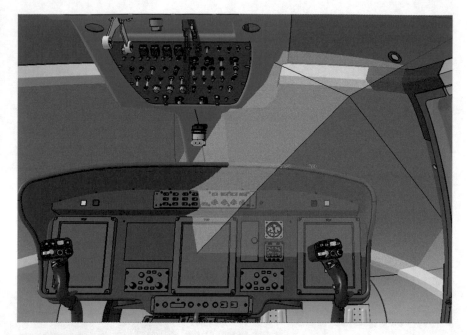

Fig. 5 The AC312E helicopter is driving the horizon at the standard eye site

an audio warning and displays it on the central warning panel. Where, the trigger conditions for ENG1(2) are ENG1(2) FAIL(engine failure) or ENG1(2) OIL LOW PRESS (engine lubricating OIL pressure is LOW) (Table 2).

For some of the more important systems, the control panel also retains the indicator light design. Such as the EC175 helicopters engine FIRE extinguishing system, when engine is on FIRE, comprehensive display have ENG FIRE warning information, at the same time the FIRE information displayed on the control panel, the execution of extinguishing FIRE program triggered a bottle, the FIRE EXTING amber 1 information displayed on the panel, and it will display amber 2 information when press it again, to alert the pilot of the corresponding FIRE extinguishing bottle is empty (Fig. 6).

The design of such devices is more concerned with the convenience of the pilot. Through the human–machine efficacy analysis of the cockpit of AC312E helicopter, it can be seen that the maximum range of the pilot's head activity is $+60°$ in the horizontal direction and $\pm50°$ in the vertical direction. The FIRE EXTING control board of EC175 helicopter is arranged in the front of the central control console, which makes it easier for the pilot to observe and operate. The AW139 helicopters also adopted the design method of the engine control board, fire board, such as landing gear control panel with display status of the control unit is decorated in the central control console front near the pilot's side, the design is still in the man–machine efficacy reasons, in order to better serve the pilot, improve flight efficiency, reduce

Table 2 List of EC175 helicopter red warning information

Serial	Name	Central alarm panel	Main message list	Audio alarm	Trigger condition
1	1 send fire	EGN 1 FIRE	ENG1 FIRE	Audio 2	Engine no. 1 detected a fire alarm
2	2 send fire	EGN 2 FIRE	FIRE ENG2	Audio 3	Engine no. 2 detected a fire alarm
3	Cargo compartment fire	CARGO FIRE	CARGO FIRE	Audio 32	The temperature detector detected a fire alarm
4	1 send failure	ENG 1	ENG1 FAIL	Audio 56	1 stop due to flameout or overturn
5	2 send failure	ENG 2	FAIL ENG2	Audio 57	2 engine stops due to flameout or overturn
6	Low oil pressure	ENG 1	ENG1 OIL LOW PRESS	Audio 62	1 grease pressure below minimum
7	Low oil pressure	ENG 2	OIL LOW PRESS ENG2	Audio 63	2 grease pressure below minimum
8	Low main lubricating oil pressure	MGB PRESS	MGB LOW PRESS	Audio 59	The main lubricating oil pressure is below the minimum
9	Battery 1 overheating	BAT 1 OVHT	BAT1 OVERHEAT	Audio 64	When the temperature exceeds the limit
10	Battery 2 overheating	BAT 2 OVHT	OVERHEAT BAT2	Audio 65	When the temperature exceeds the limit
11	Feed tank 1 low level	LOW FUEL	FUEL1 LOW	Audio 60	Feed tank 1 level below threshold
12	Feed tank 2 low level	LOW FUEL	LOW FUEL2	Audio 61	Supply tank level 2 is below threshold

(continued)

Table 2 (continued)

Serial	Name	Central alarm panel	Main message list	Audio alarm	Trigger condition
13	Failure of automatic flight control system - grip control	AP	AUTOPILOT	Audio 55	Single-axis or multi-axis autopilot is completely disabled

Fig. 6 Fire extinguishing system information display of EC175 helicopter engine

the burden of the pilot, simplifying the operation procedures, to ensure the flight safety (Fig. 7).

For some helicopters with small tonnage, such as the EC135, the design idea is also to arrange the control panel with display function within the best scope of the pilot's vision, including the warning indicator on the dashboard, the engine fire control indicator box, and the engine control panel on the central control platform. Since the mode selection switch of the engine has no information to be displayed to the pilot, it is arranged on the switch board at the top of the cockpit to achieve the purpose of reasonable space arrangement. From the perspective of the layout of an EC135 helicopter design personnel to helicopter emergency and fault disposal procedures related system with control switch clever design, it greatly convenient for the pilot's operation, this is fault execution become more convenient, disposal efficiency and success rate of disposal of the pilot will have a positive impact. Such as engine fire extinguishing procedures, the above mentioned models have a display system to alarm the pilot. At the same time, the red light of FIRE on the dashboard tells the pilot that this is an alarm program with a high warning level that needs to be dealt with immediately. The subsequent operation procedures include the operation of the engine, fuel and fire extinguishing system, which may be slightly different depending on the type, but these three systems are basically in the same area, and the operating time interval between the systems is greatly shortened, which is very beneficial for extinguishing the engine fire in time.

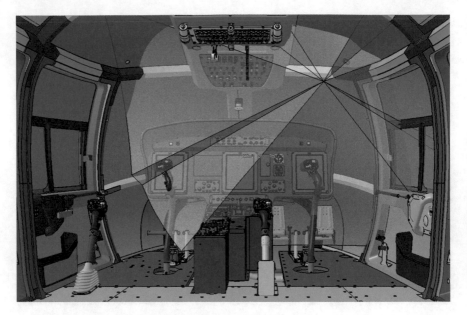

Fig. 7 Range of sight of AC312E helicopter pilot's head

4 Conclusion

Summing up the cockpit layout of the above models, and looking back at the evolution of helicopter cockpit layout, we can get some inspiration from it. Cockpit design concept is to provide timely and accurate information to the pilot, especially the failure information display, so that pilots can be the first time to master the most direct the most valuable alarm content, and the fault are the most effective treatment, which requires the helicopter at the beginning of the design will need to analyze the system of alarm information summary and classification, will be able to directly affect flight safety information in the form of red and a multiple warning told the pilot. Helicopters have always been designed with this in mind. With the development of modern information technology, the information display of each system of the helicopter is more comprehensive and complex. For such complicated system information, we need to screen it based on the determination of each system's fault mode and fault safety level. According to the common practice of modern helicopters, the information of fire, engine failure, battery overheat, main lubricating oil and fuel oil in red warning level information is displayed on the dashboard through red warning light panel. For amber warning information, the fault schematic diagram can be displayed to the pilot through the integrated display system to help the pilot understand the fault and make a correct judgment.

By means of modernization, we can do the helicopter design from simple to complex, let its all information acquisition and display, but the design of the helicopter we must by complex to simple, to classify the information and will show the necessary information to the pilot, at the same time can take multiple display mode to reduce the misjudgment on important faults, and other fault information, the use of the relative single failure mode to remind the pilot, to alleviate the burden of the pilot in order to improve efficiency, ensure flight safety. For the information collected by the system for ground maintenance, it is not necessary to show it to the pilot, only a module for ground maintenance needs to be designed to facilitate ground data download and processing.

Helicopter development make great changes have taken place in the cockpit of the helicopter, but our definition of the human–computer interaction interface and processing principle has not changed, that is to let the pilot in the first place to master the most direct and the most effective data, to implement fully controllable helicopter flight operations, the cockpit design purpose is to build a safe and efficient human–computer interaction interface.

References

1. CAAC29
2. AC-29-2C
3. GJB 1471A
4. Mintong: Calculation of the importance degree of cockpit information element based on AHP. Sci. Technol. Inf. **6**, 1–2,4 (2019)
5. Chen, D.: Research and development of helicopter situational awareness system in low visibility environment, Avionics Technol. **1**, 2–3 (2012)
6. Chen, Y.: Research on human error based on cockpit visual warning information. Sci. Technol. Innov. Appl. **25**, 1 (2014)
7. She, Q.: The integrated design of cockpit display picture based on model drive. Electron. Technol. 1–2

Aircraft-Cable Fault Location Technology Research Based on Time Domain Reflectometry

Tao Li, Danyang Wang, Jianjun Tang, Yang Liu, and Ou Chen

1 Introduction

As part of the EWIS, modern aircraft cable harnesses are assembled in cable bundles, which has very large amount of wirings with various types arranged through different aircraft bays. The Boeing 737, for instance, has a total length of 280 km cable wirings [1–3]. It is reported by FAA since late 1980s that wiring safety concerns were raised due to accidents and incidents [5, 6]. Wiring issues in the US Navy cause an average of two in-flight fires very month, more than 1,077 mission aborts, and over 100,000 lost mission hours each year [3]. Investigations into wiring issues found common degrading factors in aircraft electrical wiring systems, including design, installation, maintenance, repair and environment, together with time, that play a role in wiring degradation. This requires an electric reliable continuity test as well as a fault resolve work both in aircraft assembly stage and maintenance.

Figure 1 shows the typical cable harness bundle wirings in modern aircraft bays. It is clear that the wirings are arranged between airframe system equipment surrounding with structural parts or major section frames. As a result of the spatial constraints when cable bundles are final installed, the wirings problem like chafing and bend radius problem is not easy to be inspected, located, accessed and resolved. This leads to not only the EWIS performance issues, but also high risks in aircraft operations. Once a system fault is reported, huge work will be undertaken to find the detailed wirings fault locations in harness bundles. This time-consuming and high-cost process contains the rework of installed equipment, system components and sometimes even structure disassembly. An effective and precise fault location approach is needed to ease the cable bundles assembly and maintenance processes.

Current wirings fault location methods include: physical inspection method, electrical bridge measuring method, high-voltage pulse method, and secondary pulse method [5–7]. Due to the limited room of aircraft bays, the physical inspection

T. Li · D. Wang (✉) · J. Tang · Y. Liu · O. Chen
Chengdu Aircraft Industrial (Group) Co., Ltd, 88 Weiyi Road, Chengdu 610092, China

© The Author(s), under exclusive license to Springer Nature Switzerland AG 2021 133
D. Krob et al. (eds.), *Complex Systems Design & Management*,
https://doi.org/10.1007/978-3-030-73539-5_11

Fig. 1 Typical examples of cable bundle wirings in aircraft bays

method is only available for use at the harness ends with some open bays. If a fault is in the middle section of the bundle wrings or cannot be seen, this method would be hard to perform. By contrast, the electrical bridge measuring method requires to be operated at both ends of a wire at the same time. In industrial practice, it is actually not easy to find suitable bundle connector breaks for tester access in the complicated cable harnesses wring network. The other two methods, high-voltage pulse and secondary pulse method are usually to be used for fault inspection of the electrical power supply wirings. They are not working for low-voltage wirings especially the data bus wirings, which are widely used today on advanced aircraft avionics [8, 9].

Since the methods mentioned before are both not satisfy the requirements in wiring fault location for modern aircraft assembly and maintenance, a further investigation is taken in the field of Time Domain Reflection (TDR). This technology is to determine the fault types and locate the fault by measuring the reflected signal in the cable wires and analysing the changes in the form of wire impedance change in the reflected signal. The single-end TDR test is safe for aircraft EWIS and other devices as it is a single-based low-voltage excitation test with less dependences required from the circuit data. Currently, the TDR-based cable fault tester is demonstrated to be feasible and efficient in F-18 fighter maintenance [10]. The latest developing trend is to improve the measuring performance, minimize the tester size that fits more complicated aircraft bays, and reduce the manufacturing cost.

2 Characteristics of Aircraft Cable Wiring Fault

A modern aircraft is operated in complex environment, including vibration, high temperature, cold, moisture, and other conditions. EWIS, especially the cable harnesses are also required to be operated safely in accordance with the aircraft usage conditions. Generally, there are two aspect of reasons contributed to aircraft cable wiring degradations and fault, which are mechanical damage and environmental factors [11–13].

Mechanical damage is normally caused by physical processes such as friction with surrounding frames, devices and pipelines, unreasonable binding or laying, artificial

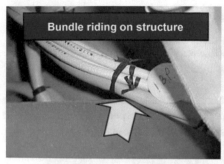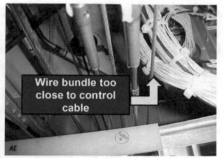

(a) Bundle riding on structure   (b) Chafing risk on moving part

Fig. 2 Examples of wire bundle mechanical damage risk [4]

stretching or trampling. The characteristics of those damage include wire chafing, fracture, looseness or damage of contact parts. It is concluded by FAA that mechanical damage is mainly resulted from improper aircraft components layout, installations and maintenance. For an advanced aircraft, the risks of mechanical damage will be greatly increased due to complex operation conditions in the air (as shown in Fig. 2). This is why a safety distance between structural surface and wires, other aircraft system components and wires are of importance in assembly and maintenance.

Besides the aircraft operation conditions in the air, the overall environmental factors include vibration, high temperature, cold, moisture, salt fog, high altitude ray particle radiation. The physical and chemical properties of cable harnesses after installation will make the EWIS performance degrading or even failure over time, including strength, insulation, electrical conductivity. What is more, in an aircraft bay, cable bundles are also affected by things like accumulation of dirt, lint, oil, grease, detergent and wastewater for a long time. This would further cause the insulation layer to be oxidized or corroded, finally resulting in the whole EWIS performance degradation.

The two aspects of fault will change the electrical characteristics of the wire itself. Generally, these cable wiring faults can be divided into open circuit faults, low resistance faults (or known as short circuit fault) and high resistance faults [3]. However, high-resistance leakage faults occur normally in medium-high voltage cables, while aircraft power grid is not included in this category. This research will only focus on open circuit fault and short circuit fault.

3 Cable Wiring Fault Location Method Based on TDR

This chapter explains the basic ideal of how TDR technology would be used for wiring fault location, and thus support to build a fault model as data baseline [15].

3.1   Fault Location Method Based on TDR

Ideally, a wire in the cable harness can be considered as a uniform transmission wire. It is therefore feasible to use distributed parameters to build a schematic model [14] as shown in Fig. 3.

In Fig. 2, R, L, C and G represent the distributed resistance, inductance, capacitance and conductance of the transmission line per unit length, respectively. According to these parameters, the wave equation can be obtained as formula (1):

$$\begin{cases} -\frac{\partial u}{\partial x} = Ri + L\frac{\partial i}{\partial t} = (R + j\omega L)i \\ -\frac{\partial i}{\partial x} = Gu + C\frac{\partial u}{\partial t} = (G + j\omega C)u \end{cases} \tag{1}$$

The characteristic impedance is the ratio of incident wave voltage to incident wave current, which can be obtained as formula (2):

$$Z_0 = \frac{u}{i} = \sqrt{\frac{R + j\omega L}{G + j\omega C}} \tag{2}$$

In the formula (2), capacitance C and inductance L are related to the dielectric constant of the cable and the cross-sectional area of the material core, which indicates that the wave impedance of different types of cables is different.

For transmission wires with small loss, the characteristic impedance Z can be simplified to $Z_0 = \sqrt{L/C}$, as $\omega L \gg R\omega C \gg G$.

The reflection coefficient can be expressed as formula (3):

$$\rho = \frac{Z_L - Z_0}{Z_L + Z_0} \tag{3}$$

In the formula (3), Z_L represents the load at the failure point. According to the time interval between the pulse's injection time and reflection time, the location of the fault point can be calculated from formula (4):

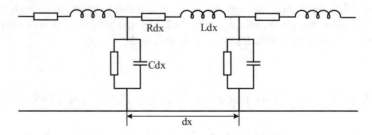

Fig. 3 The equivalent distributed parameter model of cable

$$L = \frac{1}{2}v_\rho(t_1 - t_2) \tag{4}$$

In formula (4), v_ρ is the velocity of the pulse propagating in the cable, which is related to the dielectric constant of the cable; t_1, t_2 are respectively the time of the pulse is sent and returns. When the frequency is very high, the propagation velocity of electromagnetic wave in cable tends to a constant [16].

3.2 Fault Analysis Process

In this TDR approach, the basic characteristics of impedance of cable fault can be obtained.

(1) If $Z_L = Z_0$, then $\rho = 0$. Which indicates no potential fault is found in the wire, the load impedance and the characteristic impedance are matched, according to formula (3), there is no reflection echo.

(2) If $Z_L = \infty$, then $\rho = 1$. Which indicates there is an open circuit fault in the cable, the reflection coefficient equals 1, and the incident wave and reflected wave have the same polarity, as shown in Fig. 4.

(3) If $Z_L = 0$, then $\rho = -1$. Which indicates there is a short circuit fault in the cable, the reflection coefficient equals -1, and the incident wave and the reflected wave have opposite polarity, as shown in Fig. 5.

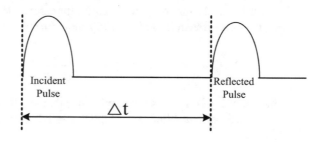

Fig. 4 Schematic diagram in case of open circuit found in a wire

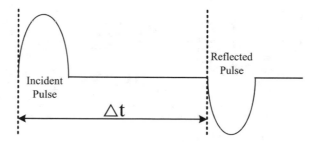

Fig. 5 Schematic diagram in case of short circuit found in a wire

(4)   If $Z_L \neq 0$ or $Z_L \neq \infty$, then $-1 < \rho < 1$. In case of wrinkling, wear and shielding damage of the insulation skin of the cable, the impedance of the cable must be inevitably affected, and the reflection coefficient equals $-1 < \rho < 1$.

The above results show that the characteristics of the reflection wave will change significantly if open circuit and short circuit are found in a wire. This is will be used as fundamentals of fault type detection.

4   Development of the TDR-Based Handheld Inspection Device

4.1   General Device Layout

The developing device is composed of a touch screen, signal generation and processing circuit, power amplifier circuit, impedance matching circuit and sampling circuit, as shown in Fig. 6.

The power amplifier circuit function is to improve the power of the transmitting pulse and make the reflected characteristic signal more obvious. The schematic circuit diagram is shown in Fig. 7.

The sampling unit collects and stores the reflected wave data, compares and calculates them with the voltage incident wave. The schematic circuit diagram is shown in Fig. 8.

In order to ensure that the pulse transmitting signal and reflected signal will not overlap, the pulse transmitting frequency should meet:

$$f \leq \frac{1}{T} = \frac{v_\rho}{L_{\max} * 2} \tag{5}$$

In the formula (5), $L_{\max}$ is the maximum cable length under test.

In order to ensure the blind area of the shortest detection distance, the pulse width should meet:

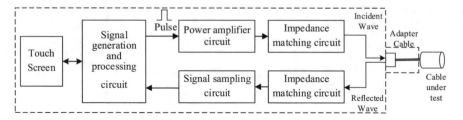

Fig. 6   General device layout

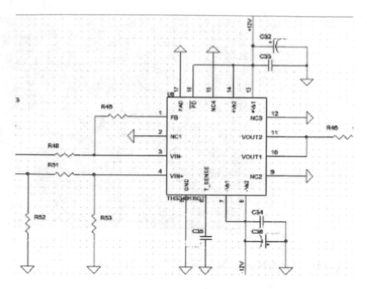

Fig. 7 The power amplifier circuit

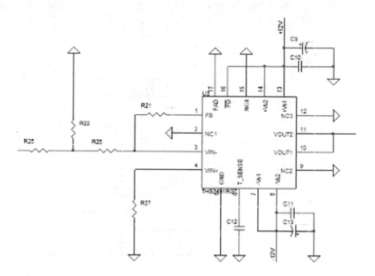

Fig. 8 The sampling circuit

$$T_{width} \leq \frac{L_{\min}}{v_\rho} \tag{6}$$

The adapter cable is shown in Fig. 9. One end of a wire is connected to the TDR device with a universal SMA interface, and the other end could adapt to common standard connectors quickly.

Fig. 9 The adapter cable for
the test

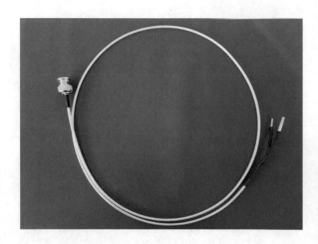

4.2 *Design of the Inspection Working Process*

Figure 10 explains the inspection working process of the developing device. After the
over-all system is powered on, the program runs automatically. Firstly, the hardware

Fig. 10 Device over-all
working process

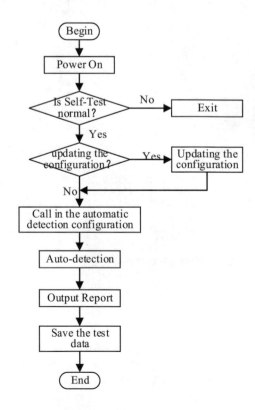

platform and communication interface are initialized, following with a self-test. If the self-test failed, the user is reminded to stop the process and check the device. If self-test passes, the device will wait for user's action. The test results are shown in the device screen and saved to the built-in database automatically.

5 Initial Implementation

The developing device is designed to test the wires of aircraft EWIS, which is part of the aircraft function verification and fault isolation process. An early developing technology demonstrator is shown in Fig. 11. In this phase, a large number of demonstration tests have been taken place to generate the initial characteristics model and prove the proposed ideas. Figures 12 and 13 show the waveform differences of short circuit and open circuit fault respectively.

According to the working process mentioned in Fig. 10 before, several fault-location tests have been done on two selected sing-core shielded cable harnesses to build the basic characteristics model for this device. As shown in Table 1, an example of wave velocity is generated from those tests.

The open circuit and short circuit faults were set on the two experimental cable harnesses respectively, and the developing device was used to detect and record the results. The experiment results are shown in Tables 2 and 3. The data presented in the two tables are then used as the initial wire characteristic baseline to support real-world fault location. It should be pointed out that, the test baseline can be iterated and studied with a large number of wires and different wire types to fasten the fault location in daily use.

The experiment results indicate that the developing device is feasible to detect the open circuit and short circuit location accurately, and the margin of error could meet

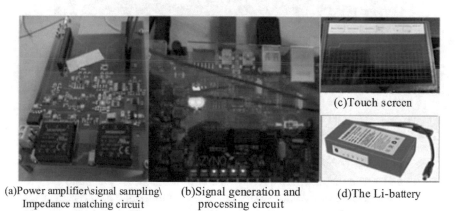

(a)Power amplifier\signal sampling\   (b)Signal generation and   (c)Touch screen
   Impedance matching circuit   processing circuit   (d)The Li-battery

Fig. 11 Technology demonstrator layout

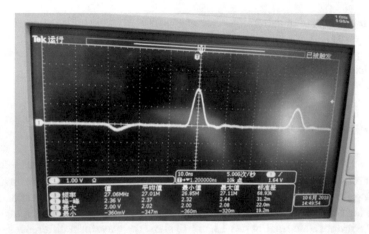

Fig. 12 Waveform result for a wire open circuit test

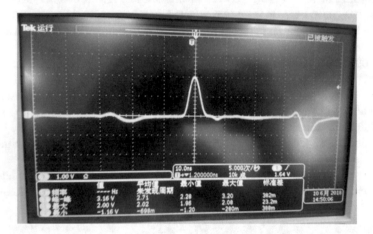

Fig. 13 Waveform result for a wire short circuit test

Table 1 Wave velocity of the two types of cables

Type	The wave speed (mm/ns)
Cable No. 1	197
Cable No. 2	202

the actual needs of the aircraft wire fault test. Otherwise, this device only weighs 820 g, and its size (as shown in Fig. 14) is also convenient for one hand to hold.

The initial technology demonstrator developed at this stage only has the capacity of testing several types of wire, due to the limited wire characteristics baseline generated. However, it still can be used to demonstrate the feasibility to implementing TDR approach in aircraft cable harness fault location. It has the potential for performance

Table 2 The cable open circuit test results

Cable type	Fault type	Reference (m)	Measured (m)	Error (m)
Cable No. 1	Open	0.52	0.48	−0.04
	Open	5.03	5.07	0.03
	Open	15.00	15.04	0.04
	Open	30.04	30.00	−0.04
Cable No. 2	Open	0.51	0.54	0.03
	Open	5.01	5.04	0.03
	Open	15.00	15.02	0.02
	Open	29.94	29.90	−0.04

Table 3 The cable short circuit test results

Cable type	Fault type	Reference (m)	Measured (m)	Error (m)
Cable No. 1	Short	0.52	0.48	−0.04
	Short	5.03	5.07	0.03
	Short	15.00	15.04	0.04
	Short	30.04	30.00	−0.04
Cable No. 2	Short	0.51	0.54	0.03
	Short	5.01	5.04	0.03
	Short	15.00	15.02	0.02
	Short	29.94	29.90	−0.04

Fig. 14 The prototype device in implementing TDR

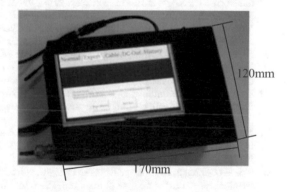

improvement. Once more types of wire characteristics are generated as a updated baseline database, this device will cover the full EWIS fault location requirements in both aircraft assembly line and maintenance stage.

6 Conclusion and Future Work

This preliminary research investigates the needs of the fast and accurate wire fault location of aircraft EWIS continuity tests. Two most found wiring fault type, the open circuit and short circuit are analysis following with the drawbacks of conventional methods. A TDR-based approach is then proposed to allow a new model-based approach by implementing TDR technology. This approach has the advantage of real-world wire fault characteristics baseline study, and thus supports a fast and more accurate fault location result. The initial technology demonstrator has been developed and tested preliminarily on aircraft assembly line, with several wire characteristics baseline generated. As a daily use device for aircraft assembly and maintenance, this device is designed to be a miniaturized device to fit the work conditions of limited aircraft bays. As discussed before, this device benefits from the wiring fault mode generated from different types of wire characteristic. To make the most use of its potential, it is suggested to have more TDR analysis data baseline of the most used wire types in the next stage work.

Acknowledgements This research is a fundamental research funded by SASTIND China under Grant JCKY2019205A004.

References

1. Zhang, J., Zhou, X., Wei, J., et al.: Method of wire insulation fault location based on wavelet transform. Trans. China Electrotech. Soc. **27**(5), 99–104 (2012)
2. Mei, Zh.P., Li, Q., Wen, J.Q., et al.: Research on optimization of wiring paths in airplane harness process. In: 2012 IEEE International Conference on Cyber Technology in Automation, Control, and Intelligent Systems, pp. 485–488. IEEE Computer Society, United States (2012)
3. Sharma, C., Madhavan, S., Sangwan, S.: Modeling of STDR system for locating faults in live aircraft wires. In: 2013 International Conference on Emerging Trends in Communication, Control, Signal Processing and Computing Applications (C2SPCA). IEEE (2013)
4. Federal Aviation Administration, Aircraft Electrical Wiring Interconnect System (EWIS) Best Practices (2010). Available online: https://www.faa.gov/training_testing/training/air_training_program/job_aids/
5. Yuan, K., Yu, Y.F.: Aircraft cable fault location system based on principle of regression analysis. In: The 5th International Conference on Computer Science and Education, pp. 1226–1229. IEEE Computer Society, United States (2010)
6. Shi, X.D., Zheng, J.Zh., Jing, T., et al.: Design of aircraft cable intelligent fault diagnosis and location system based on time domain reflection. In: 2010 8th World Congress on Intelligent Control and Automation, pp. 5856–5860. Institute of Electrical and Electronics Engineers Inc., United States (2010)
7. Nve, X., Zan, M., Yi, T.: Selection of wiring environment and failure rate comparison analysis in aircraft wiring risk assessment. Procedia Eng. **14**, 428–432 (2011)
8. Griffiths, L.A., Parak, R., Furse, C., et al. The invisible fray: a critical analysis of the use of reflectometry for fray locating. IEEE Sens. J. **6**(3), 697–706 (2006)
9. Griffiths, L.A., Parak, R., et al.: Application of phase detection frequency domain reflectometry for locating faults in am F-18 flight control harness. IEEE Trans. Electromagn. Compat. **47**(2), 327–333 (2005)

10. Zhang, J., Wei, J., Xie, H., et al.: Detection and analysis of aerospacewire insulation faults based on TDR. Acta Aeronaut. ET Astronaut. Sinca **30**(04), 706–712 (2009)
11. Zhang, Z., Wen, F., et al.: Wavelet analysis based power cable fault location. Autom. Electr. Power Syst. **27**(1), 49–53 (2003)
12. Liu, X., Yuan, K.: Design of fault diagnosis and localization system of aircraft wire based on cross-correlation algorithm. Comput. Measure Control **22**(12), 3903–3905 (2014)
13. Zhai, Y., Guo, G.: New method of data denoising and fault location for aviation cable TDR. Mod. Defence Technol. **44**(06), 128–134 (2016)
14. Cupertino, F., Lavopa, E., Zanchetta, P., et al.: Running DFT-based PLL algorithm for frequency, phase, and amplitude tracking in aircraft electrical systems. IEEE Trans. Industr. Electron. **58**(3), 1027–1035 (2011)
15. Shi, Q., Kanoun, O.: Detection and location of single cable fault by impedance spectroscopy. In: IEEE I2MTC. IEEE (2014)
16. Song, J.: Some Key Techniques of Cable Length Measurement Based on Time Domain Reflectometry, p. 23. Harbin Institute of Technology, Harbin (2010)

An MBSE Framework for Civil Aircraft Airborne System Development

Huang Xing, Yang Hong, Kang Min, and Zhang Juan

1 Introduction

Airborne System is an important part of aircraft, and their functions and performance will directly influence aircraft whole performance [1]. With the progress of science and technology, new technologies and products are used in airborne systems which contributes the complexity of system. Therefore, the possibility of design errors is increased, and these errors may lead to a system failure and even an aircraft fatal accident.

For civil aircraft, safety is an important consideration for airworthiness authorities. The certification regulation defines specific rules for system safety, such as *FAA Title 14 of the Code of Federal Regulations part 25.1309 Equipment, systems, and installations* [2]. Therefore, the Aerospace Recommended Practice (ARP) 4754A, *Guidelines for Development of Civil Aircraft and System*, is recommended by FAA & EASA as an acceptable method for establishing development assurance process to decrease those errors in requirements or design [3, 4]. To meet the objects of these guidelines, a specific development process is formed and used in airborne system industry. In this development process, requirements are an important engineering artifact. But the weaknesses of pure document-centric development approach [5],

H. Xing (✉) · Z. Juan
AVIC Xi'an Aviation Brake Technology Co., Ltd, No. 5 Keji 7th Rd, Gaoxin Zone, Xi'an, China
e-mail: Huangx004@avic.com

Y. Hong
Shanghai Aircraft Design and Research Institue, No. 5088 Jinke Rd, Shanghai, China
e-mail: yanghong@comac.cc

K. Min
Xi'an Aircraft Certification Center, Airworthiness Certification Center of CAAC, No. 35 Tangyan Road A Block the 18th Floor, Xi'an, China
e-mail: kangmin_acc@caac.gov.cn

© The Author(s), under exclusive license to Springer Nature Switzerland AG 2021
D. Krob et al. (eds.), *Complex Systems Design & Management*,
https://doi.org/10.1007/978-3-030-73539-5_12

which is traditional process in industry, are emerging as time-consuming documents reading and ambiguous literal understanding.

Nowadays, many researches [6–8] show that Model Based Systems Engineering (MBSE) is a good approach to capture and decompose requirements at top level. With SysML support, use cases, activities, sequence and other diagrams [9] are created to help designer have clear understanding about system. As airborne system design is a complex work and involve a lot of detail engineering works, how to use MBSE methodology in industry project and link to detail design works is still a topic.

Model based and test driven (MBTDD) methodology [10, 11] is already used in software industry to generate an agile development loop and enhance quality of design work. For airborne system, this method could also be used in detail design with different disciplines simulation software support, such as Matlab/Simulink, AMESim and Maxwell. And the adaptive improvement for MBTDD is the consideration of airborne system development process. Therefore, under the same development process, there is a possibility to combine MBSE and MBTDD methodologies.

In this paper, we present a new MBSE framework named as CESAM&MBTDD which combines CESAM MBSE, MBTDD methodologies and the consideration of airborne system development process. The purpose of this new MBSE framework is to improve civil aircraft airborne system design quality and efficient. We chose Wireless Tire Pressure Indication System (aliased as WTPIS) as study case to implement this framework.

The rest of this paper is organized as follows: Sect. 2 presents related information about civil aircraft airborne system development process, MBTDD and CESAM MBSE. Section 3 presents introduction of combination methodology of CESAM&MBTDD. Section 4 details the case study and introduces the results and discussions. We conclude and highlight the advantages of new framework in Sect. 5.

2   Related Information

2.1   Civil Aircraft Airborne System Development Process

To meet the objects of guidelines from ARP-4754A, civil aircraft airborne system development processes could be concluded as V-model (See Fig. 1).

This typical development process nowadays is using top-down strategy [12]. As the high complexity of system development work, the whole process is divided into several hierarchies such as system level, equipment level and implement or hardware & Software level. Aircraft level is not included in system development process, but still should be involved, because this level is the direct customer for system level and its outputs such requirements, interface is the most important inputs.

In the whole V-model, for different level process, they have similarity which could be conclude as another smaller V-chart (See Fig. 2). At each level, they will finish compliant analysis based on requirements or needs from high level (usually

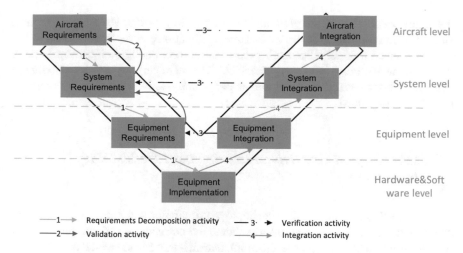

Fig. 1 V-model of airborne system development process

Fig. 2 V-chart for different level's development process

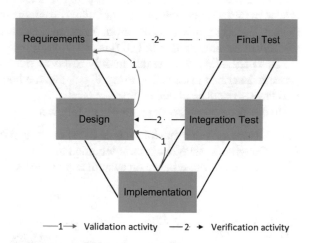

is direct customer) and after that form technical requirements which could clearly guide architecture design, logic design or mechanical design at specific domains. Implementation could be lower level requirements or drawings for mechanical equipment, schemes for electrohydraulic equipment. For airborne electrical hardware or software, even that is a kind of equipment implementation in the V-model, but the specific guidance, RTCA DO-178 and DO-254, is introduced for these development process, these parts are not included in this paper.

Validation and Verification activities are happened in whole V-model and smaller V-chart. The purpose of validation activity is ensuring lower level requirements are correct and complete or the design correct response requirements. The purpose of verification activity is confirming all the designs are properly implemented and meet the requirements. The means of validation and verification could be analysis, review, simulation, test or other efficient ways. Except validation and verification, ARP-4754 also introduce other activities such process assurance, configuration management, development assurance level assignment, etc. The main purpose of these activities is decreasing human errors in development phase and making sure the final products meet customer's needs.

2.2   Model Based Test Driven Development Methodology Introduction

During the validation phase, since the improved simulation software tools, simulation becomes a most powerful method to validate the correctness and completeness of requirements. Another big advantage of simulation is that created model in software is testable and changeable. Therefore, requirement-design-model-simulation test, this loop becomes a benign iteration and more economical and low time cost than the way to manufacture whole system or build a prototype then get test results to update design. Based on this consideration, a new methodology named model based and test driven development (MBTDD) is used for airborne system design (See Fig. 3).

This methodology is created mainly based on the typical airborne development process described in Sect. 2.1. From Fig. 3, this methodology could be divided into two level, system level and equipment level.

In the system level, 6 steps should be followed.

(1)   Create system behavior models based on system requirements.
(2)   Generate system behavior integration model.
(3)   Create test cases based on system requirements.

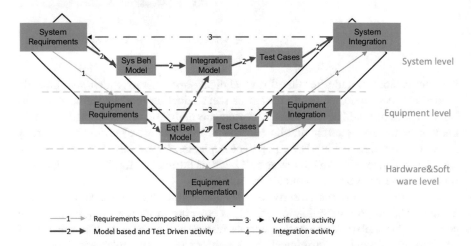

Fig. 3 Model based and test driven development for airborne system

(4) Execute test cases based on system integration model.
(5) Generate more precision integration model from equipment level.
(6) Re-execute test case based on update system integration model.

In the equipment level, 3steps should be followed.

(1) Create equipment models based on equipment requirements.
(2) Create test cases based on equipment requirements.
(3) Execute test cases based on equipment models.

The system behavior model should be as simple as possible but proper reflect main function of requirement. The cluster of these models usually base system functions but should have clear traceability with requirements. If the equipment is complex such as control unit which is comprised by software and hardware, this equipment could be regarded as sub-system and repeat the system level development process. This process of modeling is a kind of system and equipment design work, the implementation of the requirement will be checked and whole architecture will be raised during requirement modeling phase. Once the testable model is created, the test case should be generated and executed. From this simulation results, the correctness of requirement, design, test cases will be both validated. These model test cases could be reused in verification phase, meanwhile models will be calibrated since the same test cases.

The difference between system and equipment models is that lower level model will concern more implantation constrains. This strategy let the whole modeling work becomes multi-layered like the airborne system development process. Thercfore, re-executing test case based updated integration model become an essential step because more detail information is added.

Additional issue should be mentioned is model integration strategy in different levels. The simulation software in different levels will be different, such as AMESim, Simulink or Adams, the integration strategy could be co-simulation or generating C-code and operating in unique environment depend on the capability of used tools.

2.3   CESAM MBSE Methodology Introduction

CESAM is a system architecting and modeling framework, developed since 2003 [13]. It is dedicated to the working systems architects, engineers or modelers to help them to better master the complex integrated systems they are dealing with. In CESAM MBSE methodology, integrated system could be analyzed and modeled from three generic architectural visions: operational vision, functional vision and constructional vision (See Fig. 4). The model language of CESAM is SysML.

Operational vision analyses focus on the interface with environment of system of study, in this level system will be regarded as "black box". The most important work at operational vision is to understand needs of different stakeholders for the system as the main output for next vision. To support this work, operational environment,

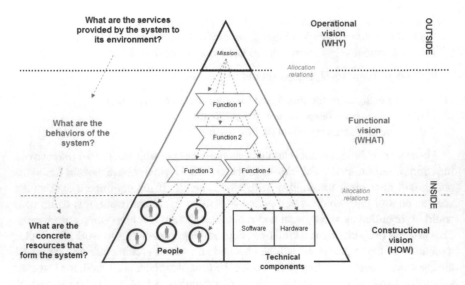

Fig. 4 The CESAM MBSE methodology

stakeholder architecture, life cycle, use case and operational scenario diagrams will be used to describe the operational architecture of system.

Functional vision analyses focus on the abstract behavior of system. In this level system is regarded as "grey box", input and output dynamics of system will be analyzed without consider which component will be used to comprise system. The main goal of this phase is to understand what the system does and form functional requirements flow down to next vision. Functional interaction, functional decomposition, functional modes and functional requirement diagrams will be used to describe the functional architecture of system.

Constructional vision analyses focus on physical components which comprise system. In this level, system is "white box". The main purpose of this phase is to understand detail information of all the related hardware, software and "human-ware" which belongs to system. Constructional interaction, constructional mode and constructional requirement diagrams will be used to describe the constructional architecture of system.

CESAM MBSE methodology could be used in different complex system domain, include airborne system. The same as airworthiness authority's requirement and the nature of civil aircraft airborne system development process, there is a principle line in CESAM which is the traceability from stakeholder's needs to functional requirements and constructional requirements. This similarity provides the possibility of the combination of CESAM and MBTDD.

3 CESAM&MBTDD Methodology Introduction

Since CESAM MBSE is more focus on top level behavior of system and the interaction between system and environment. It helps airborne system designer have a general perspective for system, decrease the risk of incomplete and poor design. Meanwhile, MBTDD could be an efficient supplement as this method will help designer to have a fast iteration and optimize the detail design work. The requirements are the link between two methods. Therefore, the general framework of combination methodology which is named as CESAM&MBTDD could be summarized in below (See Fig. 5).

From Fig. 5, the new method has three hierarchies, aircraft level; system level and equipment level.

In aircraft level. Even the name is "aircraft level", but research object is still the airborne system. The purpose of analyses and modeling work for this level is the same as operational vision described in Sect. 2.3. Stakeholders of airborne system could be airline, airport, passenger, authority, ambient environment, and other airborne systems which have interface with analyzed system. Different diagrams described in Sect. 2.3 will be used and stakeholders' needs will be generated to flow down to next level.

In system level. System functional development will be launched bidirectional. First, using CESAM method to analysis functional architecture of system and functional interaction, functional decomposition and functional modes diagrams will be modeling by SysML, the preliminary functional requirements will be documented. Second, using MBTDD method to create functional behavior models based on preliminary requirements, the purpose of behavior model is analyses function performance and implementation constrain. The test will be executed since the test cases and integrated models are prepared. A top-level test from stakeholder view is presented and integrated test cases comes from operational scenario. After that,

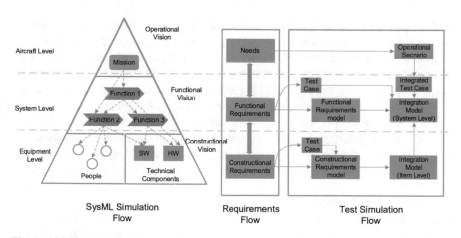

Fig. 5 CESAM&MBTDD framework

an update version of functional requirements will be documented because of added detail information or findings from tests. Last, the consistency between needs to functional requirements, SysML model to performance model should be required. Therefore, the functional requirements could release and flow down next level.

In equipment level. Similar as system level, using CESAM method, system architecture will be defined, and all functions will be allocated to different components, constructional interaction and constructional mode diagrams will be created by SysML. Then documenting primary constructional requirements and using MBTDD method to update requirements by creating functional models, documenting test cases and executing test. And all the constructional requirement should be linked to needs, functional requirements or have rationale.

The final output of this new methodology is released and validated constructional requirements to next level such as software, hardware or production phase.

4 Experiment Results and Discussion

In this section, we choose WTPIS as study case to implement CESAM&MBTDD methodology. WTPIS is installed on the aircraft nose wheels and mainwheels to measure tire pressure of wheels and send pressure information to Ground Crew and avionics system by wireless protocol.

In this specific study case, we chose *Polarion ALM* as requirements management software, *Enterprise Architecture* as SysML modeling software, *Matlab/Simulink* as functional modelling software.

4.1 Results

In this study case, 12 stakeholders, 10 use cases, 5 main operational scenarios, 9 first level functions and 16 sub-level functions are defined of WTPIS and modeled by *Enterprise Architecture*. And 56 stakeholder needs, 52 functional requirements, 82 constructional requirements, 5 integrated test cases, 50 system test cases and 56 equipment test cases are documented in *Polarion ALM* as individual work items, and traceability between them are generated as matrixes automatically.

As requirements are most important media to link CESAM and MBTDD methods, we use interface tools of *Polarion ALM* with *Enterprise Architecture* and *Matlab/Simulink* to reference SysML models and functional models. The configuration management of requirements, models or other work items is controlled in *Polarion ALM*.

Now we chose Ground Crew's need as an example to introduce specific development activities of CESAM&MBTDD methodology. As shown in Fig. 6.

In aircraft level, we identify Ground Crew as main customer of WTPIS. The use case *"WTPIS-113 To read WTPIS information by Ground Crew"* and related

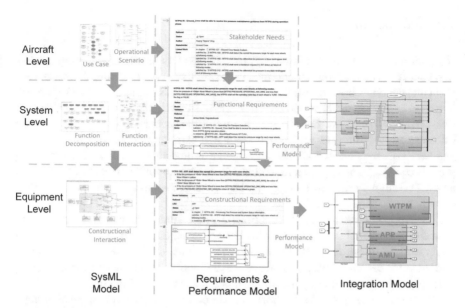

Fig. 6 WTPIS development activities architecture

operational scenario "*WTPIS-115*" are analyzed and modeled. Therefore, one of their needs "*WTPIS-76 Ground Crew shall be able to receive tire pressure maintenance guidance from WTPIS during operation phase.*" is documented in *Polarion ALM*.

In system level, from needs, use case and operational scenario diagram, two related system functions *WTPIS-167 Processing Tire Pressure and System Status Information and WTPIS-216 Processing Maintenance Guide* are identified and re-organized in functional interaction diagram and function decomposition diagram. Each functional requirement is documented in *Polarion ALM* such as "*WTPIS-150—WTPIS shall detect the normal tire pressure range for each nose wheels at following Active and degraded modes.*" Functional model and test case is created. The integration model is generated after finishing all the functional models. Each functional model and integration model are tested based on same test case to validate correctness and completeness of functional requirement. The integrate test is generate from *WTPIS-115* and used to test integration model to validate system behavior including assumption from other interaction systems.

In equipment level, functions allocation and constructional interaction diagrams are created as *WTPIS-218, WTPIS-219*. Functions *WTPIS-167, WTPIS-216* are allocated to APP which is one equipment belongs to WTPIS. From functional requirement *WTPIS-150*, constructional requirement of APP is created as "*WTPIS-365—APP shall detect the normal tire pressure range for each nose wheels.*" functional model, test case and test are similar as system level, but the differences are the boundary of integration model and test is limited in each equipment, these equipment models are integrated into system level and retested based on system test cases again.

4.2 *Discussion*

Now, let us discuss concerns from CESAM&MBTDD development activities. The first concern is about requirement line. There are three points to support the requirement line becomes the most important link between CESAM and MBTDD approach. First one is requirements are output from CESAM analyses flow down to next level and MBTDD approach. Second one is requirements are iteration point as detail information and findings was added from MBTDD approach, at same time new analyses will happen in CESAM and update requirements will be released. Last point is certification consideration, requirements traceability between different level and with design work, which is reflected by models, is convincing evidence to show compliance to stakeholders' needs.

The second concern is about functions. System functions are captured from use case and operational scenario, organized as complete architecture and flow down to equipment. In this process, the requirements and functional models will be clustered and belong to each function, so good quality of function definition and decomposition will help to build a good quality system and enhance work efficient.

The third concern, validation work is main part of airborne system development, testable models help designer to find mistakes and errors in early stage. Simulation test based on functional models and integration model validates the correctness and completeness of requirements and system designs. Re-test based on different level integration models is still needed to check system behavior along with development progress. In WTPIS case study, we create high level integration test based on operation scenario and test in *Matlab/Simulink*. New research [14] shows that SysML models could co-simulate with Simulink models, this provide a higher-level test from stakeholder view and integrate system design details.

The fourth concern, this new methodology will help to promote verification work and build model repositories which will benefit future projects. Design risk and verification test workload are decreased thanks to simulation test and test cases reusing. Meanwhile, in verification phase, test cases and different level's models which are generated in early stage will be optimized and calibrated system test and equipment test are executed based on real product. After this step, models (include SysML and performance models) and test case repositories are created. Therefore, in future project, same system design or equipment will be called from repositories as model form to shorten development time.

5 Conclusions

In this paper, a new combination methodology CESAM&MBTDD used in airborne system development V-model process is presented, and WTPIS is chosen as an example airborne system to implement this methodology, the detail development activities and related model tooling chain are introduced. From the results and discussion concerns, the advantages of this new airborne system development methodology could be concluded as below:

(a) *Optimized requirements in different hierarchy.*
(b) *Fast iteration loop for validation work.*
(c) *Clear and understandable system design thanks to model support.*
(d) *Reusable achievements thanks to repositories creation.*

In future work, we intend to add more detail procedures of development activities based on a real and specific airborne system project and build a complete tooling chain include more disciplines. Safety analyses which is an important consideration for certification will be also added in this new methodology.

References

1. Compile group of an introduction to aviation, M.: an introduction to aviation, Aviation Industry Press, Beijing, China, April 2010
2. FAA: Electronic code of federal regulations: Part 25—Airworthiness standards: transport category airplanes. https://www.ecfr.gov/cgi-bin/retrieveECFR?gp=&SID=65aae5adca38203897808925d41fa6c2&mc=true&n=pt14.1.25&r=PART&ty=HTML
3. AC 20-174: Development of civil aircraft and systems. https://www.faa.gov/documentLibrary/media/Advisory_Circular/AC_20-174.pdf
4. CS-25: Certification specification and acceptable means of compliance for large aeroplanes CS-25. https://www.easa.europa.eu/document-library/certification-specifications/cs-25-amendment-23
5. Wang, Y., Zhang, A., Li, D., et al.: Research on Civil Aircraft Design Based on MBSE. Aeronautical Computing Technique (2019)
6. Li, L., Soskin, N.L., Jbara, A., et al.: Model-based systems engineering for aircraft design with dynamic landing constraints using object-process methodology. IEEE Access **7**, 61494–61511
7. Zhu, S., Tang J, Gauthier J M, et al. J.: A formal approach using SysML for capturing functional requirements in avionics domain. Chin. J. Aeronaut. (2019)
8. Yang, H., Xiao, Y., Li, B.J.: MBSE application in civil aircraft brake system requirements analysis. Civil Aircraft Des. Res. **131**(04), 110–114 (2018)
9. Delligatti, L.M.: SysML Distilled: A Brief Guide to the Systems Modeling Language. Addison-Wesley (2013)
10. Mou, D, Ratiu, D.C.: Binding requirements and component architecture by using model-based test-driven development. In: 2012 First IEEE International Workshop on the Twin Peaks of Requirements and Architecture (TwinPeaks), pp. 27–30. IEEE (2012)
11. Sadeghi, A., Mirian-Hosseinabadi, S.H.J.: MBTDD: model based test driven development. Int. J. Software Eng. Knowl. Eng. **22**(08), 1085–1102 (2012)
12. ARP4754A S A E. J.: Guidelines for Development of Civil Aircraft and Systems. SAE International (2010)
13. CESAM: CESAMS systems architecting method. A pocket guide. CESAM Community (2017). https://www.cesames.net/wp-content/uploads/2017/05/CESAM-guide.pdf.
14. Hai, X., Zhang, S., Xu, X.C.: Civil aircraft landing gear brake system development and evaluation using model based system engineering. In: 2017 36th Chinese Control Conference (CCC), pp. 10192–10197. IEEE (2017)

An Optimization Method for Calibrating Wireline Conveyance Tension

Can Jin, Xin Peng, Qing Liu, Xinhan Ye, Lang Chen, Qiuyue Yuan,
Xiang-Sun Zhang, Ling-Yun Wu, and Yong Wang

1 Introduction

A good wireline conveyance tension prediction benefits both in wireline job planning and logging operation phase. On one hand, tt can improve logging efficiency from better cable type and weak point type selection in plan which leads to fewer runs. On the other hand, it is a better reference to real-time logging data which help identify abnormal trend earlier to avoid stuck and reduce the risk of fishing operations and associated non-productive time.

For one wireline job, the depth-based conveyance tension profile can be predicted by a physical model with borehole, formation, and mud information, surface equipment and tool string information and some other parameters as input [1, 2]. The challenge of applying a numerical model to predict the tension is that there are input parameters such as cable friction coefficient and tool friction coefficient which are difficult to measure. This is due to the complex situation underground, together with some uncertainties from the formation and mud.

Can Jin, Xin Peng: Authors contributed equally to this work.

C. Jin · Q. Liu (✉)
Schlumberger Technologies (Beijing) Ltd., Beijing 100049, China
e-mail: QLiu10@slb.com

X. Peng · X. Ye · L. Chen · Q. Yuan · X.-S. Zhang · L.-Y. Wu · Y. Wang
CEMS, NCMIS, MDIS, Academy of Mathematics and Systems Science, Chinese Academy of Sciences, Beijing 100080, China

X. Peng · X. Ye · L. Chen · Q. Yuan
School of Mathematical Sciences, University of Chinese Academy of Sciences, Beijing 100049, China

D. Krob et al. (eds.), *Complex Systems Design & Management*,
https://doi.org/10.1007/978-3-030-73539-5_13

159

In this paper, we propose to utilize measured tension data to calibrate the unmeasurable parameters (mainly the cable and tool friction) in physical model. The motivation is that oil industry measures and collects large scale data for every well and daily jobs [3–5], which allows us to leverage the concept and technologies from data science, machine learning, and artificial intelligence to help decision making and automation [6, 7]. We expect that the delicate integration of physical model with measured data leads to a well calibrated model, which can then be applied for predicting next wireline job in the same borehole or nearby.

Formally, we formulate this calibration task as an optimization problem to obtain the 'best fit' parameters for the prediction. Decision variables are the unmeasurable input parameters in the physical model. The objective function to be minimized is the difference between predicted tension and measured data. A new stochastic direction descent (SDD) search strategy is proposed to solve this optimization problem. Field case studies show that this optimization model and strategy is efficient to calibrate the physical model.

2 Optimization Model

2.1 Model 1

Our goal is to calibrate parameters via minimizing the distance between predicted tension profile and measured data. The model is expressed as

$$\min_{f_c, f_t, d} D\big(F(f_c, f_t) + d, \hat{y}\big) \tag{1}$$

where f_c and f_t are averaged cable and tool friction coefficient respectively, and d value is offset or residue and $\hat{y}$ is the measured tension along hole. The function F is the conveyance tension model as in [1]. D is the distance function that may have various forms. In this paper, we use L_2 norm as follows.

$$D(x_1, x_2) = \|x_1 - x_2\|_2^2 \tag{2}$$

Model 1 is an unconstrained oracle optimization problem, with three decision variables f_c, f_t and d. f_c and f_t are the unmeasured cable and tool friction parameters in the physical model. They are usually bounded by its physical meaning. d is introduced as a residue term. It can partially absorb the effect of uncertainty that is not considered in the physical model. This allows us to relax the upper and lower bounds of f_c, f_t mathematically in the optimization model. We further notice that optimization problem $\min_d D\big(F(f_c, f_t) + d, \hat{y}\big)$ is quadratic programming problem (when D is L_2 norm) and can get the exact optimal solution directly. We introduce an equivalent model as below.

2.2 Model 2

We equally transform Model 1 into a bilevel optimization problem: the lower-level optimization task is a quadratic optimization $\min_d D(F(f_c, f_t) + d, \hat{y})$ with single variable d, and the upper-level optimization task is $\min_{f_c, f_t} G(f_c, f_t)$ with two decision variables f_c, f_t. In this way, we have the following optimization model:

$$\min_{f_c, f_t} G(f_c, f_t)$$
$$\text{s.t. } G(f_c, f_t) = \min_d D(F(f_c, f_t) + d, \hat{y}) \tag{3}$$

3 SDD Search Method

Model (3) is an oracle optimization problem since evaluation of objective function $G(f_c, f_t)$ relies on a numerical computing engine F, which simulates the downhole context with given cable and tool friction coefficients. The computation complexity is known as NP-hard given an oracle representation of the function [8]. There exist several challenges in efficiently solving model (3). The objective function is in the low-dimensional unconstrained setting but non-convex and one needs to avoid local optima. In addition, since it's not possible to obtain gradient (first-order) and Hessian (second-order) information, one has to relies on the function queries (zeroth-order). Here, we propose a stochastic zeroth-order algorithm for model (3) inspired by the decades of work in zeroth-order optimization literature and primarily motivated by contemporary statistical machine learning problems.

SDD is a stochastic search method proposed specifically for the oracle optimization problem (3). At each iteration, we will generate a random direction, and search the descent of objective function in this direction. A quadratic function will be fitted in the search direction to accelerate the search of optimal solution. Formally, we maximize an expensive function $G : \Omega \to R$ to obtain

$$x_{opt} = \operatorname*{argmax}_{x \in \Omega} G(x)$$

Within a domain $\Omega \subset R^k$ which is a bounding box and k is typically small (k = 2 or 4 in our case). SDD generates the iterative point by the following search strategy,

$$x_{star} = \operatorname*{argmin}_{x \in \aleph(P)} Q(x)$$

where P is a randomly generated unit vector and $\aleph(P)$ is the linear space expanded by vector P. Q is a quadratic function by fitting three points x_0, x_1, x_2. x_0 is the initial point. $x_1, x_2,$ are the two random points along vector P, i.e., $x_1 = x_0 + \lambda_1 P, x_2 =$

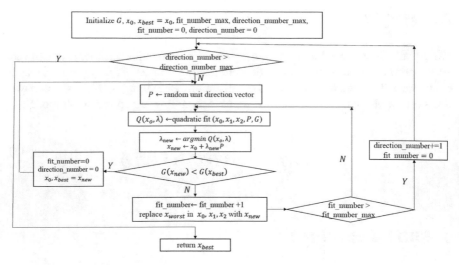

Fig. 1 The general workflow of SDD search method

$x_0 + \lambda_2 P$. λ_1, λ_2 are two parameters determined by the value of $G(x_0), G(x_1), G(x_2)$. The algorithm workflow is shown in Fig. 1 and the details can be seen in Appendix.

Our motivation for SDD method is that the deterministic search methods such as simplex method often stuck at the local optima or the points with vanishingly small gradient for this calibration problem. By introducing stochastic mechanism, the SDD method can help the solution escape from these traps.

4 Datasets and Measurement of Prediction Quality

4.1 Datasets

We employed five typical scenarios to test the above optimization models and searching strategy. The five cases are described in Table 1 and trajectories and casings are shown in Fig. 2. We compare SDD with other popular search methods (Grid,

Table 1 Field cases under typical scenarios

Data ID	Well trajectory profile	Type
Case 1	Vertical	Half casing, half open hole
Case 2	Vertical	Half casing, half open hole
Case 3	S-shape well	Casing
Case 4	J-shape well	Half casing, half open hole
Case 5	J-shape well	Casing

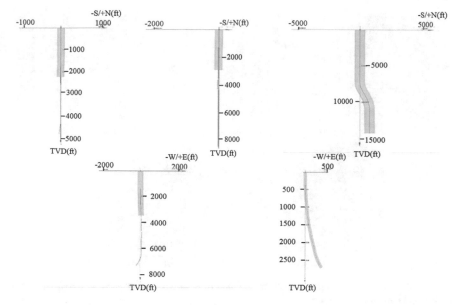

Fig. 2 Wellbore trajectories and casings for cases 1–5

Bisection, Simplex, the details are described in Appendix).

4.2 Measurement of Prediction Quality

In this paper, we use two well-known evaluation indicators to evaluate calibration performance. One is root mean square error (RMSE). Another is mean absolute percentage error (MAPE). They are formulated as follows:

$$\text{RMSE} = \sqrt{\sum_{i=1}^{n}(y_i - \hat{y}_i)^2 \bigg/ n}$$

$$\text{MAPE} = \left(\sum_{i=1}^{n}\left|\frac{y_i - \hat{y}_i}{\hat{y}_i}\right| \times 100\%\right)\bigg/ n$$

where n is the dimension of $\hat{y}$.

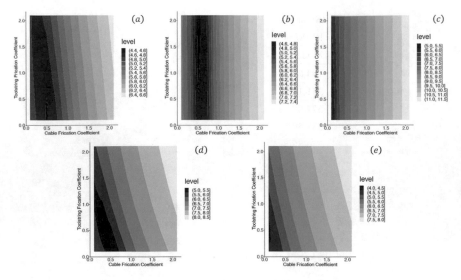

Fig. 3 The approximate $\log_{10}$-transformed function landscape of datasets **a** Case1, **b** Case2, **c** Case3, **d** Case4, and **e** Case5

5 Numerical Results

5.1 Objective Function Landscape by Grid Method

First, we shape function landscape through grid method. It is worth noting here that we assume that the friction coefficients f_c, f_t are constant throughout the whole well regardless of the well type. We set f_c and f_t from 0.1 to 2.1 with an interval of 0.2. The heatmap of $\log_{10}$-transformed mean square error (MSE) in five cases are plotted in Fig. 3.

From Fig. 3 we can see that the objective function with friction coefficients in five cases approximately have unimodal structure. In addition, the loss function is much sensitive to the tool's friction coefficient than the table's friction coefficient, especially in case 1, case 2, and case 3.

5.2 Model 2 is More Effective Than Model 1

We applied the simplex method to Model 1 and Model 2. Table 2 summarizes RMSE and MAPE results based on same simplex iterations in five cases.

Table 2 Comparison of RMSE and MAPE based on Model 1 and Model 2

Data ID	RMSE/MAPE	
	Model 1	Model 2
Case 1	194.03/2.998%	185.10/2.881%
Case 2	341.12/2.785%	248.26/2.086%
Case 3	577.53/3.449%	502.64/3.172%
Case 4	331.59/1.959%	341.67/1.997%
Case 5	344.86/5.357%	150.07/1.927%

As shown in Table 2, model simplification exhibits a significantly decrease in RMSE and MAPE in most cases. Especially in case 5, the RMSE decreases from 344.86 to 150.07 while MAPE decreases from 5.357 to 1.927%. The predicted tension curves of simplex solution based on Model 1 and Model 2 in case 5 are visualized in Fig. 4 (The results in other four cases can be found in appendix Figs. 6, 7, 8 and 9).

In addition, Table 2 shows that the MAPE based on Model 2 are smaller than 5% in all five cases. This demonstrates the power of model to predict surface tension by calibrating parameters. Next, we will compare the performance of different search methods.

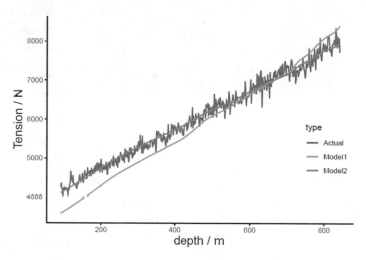

Fig. 4 The predicted tension curves of simplex solution based on Model 1 and Model 2 in case 5

5.3 SDD Outperforms Gird, Bisection, and Simplex Methods

We observed that the above search method doesn't converge to a good solution in some difficult cases, based on Model 2. Next we apply our new search method SDD in five cases and compare with three existing search methods (grid, bisection, and simplex). The results are summarized in Table 3.

As shown in Table 3, the SDD method outperforms other three search methods in five cases. Especially in case 1 and case 2, the RMSE and MAPE decrease significantly. Importantly our new search method overcomes several limitations of the existing methods. The grid search method is simple but has high complexity, the bisection search method converges quickly but is difficult to reach an optimal solution, and the simplex search method is efficient compared with grid and bisection method but is hard to converge in some difficult cases. In contrast, SDD converges slightly faster and usually reaches a local optimum. The predicted tension curves of simplex and SDD solution based on Model 2 in case 1 are visualized in Fig. 5 (The

Table 3 Comparison of RMSE and MAPE based on different search methods

	RMSE/MAPE			
	Grid	Bisection	Simplex	SDD
Case 1	194.23/3.014%	211.54/3.386%	185.10/2.881%	85.898/1.046%
Case 2	248.26/2.086%	248.26/2.086%	183.77/1.402%	99.342/0.672%
Case 3	502.64/3.172%	502.64/3.172%	428.53/3.006%	423.86/3.036%
Case 4	332.37/1.941%	337.88/1.897%	341.67/1.997%	308.49/1.743%
Case 5	148.40/1.902%	150.09/1.943%	150.07/1.927%	148.19/1.894%

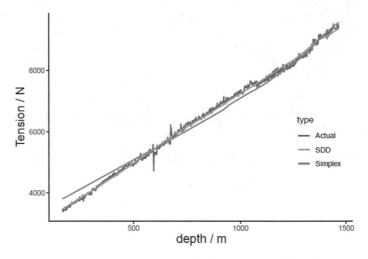

Fig. 5 The predicted tension curves of simplex and SDD solution based on Model 2 in case 1

results can be found in appendix for other four cases Figs. 10, 11, 12 and 13).

In case 1, the RMSE decreases from 185.10 to 85.898 and MAPE decreases from 2.881 to 1.046% by SDD compared with simplex search method. The error at about 300 and 1000 m decrease significantly (Fig. 5). Since SDD is a random search algorithm, we next test the sensitivity of SDD method to the initial solution in case 3 and case 5.

5.4 SDD Has Good Stability to the Initial Solution and High Efficiency

To test SDD method's robustness to the initial solution, we use two cases and four or five different initial solutions. The results are shown in Tables 4 and 5 respectively.

From Tables 4 and 5 we can see that using different initial values, the final solutions are all in a close area. This demonstrates that SDD has good stability to the initial solution. At the same time, the number of search points are close to simplex method (about 500). In general, SDD is good choice to solve oracle optimization problems.

Table 4 Comparison of SDD results in different initial values in case 5

Initial point	Solution	MSE	#search point
fc = 0.5, ft = 0.1	fc = 0.144, ft = 0.351	21,960.6	740
fc = 0.5, ft = 0.5	fc = 0.144, ft = 0.348	21,989.8	680
fc = 2.0, ft = 1.0	fc = 0.143, ft = 0.349	21,959.8	685
fc = 0.14, ft − 0.35	fc = 0.144, ft = 0.347	21,959.8	443

Table 5 Comparison of SDD results in different initial values in case 3

Initial point	Solution	MSE	#search point
fc = 1, ft = 0.1	fc = 0.076, ft = −0.13	179,655.5	673
fc = 0.5, ft = 0.5	fc = 0.076, ft = −0.08	179,531.6	659
fc = 2.0, ft = 1.0	fc = 0.076, ft = −0.12	179,556.3	574
fc = 0.1, ft = 1.0	fc = 0.076, ft = −0.07	179,595.2	457
fc = 10, ft = −10	fc = 0.076, ft = −0.13	179,632.3	984

6   Discussion

In this paper, we rely on the measured data and introduce the optimization model to calibrate wireline conveyance tension prediction in well logging. The challenge is that the objective function doesn't have a closed form in mathematics and its value is passed by an oracle (numerical computing engine). Firstly, we establish two equivalent models and implement simplex method to compare their power by RMSE/MAPE. The results show that Model 2 is more efficient than Model 1. Secondly, we propose a new search method SDD and compare with three existing search methods. The results show that SDD outperforms other three search methods in all five cases. Finally, we test the sensitivity of SDD method to the initial solutions and show that SDD is robust for the initial solution and its efficiency can be comparable to simplex search method.

In the end, the RMSE in case 3 and case 4 are 423.86 and 308.49 and there are much room for improvement compared to other three cases. A very important reason may be the constant assumption of the coefficient friction along the hole. Because of complexity of formation structure and well shape, it is unreasonable to suppose the friction coefficient f_c, f_t are constant in casing/open section of well. We can update the computing engine to allow multi-segment friction coefficients and implement SDD methods in our future work.

Appendinx

Search Methods

Grid search. The grid method is one of the simplest search method that explores the parameter settings using a grid. It can help us visualize function lanscape and is currently the most widely used method for global optimization.

Bisection search. The bisection method is one of reliable, easy to implement, and convergence method. It is well-known in finding real root of non-linear equations and can be extended to solve optimization problem.

Simplex search. Simplex method is a classical derivative-free optimization algorithm which depends on the comparison of function values at a general simplex, followed by the replacement of the worst vertex by another point [8, 9]. It is more efficient than the grid method and bisection method [10–12].

SDD method

Input: G, x_0, $x_{best} \leftarrow x_0$, fit_number_max, direction_number_max,
fit_number $\leftarrow 0$, direction_number $\leftarrow 0$
While True
  if direction_number > direction_number_max
    break
  else
   $P \leftarrow$ random unit direction vector
   x_1 , $x_2 \leftarrow$ $x_0 + \lambda_1 P$, $x_0 + \lambda_2 P$, satisfied $0 < \lambda_1 < \lambda_2$
and $G(x_0), G(x_2) >$ $G(x_1)$ or satisfied $\lambda_1 < 0 < \lambda_2$ and
$G(x_1), G(x_2) > G(x_0)$ or satisfied $\lambda_1 < \lambda_2 < 0$ and
$G(x_1), G(x_0) > G(x_2)$
    While True
      $Q(x_0, \lambda)$ quadratic fit (x_0, x_2, x_3, P, G)
      $\lambda_{new} \leftarrow argmin\ Q(x_0, \lambda)$, $x_{new} \leftarrow x_0 + \lambda_{new} P$
      If $G(x_{new}) < G(x_{best})$
        fit_number $\leftarrow 0$
        direction_number $\leftarrow 0$
        $x_{best} \leftarrow x_{new}$
        break
      else
        replace x_{worst} in x_0, x_2, x_3 with x_{new}
        fit_number fit_number + 1
        if fit_number > fit_number_max
          fit_number = 0
          direction_number $\leftarrow$ direction_number + 1
          break
return x_{best}

Supplementary Figures

See Figs. 6, 7, 8, 9, 10, 11, 12 and 13.

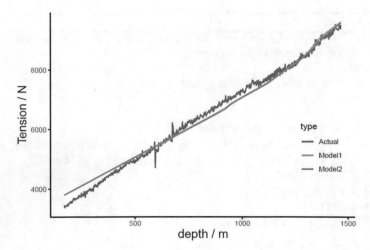

Fig. 6 The plan tension curves of simplex solution based on model 1 and model 2 in case 1

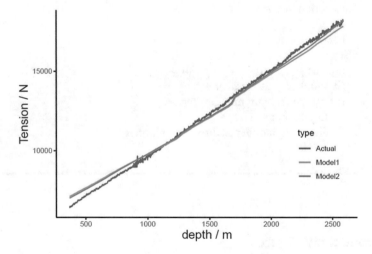

Fig. 7 The plan tension curves of simplex solution based on model 1 and model 2 in case 2

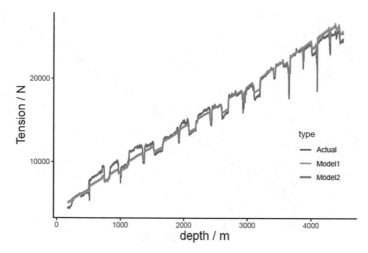

Fig. 8 The plan tension curves of simplex solution based on model 1 and model 2 in case 3

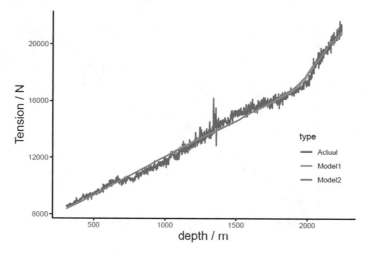

Fig. 9 The plan tension curves of simplex solution based on model 1 and model 2 in case 4

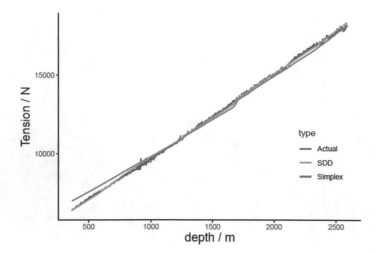

Fig. 10 The plan tension curves of simplex and SDD solution based on model 2 in case 2

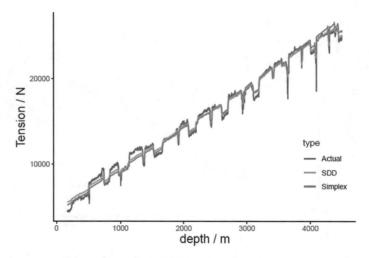

Fig. 11 The plan tension curves of simplex and SDD solution based on model 2 in case 3

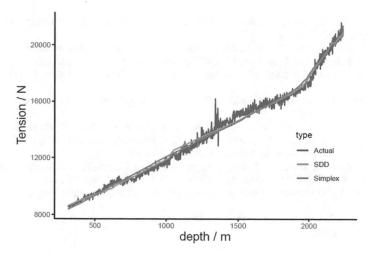

Fig. 12 The plan tension curves of simplex and SDD solution based on model 2 in case 4

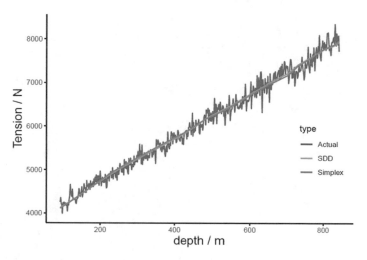

Fig. 13 The plan tension curves of simplex and SDD solution based on model 2 in case 5

References

1. Underhill, W.B, Moore, L., et al.: Model-based sticking risk assessment for wireline formation testing tools in the U.S. Gulf Coast. Well Testing (1998)
2. Kolda, T.G., Lewis, R.M., Torczon, V.: Cable-conveyed well intervention. Paper SPE 71560 presented at the SPE Annual Technical Conference and Exhibition held in New Orleans, Louisiana (2001)
3. Mohammadpoor, M., Torabi, F.: Big Data analytics in oil and gas industry: an emerging trend. Petroleum (2018)

4. Lu, H., Guo, L., Azimi, M., Huang, K.: Oil and Gas 4.0 era: a systematic review and outlook. Comput. Ind. **111** (2019)
5. Wu, T., Mao, Y., Zhao, G.: A model designed for HSE big data analysis in petroleum industry. In: International Petroleum Technology Conference 2019 (IPTC 2019) (2019)
6. Nelder, J.A., Mead, R.: A simplex method for function minimization. Comput. J. **7**(4), 308–313 (1965)
7. Mckinnon, K.I.M.: Convergence of the Nelder-Mead simplex method to a nonstationary point. SIAM J. Optim. **9**(1)
8. Boyd, B., Vandenberghe, L.: Convex optimization. Cambridge University Press (2004)
9. Beltran, C., Edwards, N.R., Haurie, A.B., et al.: Oracle-based optimization applied to climate model calibration. Environ. Model. Assess. **11**(1), 31–43 (2006)
10. Lewis, R.M, Shepherd, A., Torczon, V.: Implementing Generating Set Search Methods for Linearly Constrained Minimization. Society for Industrial and Applied Mathematics (2005)
11. Powell, M.J.D.: On search directions for minimization algorithms. Math. Program. **4**(1), 193–201 (1973)
12. Powell, M.J.D.: The BOBYQA algorithm for bound constrained optimization without derivatives (2009)

An Overview of Complex Enterprise Systems Engineering: Evolution and Challenges

Xinguo Zhang, Chen Wang, Lefei Li, and Pidong Wang

1 Introduction

With the integration of the digital, physical and human domains, industries across the world have undergone profound transformations. The fourth industrial revolution, Internet of Things, the industrial internet, and the big data era are altering the approaches toward designing, developing, manufacturing, and providing a product (or service) [1, 2]. The transformation has resulted in increasing complexity of systems, particularly in the aerospace, energy, and automotive industries etc. [3]. The complexity of a system increases as the number of interconnected parts grows, resulting in abundant possibilities of emergent behaviors. The complexity of a purely mechanical system is relatively low, as compared to a mechatronic system and an integrated system consisting of mechanical, electronic, software, network (M/E/S/N) and other components. The levels of system complexity growth are presented in Fig. 1.

Recently, Enterprise Systems (ES) have drawn much attention with the growing recognition of the significance of organizational developments and operations. As "a complex socio-technical system that comprises interdependent resources of people, information, products and technology that must interact with each other and their environment in support of a common mission" [4], ES is a purposeful integration of interdependent resources (including people, processes, organizations, products, technology, and funding) to achieve business and operational goals [5]. Because

X. Zhang (✉) · C. Wang
Tsinghua University, No. 4 Room 1107, Shuangqing Building, No. 77 Shuangqing Road 100084, Beijing, China
e-mail: zhangxinguo@tsinghua.edu.cn

L. Li · P. Wang
Tsinghua University, No. 4 Room 1103, Shuangqing Building, No. 77 Shuangqing Road, Beijing 100084, China
e-mail: lefeili@tsinghua.edu.cn

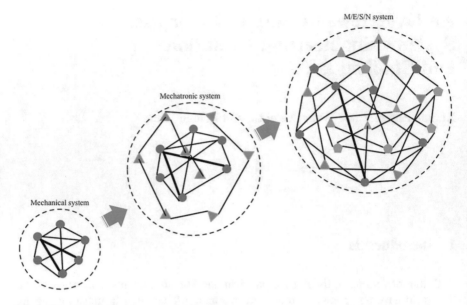

Fig. 1 The growth of system complexity

the performance of decision makers within the enterprise, the workflow and business processes, and all technical systems to support decision have to be considered [6], it is challenging to evaluate the operational capability of an enterprise. Despite arising as an interdisciplinary and collaborative approach to deal with these challenges [7], Systems Engineering (SE) approach is not suitable for ES. It is ineffective for handling the complexity of ES [8], considering the operational environment of the system under development, as well as its interactions with other systems to complete an overall operational mission. The emergence, derived from components and architectures of ES, along with the ever-growing interconnectivity and interoperability, makes it difficult to make good design choices [9]. This trend has weakened the ability of system engineers to predict the results of their design decisions [10]. Owing to high interdependency between the constituent systems, the development team should balance the performance of them [11].

In this paper, we first review recent research works on Enterprise Systems Engineering (ESE) and Enterprise Architecture (EA), and highlight the new trend of model-based approach. The approach builds on the principles of model-based on ESE [12] and present challenges of future research on ESE.

2 Background

2.1 Enterprise Systems Engineering

In ES, the involvement of human and organizational behaviors imparts complexity and unprecedented challenges to the SE approaches [13]. Hence, Enterprise Systems Engineering (ESE) is the application of System of Systems Engineering to the planning, design, improvement, and operation of an enterprise [14], as "the body of knowledge principles and practices to design an enterprise" [15]. ESE focuses on architecture, methods, and tools to solve problems and addresses inherent complexity of an enterprise [16]. It enables top management to rapidly implement new strategies and control key business parameters to gain a competitive advantage [4]. In addition to solving problems, ESE also addresses the creation of opportunities for better ways to achieve the enterprise's goals [8], such as optimizing costs, quality and speed of delivery. The fundamental concept of ESE is rooted in the theories of systems thinking and complex systems science, and incorporates SE principles, concepts, and methods into the enterprise context [14]. The evolution of ESE and Complex Systems Engineering (CSE) are shown in Fig. 2. ESE is confronting a new era of digitalization enterprise: the development of a broader areas of digital technology and information, the innovation of more-integrated business and manufacturing, and a need for much faster adaptability [4]. The development poses new challenges for SE to address enterprise architecture (EA), strategic technical planning, enterprise analysis and organizational behavior.

Considering the life cycle of many ESE plans, the initial EA may no longer be applicable when new business processes and manufacturing systems are brought out. Apart from the SE roles in the acquisition and development of product systems,

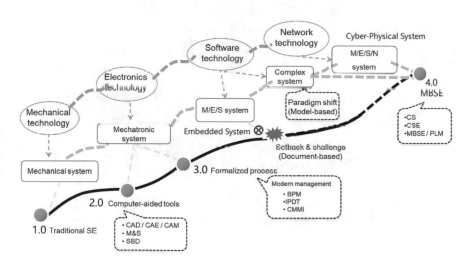

Fig. 2 The evolution of complex systems engineering (CSE)

ESE extends to the development of strategic goals, EA, and enterprise technology capabilities [17]. It focuses on defining the business progresses of an enterprise, including the requirement definition concerning business progress, design synthesis, and progress verification and validation [18].

2.2 Enterprise Architecture

IBM develops EA that "defines and maintains the architecture models, governance and transition initiatives needed to effectively coordinate semi-autonomous groups towards common business and/or IT goals [19]." However, EA has been conceptualized at a high level, there is no accepted definition of EA in the world at present. In general, for an industrial enterprise, EA can be considered as a structured and consistent set of plans for the integrated representation of the business and information technology (IT) domains of the enterprise in past, current, and future states [20]. It applies architectural principles to conduct enterprise analysis, design, planning and implementation, and presents an integrated view of organization, business and technology for the development and execution of the enterprise strategy. Since EA covers the activities of the whole organization from business to technology domains systematically, EA is no longer only a method for information management system and system design. The activities of EA are delivered through domains including business, information, technical and application architecture [21]. Each domain has its own deliverables, which determines its implementations. As shown in Fig. 3, the EA lays basis for the business model and supports business processes.

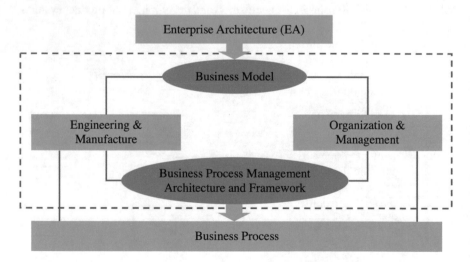

Fig. 3 Architecture-centric transformation

In Fig. 3 the architecture-centric transformation requires engineering management to extend SE practices to deal with people, process, organization, technology and funding. These elements are closely connected to each other. As a result, engineers and managers have to coordinate functions, share information, build working platforms, allocate resources and make adaptive decisions. Hence, these domains are unique architecture disciplines, but not independent of each other [22], interrelated for the overall realization of goals and objectives of ES [23]. It brings the frameworks challenging and complex when mapping the domains with activities of ES and deliverables. Which based on how EA defined in an ES [24]. Implementing EA requires a lot of investment, which is not always feasible in many organizations.

Several literature reviews on EA have been conducted [25]. There are different methods of implementing EA [26]. Many EA frameworks or methodologies have emerged, in recent years [27]. In order to increase opportunities for successful implementation of sustainability and competitiveness, different ES have adopted different approaches including implementations of different frameworks, such as Gartner Inc., Enterprise Architecture Planning (EAP), Forester and The Open Group Architecture Framework (TOGAF), Department of Defense Architecture Framework (DODAF) for EA implementation [28, 29]. With context and uniqueness requirements, it is hard for some frameworks to implement in various ES. Some frameworks are so general that it cannot be applied to specific environments [30]. In addition, due to the complexity of ES, some frameworks are oriented to the enterprise level, while some frameworks are only for the development of information systems or technology systems. It makes the differences between these frameworks and has an impact on their implementation [31]. With the lack of compatibility, the adoption of some frameworks has declined.

Therefore, only a few frameworks (such as Gartner Enterprise Architecture, Federal Enterprise Architecture Framework (FEA) and TOGAF, Zachman Enterprise Architecture Framework (ZEAF)) are currently available [32]. Among these frameworks, the ZEAF was the most popular [33]. Nevertheless, ZEAF still presents some challenges for architects and ES [34]. Models and approaches have been proposed in recent years [35, 36], such as models for implementing the Zachman Framework through methodology based on action research [37], a knowledge architecture model [38], and a reference architecture model for collaborative activities [24]. Some scholars argue that some shortfall of Zachman Framework come from both business and information system/information technology domains [39]. An action research by using the Zachman Framework to assist in the implementation of the EA framework has been conducted, in which they focused on understanding the data of enterprise [30]. It brings the frameworks challenging and complex when mapping the domains with activities of ES and deliverables. Which based on how EA defined in an ES. The major contribution of this paper is to establish an updated ontology of EA framework.

3 A New Paradigm for Complex Enterprise Systems with EA Modeling and Enterprise Ontology

3.1 Enterprise Ontology

The Zachman Framework provides an enterprise ontology (EO) with which we can view and analyze an enterprise system systematically. The Zachman Enterprise Architecture Framework (ZEAF) is a two-dimensional schema. One of the dimensions specifies five levels of perspectives including contextual, conceptual, logical, physical, and detailed. Meanwhile, the other dimension adopts the 5W1H approach comprising why, how, what, who, where, and when [40]. We update the EO of ZEAF to make it suitable for an enterprise that provides technology products (or services) to satisfy stakeholders' expectations. The details are shown in Fig. 4. In particular, the high-level leadership perspective defines the tasks of a business planner including engineering and management, i.e., to identify the business scope, target, background, and environment.

The high-level leadership is confronted with guiding enterprises that exhibit high complexity and uncertainty. In such conditions, the present methods are likely to fail. They also have to be able to establish the "right" partnerships and build the success of their organizations on inter-organizational trust and openness because isolation will cause failure. The business management perspective defines the tasks of a business concept owner, who is responsible for the design, operation, and improvement of the organizational model. At a level below, the architect's perspective defines the tasks of a business logic designer, who pays more attention to the logic of modeling a system. Meanwhile, the engineers focus on the technical model at the physical level.

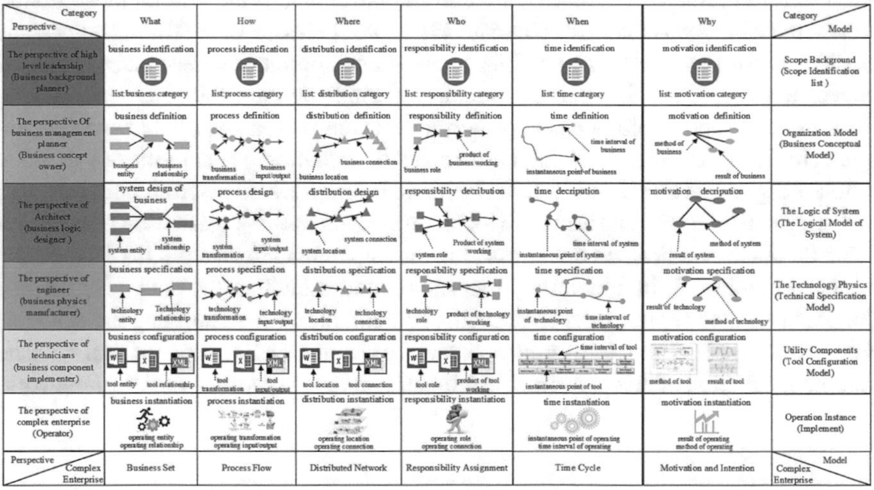

Perspective \ Category	What	How	Where	Who	When	Why	Category \ Model
The perspective of high level leadership (Business background planners)	business identification / list business category	process identification / list process category	distribution identification / list distribution category	responsibility identification / list responsibility category	time identification / list time category	motivation identification / list motivation category	Scope Background (Scope Identification list)
The perspective Of business management planner (Business concept owner)	business definition	process definition	distribution definition	responsibility definition	time definition	motivation definition	Orgnization Model (Business Conceptual Model)
The perspective of Architect (business logic designer)	system design of business	process design	distribution design	responsibility decribution	time decription	motivation decription	The Logic of System (The Logical Model of System)
The perspective of engineer (business physics manufacturer)	business specification	process specification	distribution specification	responsibility specification	time specification	motivation specification	The Technology Physics (Technical Specification Model)
The perspective of technicians (business component implementer)	business configuration	process configuration	distribution configuration	responsibility configuration	time configuration	motivation configuration	Utility Components (Tool Configuration Model)
The perspective of complex enterprise (Operator)	business instantiation	process instantiation	distribution instantiation	responsibility instantiation	time instantiation	motivation instantiation	Operation Instance (Implement)
Perspective \ Complex Enterprise	Business Set	Process Flow	Distributed Network	Responsibility Assignment	Time Cycle	Motivation and Intention	Complex Enterprise \ Model

Fig. 4 An ontology of the enterprise architecture

The perspective of a complex enterprise is to provide use-cases of business, process, distribution, responsibility, and time to all the relevant users in an enterprise.

The proposed EA framework has thirty-six core elements covering the conceptual issues within an enterprise system. The framework establishes a common language for the managers and the engineers from different disciplines and enables effective communication towards shared organizational goals. For example, different rows (and perspectives) correspond to the missions of different roles in an enterprise. The "what" column is updated with respect to the business sets that enterprises must track and manage and that may have multiple uses. These include input/output of system design processes, input/output of manufacturing processes, input/output of organizational (management) processes, modes for execution of system design, and manufacturing or organizational processes. The "what" column may transform rapidly and even not be under complete control of the enterprise [23]. The "when" column specifies all important lifecycle considerations of the enterprise and of the product (or service) systems it generates. Enterprises have to integrate, synthesize and analyze cross-data sets including external and internal data in real time in order to obtain information from the network of interrelated data items. The challenge for the enterprises is how to effectively understand large data in real time and take corresponding actions, instead of understanding data from isolated sources. Therefore, due to the complexity of EA, it is suggested that the interpretable EA description is important for complex enterprise systems. If the architecture description can be understood and interpreted by computers, it will allow answers such as "What processes will a replacement application affect?" "What roles are involved in the process?" "Why do we decide to customize for this particular application?".

According to Hinkelmann, EO can meet this requirement. One advantage of having an ontological description of an EA is machine-interpretable [4]. Moreover, EO can be used for realizing interoperability among systems, between systems and humans, or among humans. They can be used for uniquely identifying and disambiguating concepts through formal semantics, thereby facilitating knowledge transfer. They enable the determination of implicit facts and can be used for analyzing and detecting logical inconsistencies. Moreover, they enable systematic domain descriptions and the reuse of knowledge models in new applications [41]. EO has many representation formalisms interpreted by machines [4]. But, there is no 'right' language to formally describe an enterprise ontology. The "choice of the language to use in a system or analysis will ultimately depend on what types of facts and conclusions are most important for the application" [4, 42]. Hence, we add the rightmost column to the framework by introducing appropriate models developed under the recent trend of Model-based Systems Engineering (MBSE) and Model-based engineering (MBe). We further elaborate on the concept of Model-based Enterprise (MBE) in the following.

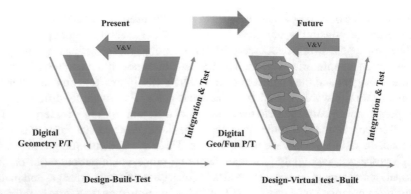

Fig. 5 Paradigm shifting of Industry Engineering 4.0

3.2 The Model-Based Enterprise Systems Engineering

The paradigm of model-driven engineering to Industry Engineering 4.0 has shifted as shown in Fig. 5: Tests and reworks are conducted on models rather than on entities, which significantly reduces the cost and time incurred. Models are used for designing and adapting enterprises and manufacturing systems before they are altered in reality. Therefore, all the likely problems are revealed in an early stage of design. The challenge is how to rapidly respond to transformations in business intelligence and manufacturing intelligence.

There are many attentions and in-depth research on model-based techniques as means to deal with the challenges of ESE [43]. Engineers can use models of EA, constituent systems, infrastructure, and environment to explore alternative designs before making commitments through detailed design and prototyping. The models also enable the validation of quality characteristics such as reliability, resilience and security. Furthermore, if they have formal semantics, it becomes feasible to use computer aided engineering to identify potential defects.

It is still difficult to apply model-based methods to ESE. As shown in Fig. 5, three main phases constitute the traditional V-model development process at present: design in the virtual domain, and build and test in the physical domain. Under the new model-based industrial paradigm, we emphasize the iterations between digital design, virtual integration and testing, and digital manufacturing, all in the virtual domain. The product system is physically built and tested in the physical domain only after the design is sufficiently verified and validated. It is a paradigm shift that significantly reduces the cost and time incurred by tests and reworks on physical prototypes. In the enterprise context, the concept of MBE (by SIEMENS) is proposed to describe the enterprise-level application of model-based methods. An MBE is an organization that utilizes modeling and simulation technologies to integrate and manage its technical and business processes related to production, support, and product disposal [44].

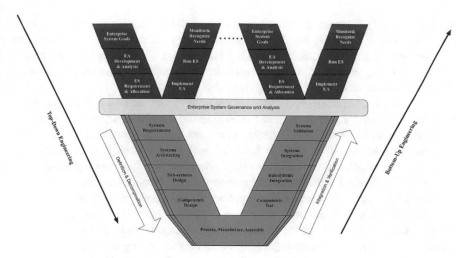

Fig. 6 The ES "Vee" model

According to the System of Systems Engineering and Integration (SOSE&I) methodology and "Vee" model [45], the MBE methodology can be demonstrated through a ES "Vee" process model (V model) (shown in Fig. 6).

The bottom of the V model is made up of multiple V model of each individual system, including the traditional SE activities. The coexistence of these models indicate that many systems are being developed and managed in parallel, with each system being an independent time and space point in its life cycle. The upper side of the V model represents the activities cycle including establishing/adjusting ES goals, developing and analyzing EA, allocating ES requirement, implementing and running ES, monitoring the status of ES and recognizing adaptation needs. If a need is recognized, the cycle starts again. These V models from end to end are established to describe the process of the ongoing iteration on design and redesign in engineering the evolving ES. And the governance-policy processes covers all the areas of the "V" are represented in the middle area of "V" [45]. A top-down engineering approach consists of the activities on the left hand side of the "V", whereas a bottom-up one are represented by the activities on the right-hand side. Thus, the ES requirements developed on the leftmost-side of the "V" are demonstrated and certified on the rightmost-hand side.

A strong governance process must be established to communicate and manage changes during the life cycle, simultaneously maximizing the ES and mission effectiveness. If the framework is established to coordinate among the individual systems and programs better, the ES will benefit from it. Governance in this model implies the set of rules, policies, and decision-making criteria. They can guide the ES toward completing its goals and objectives [46]. A dependable governance plan is critical to reduce the time to deploy; increase the qualities of constituent systems and the ES; increase flexibility of ES to rapidly adapt to the changing environments and threats; and identify technical problems in reducing mission efficiency or increasing the ES cost [47]. Before the development of the governance architecture, the type of ES must be understood deeply, due to "one size does not fit all." However, the effective governance is required to identify clear scope of authority for most stakeholders, ensure greater acceptance of newly implemented ES governance authority, and ensure consistency and transparency in decision-making.

To ensure a successful ES, there are two important activities must be done. First, the ES team must have in-depth knowledge of development, sustainment, and management in the constituent systems to ensure that the systems are compatible with the ES. Second, it is critical that the constituent-system functionalities and programmatic problems must be understood, because the constituent systems in an ES are interdependent with each other to complete ES mission success.

Innovative approach must be employed to evaluate and predict performance of the ES, such as MBSE to simulate and predict performance and the evaluation of ES performance during demonstrations. In most cases, the emergence demands new functions including the control of the adoption of new technologies and interactions in a system [45]. New hazards that frequently occur in an ES are usually the result of new function or changes brought in by the integration of constituent systems into the ES. Further research needs to focus on how the processes provide appropriate guidance on the inspection and control of safety, reliability, or hazards [45]. Moreover, there are two challenges for the modeling procedures: (1) extend with phases of machine-interpretation, which has been realized for only certain specialized procedures [4]; (2) keep the connection between human- and machine-interpretable models [4].

4 Conclusions

With the growing complexity and scales of systems, system engineering's horizons have expanded substantially. With regard to ESE, it is critical to ensure effective interactions among interdependent resources including people, information, and technology to achieve the overall mission of an enterprise. We identify the following challenges in ESE research.

(1) ES requirements should focus on the capacity, performance, efficiency, cost, and reliability of the overall ES, rather than specifying a particular design approach or implementation. Until now, the factors causing success or failure of ES remains unclear [13]. Future work needs to investigate the contributors of enterprise successes and failures, including organizational and cultural factors.

(2) The independence of constituent systems in an ES is the source of a number of technical issues confronted by ESE. This complexity can result in unanticipated effects at the ES level and result in unexpected or unpredictable behavior in an ES, despite a clear understanding of the behavior of the constituent systems. System synthesis should focus on providing holistic solutions with the following properties: (1) desired functions; (2) emergence in the presence of

multi-interactions at different system levels; (3) physical separability of solutions; (4) appropriate balance between purposes and methods; (5) adequate commonality and flexibility for both products and processes; (6) interoperability of ESs; and (7) standardization and unification of development processes and tools.

(3)   The modeling effort needs to focus on the deviation between the virtual ES and real ES. It implies that apart from the ES, its operators and operating environment should also be considered. The accuracy and efficiency of the simulation models should be studied to support an analysis of the system behaviors under normal and abnormal conditions in practice. Moreover, the data for model-based design and analysis should be available throughout the product lifecycle.

(4)   In the ES integration process, we need to optimize system performance as well as consider the size, life and lifecycle cost. It is effective to incorporate high software intelligence with simple operation and maintenance when developing a complex ES. Reusability at both the component and system levels is also an important design factor. Moreover, reliability plays a vital role in ESE. Early consideration of reliability generally results in better product quality and reduced cost and time. Finally, when human behavior is involved, we encounter epistemic uncertainty caused by ambiguity, unknown information, ignorance, and man-made mistakes [48], many of which cannot be modeled as pure randomness. Uncertain factors result in the emergence of uncertain behavior at the enterprise level.

(5)   The stability, robustness, and reliability of ES should be demonstrated from a systemic perspective. The validation should focus on comparing the model with the real system, rather than depending only on expert assessments. ESs composed of constituent systems that are independent of the ES poses challenges in conducting end-to-end ES testing typically carried out in system design and development. In particular, with large ESs, traditional testing of the entire ES following each change in the constituent systems may not be affordable or feasible. This implies that ESE needs to consider a wider range of options for collecting data to assess the risks to the ES. This includes data collection from actual operations or through estimates based on modeling, simulation, and analysis. Despite these challenges, it is the ESE team's responsibility to ensure the continuity of operation and performance of the ES.

Acknowledgements This study was co-supported by "National Key R&D Program of China" (No. 2017YFF0209400) and "Key Laboratory of Quality Infrastructure Efficacy Research Funding" (No. KF20180401).

References

1. Gorod, A., White, B.E., Ireland, V., et al.: Case Studies in System of Systems, Enterprise Systems, and Complex Systems Engineering, 1st edn. CRC Press, Boca Raton, USA (2014)
2. Zhong, R.Y., Xu, X., Klotz, E., et al.: Intelligent manufacturing in the context of Industry 4.0: a review. Engineering (3), 630 (2017)
3. Norman, D.O., Kuras, M.L.: Engineering complex systems. In: Complex Engineered Systems. Springer, Berlin, Heidelberg (2006)
4. Hinkelmann, K., Gerbe, A., Karagiannis, D., et al.: A New paradigm for the continuous alignment of business and IT: combining enterprise architecture modelling and enterprise ontology. Comput. Ind. **79**, 77–86 (2016)
5. Teoh, S.Y., Yeoh, W., Zadeh, H.S.: Towards a resilience management framework for complex enterprise systems upgrade implementation. Enterprise Inf. Syst. **11**(5), 694–718 (2017)
6. Kotzé, P., Smuts, H.: Applying the foundations of enterprise systems engineering in complex real-world environments: lessons learnt from real-world project examples. In: 2016 4th International Conference on Enterprise Systems (ES), pp. 1–12, Melbourne, VIC, Australia, March (2017)
7. Petter, K., Pereira, L.: Systems engineering: an interdisciplinary challenge. In: 30th Congress of the International Council of the Aeronautical Sciences, pp. 1–8, Seoul, Korea, August (2016)
8. Keating, C., Rogers, R., Unal, R., Dryer, D., Sousa-Poza, A., Safford, R., Peterson, W., Rabadi, G.: System of systems engineering. EMI-Eng. Manage. J. **15**(3), 36 (2003)
9. Ambrosio, J.: Systems engineering challenges and MBSE opportunities for automotive system design. In: 2017 IEEE International Conference on Systems, Man, and Cybernetics (SMC), pp. 2075–2080. Banff Center, Banff, Canada, 5–8 Oct 2017
10. Pennock, M.J., Wade, J.P.: The top 10 illusions of systems engineering: a research agenda. Procedia Comput. Sci. **44**, 147–154 (2015)
11. Madni, A.M., Sievers, M.: System of systems integration: key considerations and challenges. Syst. Eng. **17**(3), 330–347 (2014)
12. Iris, G., Yang, X.: Interdisciplinary development of production systems using systems engineering. Procedia CIRP **50**, 653–658 (2016)
13. Sauser, B., Boardmen, J., Corod, A.: System of Systems Management. System of Systems Engineering. Wiley, Hoboken, NJ, USA (2008)
14. BKCASE Editorial Board, Enterprise Systems Engineering. In: Guide to the Systems Engineering Body of Knowledge (SEBoK) V1.6, The Trustees of the Stevens Institute of Technology (2016)
15. Giachetti, R.E.: Design of Enterprise Systems: Theory, Architecture, and Methods, Boca Raton, FL. CRC Press, Taylor and Francis Group, Boca Raton, FL, USA (2010)
16. Stanford, N.: Guide to Organisation Design: Creating High-performing and Adaptable Enterprises. The Economist, Profile Books, London (2015)
17. Brook, P., Riley, T.: 8.2.1 Enterprise systems engineering—practical challenges and emerging solutions. INCOSE Int. Symp. **22**(1), 1085–1101 (2012)
18. Chen, P., Zhang, C.Y.: Data-intensive applications, challenges, techniques and technologies: a survey on big data. Inf. Sci. **275**, 314–347 (2014)
19. Isom, P.K., Miller-Sylvia, S.L., Vaidya, S.: Intelligent enterprise architecture. IBM J. Res. Dev. **54**(4), 1–2 (2010)
20. Saint-Louis, P., Morency, M.C., Lapalme, J.: Defining enterprise architecture: a systematic literature review. In: 2017 IEEE 21st International Enterprise Distributed Object Computing Workshop (EDOCW) (2017)
21. Safari, H., Faraji, Z., Majidian, S.: Identifying and evaluating enterprise architecture risks using FMEA and fuzzy VIKOR. J. Intell. Manuf. **27**(2), 475–486 (2016)
22. Niemi, E., Pekkola, S.: Using enterprise architecture artefacts in an organisation. Enterprise Inf. Syst. **11**(3), 313–338 (2017)
23. Lapalme, J., Gerber, A., Alta, V.D.M., et al.: Exploring the Future of Enterprise Architecture: A Zachman Perspective. Elsevier Science Publishers B. V. (2016)

24. Hernández, J.E., Lyons, A.C., Poler, R., Mula, J., Goncalves, R.: A reference architecture for the collaborative planning modelling process in multitier supply chain networks: a Zachman-based approach. Prod. Plan. Control **25**(13/14), 1118–1134 (2014)
25. Dang, D., Pekkola, S.: Systematic literature review on enterprise architecture in the public sector. Electron. J. e-Govt. **5**(2), 132–154 (2017)
26. Rouhani, B.D., Mahrin, M.N., Nikpay, F., Rouhani, B.D.: Current issues on enterprise architecture implementation methodology. In: New Perspectives in Information Systems and Technologies, vol. 2, pp. 239–246. Springer International Publishing, Berlin
27. Lim, N., Lee, T.G., Park, S.G.: A comparative analysis of enterprise architecture frameworks based on EA quality attributes. In: 10th ACIS International Conference on Software Engineering, Artificial Intelligences, Networking and Parallel/Distributed Computing, pp. 283–288. IEEE (2009)
28. Rouhani, B.D., Mahrin, M.N., Nikpay, F., Nikfard, P.: A comparison enterprise architecture implementation methodologies. In: International Conference on Informatics and Creative Multimedia (ICICM), pp. 1–6. IEEE, Kuala Lumpur (2013)
29. Alwadain, A., Fielt, E., Korthaus, A., Rosemann, M.: Empirical insights into the development of a service-oriented enterprise architecture. Data Knowl. Eng. **105**, 39–52 (2016)
30. Panetto, H., Cecil, J.: Information systems for enterprise integration, interoperability and networking: theory and applications. Enterprise Inf. Syst. **7**(1), 1–6 (2013)
31. Iyamu, T.: Understanding the Complexities of enterprise architecture through structuration theory. J. Comput. Inf. Syst., 1–9 (2017)
32. Simon, D., Fischbach, K., Schoder, D.: Enterprise architecture management and its role in corporate strategic management. IseB **12**(1), 5–42 (2014)
33. Benkamoun, N., ElMaraghy, W., Huyet, A.L., Kouiss, K.: Architecture framework for manufacturing system design. Procedia CIRP **17**, 88–93 (2014)
34. Löhe, J., Legner, C.: Overcoming implementation challenges in enterprise architecture management: a design theory for architecture-driven IT Management (ADRIMA). Inf. Syst. e-Bus. Manage. **12**(1), 101–137 (2014)
35. Tiko, I.: Implementation of the enterprise architecture through the Zachman Framework. J. Syst. Inf. Technol. **20**(1), 2–18 (2018)
36. Pereira, C.M., Sousa, P.: A method to define an enterprise architecture using the Zachman framework. In: Proceedings of the 2004 ACM Symposium on Applied Computing, pp. 1366–1371. ACM (2004)
37. Nogueira, J.M., Romero, D., Espadas, J., Molina, A.: Leveraging the Zachman framework implementation using action–research methodology—a case study: aligning the enterprise architecture and the business goals. Enterprise Inf. Syst. **7**(1), 100–132 (2013)
38. Jafari, M., Akhavan, P., Nouranipour, E.: Developing an architecture model for enterprise knowledge: an empirical study based on the Zachman framework in Iran. Manag. Decis. **47**(5), 730–759 (2009)
39. Robertson-Dunn, B.: Beyond the Zachman framework: problem-oriented system architecture. IBM J. Res. Dev. **56**(5), 10–19 (2012)
40. Pereira, C.M., Sousa, P.A.: Method to define an enterprise architecture using the Zachman framework. In: ACM Symposium on Applied Computing, pp. 1366–1371 (2004)
41. Tobias, B., Simperl, E.: Measuring the benefits of ontologies. In: OTM Confederated International Workshops & Posters on the Move to Meaningful Internet Systems: Workshops: Adi. Springer-Verlag (2008)
42. Brachman, R.J.: Knowledge Representation and Reasoning an Introduction to Knowledge Engineering. Springer, London (2007)
43. John, F.: Comprehensive model-based engineering for systems of systems. Insight **19**(3), 59–62 (2016)
44. Frechette. S.: Model based enterprise for manufacturing. In: 44th CIRP International Conference on Manufacturing Systems, At Madison, WI, United States (2011)
45. Vaneman, W.K.: The system of systems engineering and integration. In: Systems Conference. IEEE (2016)

46. Vaneman, W.K., Jaskot, R.D.: A criteria-based framework for establishing system of systems governance. In: Systems Conference (2013)
47. Warren, K.V.P.D., Budka, R.: Defining a system of systems engineering and integration approach to address the Navy's Information Technology Technical Authority. Incose Int. Symp. **23**(1), 1202–1214 (2013)
48. Baoding, L.: Uncertainty Theory: A Branch of Mathematics for Modeling Human Uncertainty. DBLP (2010)

Boost System of Systems Modeling via UPDM: A Case Study from Air Traffic Management

Huanhuan Shen, Weijie Zhu, and Aishan Liu

1 Introduction

Systems of Systems (SoS) are generally large complex systems, with varying degrees of operational independence, managerial independence, evolutionary development, geographical distribution, and lifecycle independence. These independent systems are integrated into an SoS that delivers unique capabilities. Both individual systems and SoS conform to the accepted definition of a system in that each consists of parts, relationships, and a whole that is greater than the sum of the parts. An SoS can be a logical configuration of existing and new systems, where the systems retain their identity, and management and engineering continue in the systems concurrently with the SoS.

Meanwhile, SoS is widely witnessed in our current society with plenty of system elements and complex interactions within the big system in different fields, e.g., air traffic management systems. With the increase of global warming, the European Union has proposed the Clean Sky Plan to meet new environmental requirements. The plan was designed to reduce greenhouse gas emissions and the environmental impacts of air transport through different measures. To achieve the goal, billions of Euros have been adopted to establish an innovative and competitive European air transport system. Obviously, building an efficient air traffic management system would for certain save more energy and reduce more pollution than many elaborately designed airplane techniques. Inspired by that, from the perspective of the air traffic management system, this paper tries to analyze and understand the functions, behaviors, and interactions within the SoS. With this case study, we believe, it would be very useful and beneficial to obtain deeper insights into the SoS, which in turn

H. Shen · W. Zhu · A. Liu (✉)
AVIC-Digital, No. 7 Jingshun Road, Beijing, China

H. Shen
e-mail: shenhh@avic-digital.com

© The Author(s), under exclusive license to Springer Nature Switzerland AG 2021
D. Krob et al. (eds.), *Complex Systems Design & Management*,
https://doi.org/10.1007/978-3-030-73539-5_15

helps to further improve efficiency, save energy and reduce pollution for the whole air traffic management system.

1.1 SoS Modeling and Engineering

Systems of systems modeling and engineering deals with planning, analyzing, organizing, and integrating the capabilities of new and existing systems into an SoS capability greater than the sum of the capabilities of its constituent parts. In contrast to a system, SoS may deliver capabilities by combining multiple collaborative and independent-yet-interacting systems. The mix of systems may include existing, partially developed, and yet-to-be-designed independent systems. As these outcomes become more complex and the associated systems more complex, the management, modeling, and simulation of these SoS become equally challenging in both civilian and military systems. Evaluation at the level of individual requirements is too low level, time-consuming and complex to determine how to assemble the SoS.

1.2 SoS Architecture

Organizations are changing their emphasis from a platform focus to an emphasis on capabilities. In general, this is a change from "We need a new system" to "We need to achieve a specific outcome." The Department of Defense Architecture Framework (DoDAF) [1] and Ministry of Defence Architecture Framework (MODAF) [2] were developed for modeling enterprise architectures from capabilities to detailed components, and the technical and operational management of these architectures as they evolve.

The frameworks define a standard way to organize an enterprise architecture (EA) or systems architecture into complementary and consistent views. DoDAF V1.0 contained four basic views: the overarching All Views (AV), Operational View (OV), Systems View (SV), and the Technical Standards View (TV/StdV). Each view is aimed at different stakeholders, and it is possible to create cross-references between the views. Although they were originally created for military systems, they are commonly used by the private, public and voluntary sectors around the world, to model SoS. Obviously, they are aiming to improve the planning, organization, procurement, and management of these complex systems.

Besides, the Object Management Group (OMG) Unified Profile for DoDAF and MODAF (UPDM) [3] was created by INCOSE and OMG to define a consistent, standardized means to describe DoDAF V 2.02 and MODAF architectures using the Systems Engineering Language (SysML) [4]. The goals of UPDM are to significantly enhance the quality, productivity, and effectiveness associated with enterprise and system of systems architecture modeling, promote architecture model reuse and

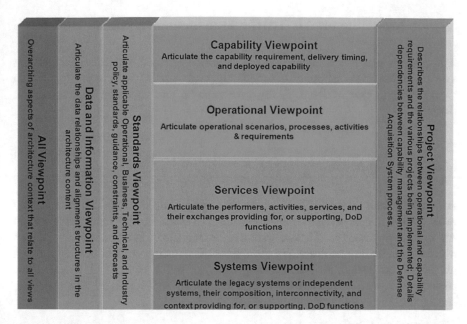

Fig. 1 UPDM views

maintainability, improve tool interoperability and communications between stake-holders, and reduce training impacts due to different tool implementations and semantics (Fig. 1).

1.3 Leverage SysML

OMG SysML is a visual modeling language that extends UML 2 in order to support the specification, analysis, design, verification, and validation of systems that include components for hardware, software, data, personnel, procedures, and facilities. SysML is intended to be used with different methodologies including structured analysis, object orientation, and others. SysML reuses a subset of UML 2 concepts and diagrams and augments them with some new diagrams and constructs appropriate for systems modeling. In particular, the language provides graphical representations with a semantic foundation for modeling system requirements, behavior, structure, and parametric, which is used to integrate with other engineering analysis models. Consequently, UPDM utilizes features from SysML to provide a robust modeling capability [5–8].

2 SoS Modeling on Air Traffic Management System

This chapter will first demonstrate the definition of the Air Traffic Management system (ATM), then elaborates our modeling process on ATM from the operational view to system view in detail, finally, give some insights about the implementation details.

2.1 Air Traffic Management

Air Traffic Management (ATM) is an aviation system encompassing all systems that assist aircraft to depart from an aerodrome, transit airspace, and land at a destination aerodrome. Apparently, ATM is a system of systems (SoS), in which systems include air traffic control, air traffic safety electronics personnel, aeronautical meteorology, air navigation system, air space management, air traffic service, Air Traffic Flow Management, etc., interacts with each other. As an SoS, however, ATM contains several challenges and problems to be solved.

Challenges and Problems

- The growing complexity of the ATM system. More and more integrated function-alities for different services in the system, e.g., more planes, passenger services, collision avoidance, etc.
- The growing complexity of systems within the ATM system. Different factors are increasing the complexity of the systems in ATM. Technical complexity growth by increased electronic and software-driven functionality of the subsystems. Increased performance requirements lead to more sophisticated and precise control of the subsystems.
- Strong uncertainties. With such big and huge capacities, ATM contains a lot of uncertainties, e.g., meteorology uncertainty, prediction uncertainty, environmental uncertainty, etc.

Since the complexity of ATM is extremely high, in order to be successful, even as a modernization project, ATM is required a system architecture that defines how the pieces of the system fit together and allow for modeling and reasoning about possible futures. To manage the complexity and risk in the ATM system, system architecture and model are essential tools (Fig. 2).

2.2 Overall Development Process

Our main motivation comes from the fact that to make comprehensive investigation and insights into an SoS, we need to start from goals to system functions. To make the requirements and models consistent, verifiable, and traceable, the UPDM framework

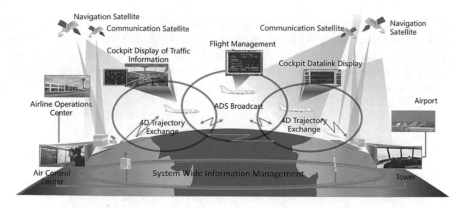

Fig. 2 Air traffic management system

is utilized to analyze the system from different viewpoints with SysML, in which models are consistent, unambiguous, and executable.

Given our objective as making a more efficient air traffic management system, it is intuitive for us to define SoS level operational activities which could be beneficial and helpful for us to further study the functions. By extracting the functions of the SoS, we could better understand the "what should the system do" in detail. Then, we need to decompose the SoS into systems and further perform analyses and investigations at the system level to give deeper insights into the whole lifecycle of the SoS.

Specifically, as shown in Fig. 3, with the concept of operations as inputs of our whole development process, we mainly analyze ATM from two viewpoints, i.e., Operational Viewpoint and System Viewpoint. From the mission-oriented Operational Viewpoint, different aspects of the SoS are studied including capabilities, missions, tasks, and requirements to determine 'what to be done'. However, from the system-oriented System Viewpoint, system-level investigations are conducted,

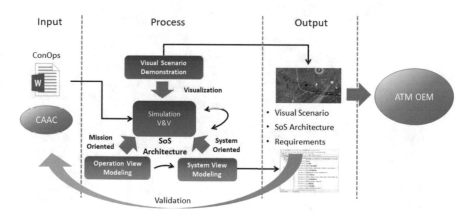

Fig. 3 Overall development process

e.g., how the systems operate coordinately, to determine 'how to be done'. Based on the previous analyses, the iteration process is performed to verify our generated models with visual scenario simulations. Further, we output the analytical results including requirements and SoS architecture, which could be used to both validate initial ConOps and motivate ATM development.

2.3 Operational Viewpoint Modeling

The Operational Viewpoint (OV) describes organizations, activities they perform, and resources they exchange to fulfill missions. This viewpoint includes the types of information exchanged, the frequency of such exchanges, the activities supported by information exchanges, and the nature of information exchanges.

OV-1: Define High-Level Operational Concept (OpsCon)

In this view, we try to give an overview of the concepts of operations in ATM. In order to motivate deeper investigation and put the operational situation into the context, we depict missions, scenarios, operational concepts, interactions between the architecture and the environment, etc.

Combining 3D trajectory with time, 4D trajectory (4DT) based operation in ATM aims to acquire accurate, continuous and real-time trajectories of flights. By introducing digital and automation technique, 4DT based operation in ATM shall enable dynamic sharing, synchronization, and negotiation of 4D trajectories, improve coordination ability of aircraft and ATM-related systems, improve precision of air traffic monitoring and accuracy of flight trajectory prediction, finally achieve more effective, efficient, environmental friendly and safer air traffic operations. The illustration of the high-level OpsCon of ATM can be viewed in Fig. 4.

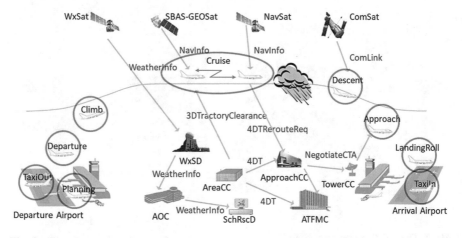

Fig. 4 4D-trajectory based operation of ATM

OV-5: Define Operational Activities and Activity Flows

With the results from OV-1, we further define and extract the activities and hierarchical structures of activities carried out within the scope of ATM. During the process, activities, rules, and conditions shall be modeled. Meanwhile, the hierarchical structure of activities may be modeled using whole-part relationships. Also, activities may be categorized using super-sub type relationships. The illustration of our activity diagrams can be found in Fig. 5.

OV-6c: Define Operational Interactions between Operation Nodes

After analyzing the hierarchical structure of activities, the behavior of the system should be paid attention to. This model identifies and describes a sequence of activities within a described architecture. The sequence diagrams are illustrated in Fig. 6.

Fig. 5 OV-5 activity diagrams

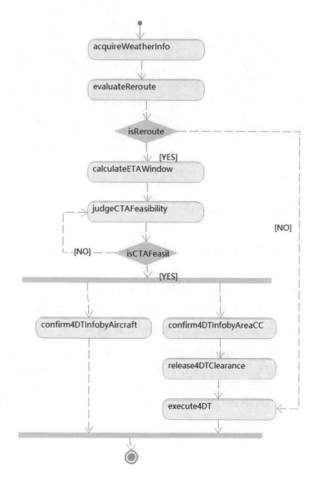

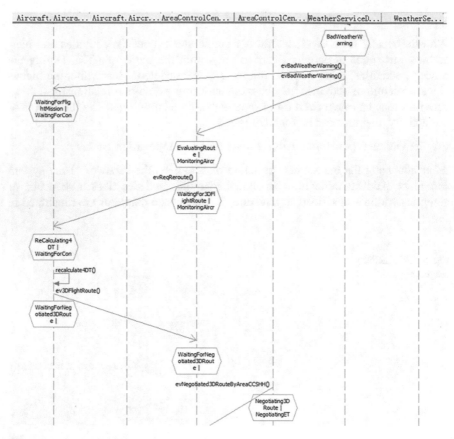

Fig. 6 OV-6c sequence diagrams

OV-2: Define Interfaces Based on Interactions

Moreover, we model the dependency between systems, which can be described as the flow of information from one system to another. The illustration of the interfaces between systems in ATM can be viewed in Fig. 7.

OV-6b: Build State Machine Models

Finally, we identify and describe changes in the states of the systems contained in ATM. In this part, we emphasize certain systems within ATM, specifically those systems that are critical to identifying the behaviors and interactions between elements in the system across its whole life cycle (Fig. 8).

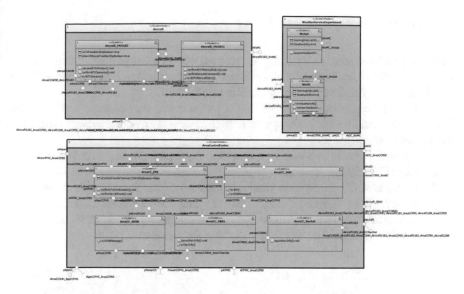

Fig. 7 OV-2 interface diagrams

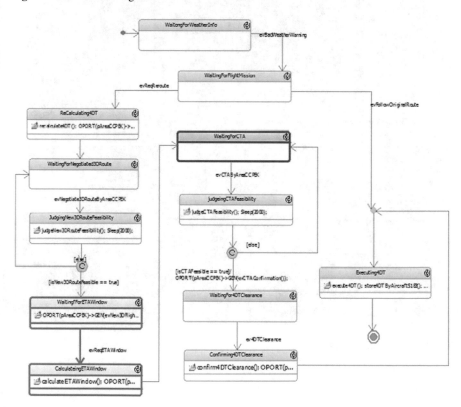

Fig. 8 OV-6b state machine diagrams

2.4   System Viewpoint Modeling

The Systems Viewpoint (SV) describes system activities and resources that support operational activities. This viewpoint traces system activities and resources to the requirements established by the Operational Viewpoint.

Similar to the process in the Operational Viewpoint, we decompose the system into elements and identify these systems. Then, we, in turn, define system functions, interactions, and behaviors.

Firstly, with the help of SV-1, system composition and interface identification can be achieved (Fig. 9).

Then, we analyze the system functions and relate the functions to operational activities in SV-4 by defining the hierarchical structures of system activities. Moreover, as illustrated in Fig. 10, we define system operational interactions in SV-10c and study system behaviors with state machine diagrams in SV-10b.

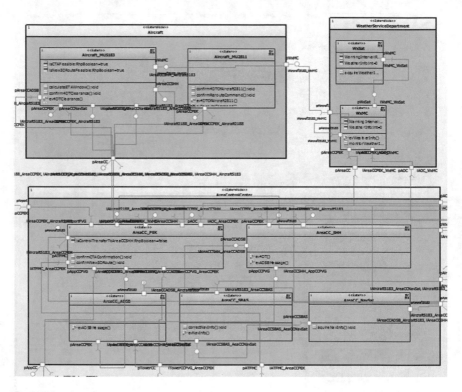

Fig. 9 SV-1 interface diagrams

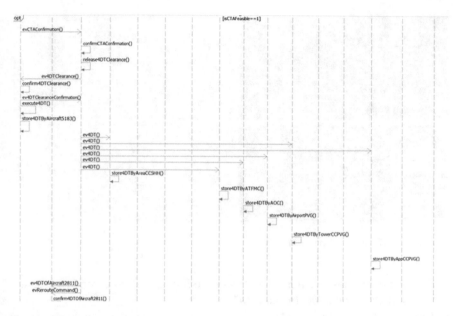

Fig. 10 SV-10b sequence diagrams

2.5 Simulation and Verification

In order to make an accurate investigation and locate errors ahead of time, with the analytical results before, the simulation and verification process is employed. At different levels, i.e., SoS and system, respectively, we perform visual scenario simulation to verify our generated requirements, functions, and abilities. Since the models in SysML are executable, with elaborately designed triggers and events, the models are able to drive the visual simulation system to perform co-simulation as shown in Fig. 11.

3 The Advantages of Our Development Process

- **Cope with the Complexity of SoS**
  Since the complexities of SoS are becoming extremely high, it is for certain very difficult for engineers to analyze them comprehensively. With our development process, SoS is investigated from different views at different levels, which could be significantly helpful for us to make a deeper understanding and insights about the SoS.
- **Enhance System Decomposition and Explore Model Details**
  With the standardized process, we can decompose the SoS into systems accordingly and make the process more distinct and clear. By investigating both static and

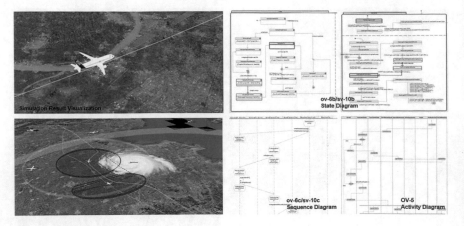

Fig. 11 Simulation and verification

dynamic properties and characteristics of the SoS as well as contained systems, we can better explore the model details.

- **Keep Consistency in Models**
  In our development process, behaviors, functions, and interactions of the systems are modeled with SysML, which is traceable and consistent. In addition, we further confirm the consistency of the simulation-based iteration and verification process.
- **Understand the Emergence**
  The emergence is one of the key factors of a system. It is much critical though difficult to understand emergent behaviors of the SoS, as there exist more systems and interactions within. By analyzing the SoS in different levels, we could better understand system behaviors when they are decomposed and combined together.

4 Conclusion

SoS modeling is becoming increasingly important since the numbers of system elements and interactions within a system increase significantly. Traditional strategies fail to manage the complexities in SoS leading to inconsistent and incomplete analyses. In this paper, we proposed an SoS modeling case study on the Air Traffic Management system (ATM) with UPDM, a unified profile to model enterprise architectures from capabilities to components. From the Operational Viewpoint, capabilities and missions are studied to determine what the SoS should be done, which further motivates the System View analysis. From the System Viewpoint, more detailed analyses are conducted by decomposing SoS into elements to investigate the functions, operations, and interactions. By iterative analyses and simulation, models are consistent, verifiable, and traceable leading to better SoS modeling and engineering [9].

References

1. Hause, M.C.: SOS for SoS: a new paradigm for system of systems modeling. In: 2014 IEEE Aerospace Conference, pp. 1–12. IEEE (2014)
2. DoD CIO: DoD Architecture Framework Version 2.02, DoD Deputy Chief Information Officer. Available online at https://dodcio.defense.gov/dodaf20/dodaf20_pes.aspx. Accessed Nov 2013
3. Friedenthal, S., Moore, A., Steiner, R.: Practical guide to SysML. In: The Systems Modeling Language, 2nd edn. Morgan Kaufman (2011)
4. MOD Architectural Framework, Version 1.2.004, Office of Public Sector Information, https://www.gov.uk/government/uploads/system/uploads/attachment_data/file/36757/201 00602MODAFDownload12004.pdf
5. Object Management Group (OMG), June, 2012, OMG Systems Modeling Language (OMG SysML™), V1.3, OMG Document Number: formal/2012-06-01, https://www.omg.org/spec/ SysML/1.3/PDF/. Accessed Nov 2013
6. UPDM, 2013, Object Management Group (OMG), 2013, Unified Profile for DoDAF/MODAF (UPDM) 2.1, formal/2013-08-04. Available at https://www.omg.org/spec/UPDM/2.1/PDF
7. Hause, M.: Model based system of systems engineering with UPDM. INCOSE Int. Symp. **20**(1), 580–594 (2010)
8. Ferrogalini, M., Linke, T., Schweiger, U.: How to boost the extended enterprise approach in engineering using MBSE—a case study from the railway business. In: International Conference on Complex Systems Design & Management, pp. 79–96. Springer, Cham (2018)
9. Tang, J., Zhu, S., Faudou, R., et al.: An MBSE framework to support agile functional definition of an avionics system. In: International Conference on Complex Systems Design & Management, pp. 168–178. Springer, Cham (2018)

Closed-Loop Systems Engineering—Supporting Smart System Design Adaption by Integrating MBSE and IoT

Thomas Dickopf, Sven Forte, Christo Apostolov, and Jens C. Göbel

1 Introduction

By demanding innovative products, customers pressure companies to shorten their development cycles for new innovative products. Nowadays, product innovation is more and more information technology-driven; those products evolve to highly digitized products and services, so-called smart products, and services [1]. These smart products and services also contribute to the vision of an open and autonomous system of systems (SoS) but come at the price of increased engineering complexity. According to a survey by the Aberdeen Group, complexity was the top systems engineering challenge already in 2014, with 51% of respondents ranking it accordingly and an increase of 24% over just four years [2]. In order to address the increasing complexity, researchers are investigating systems-engineering-based approaches to master the transition from traditional products to digital, interactive smart products and services. While Model-Based Systems Engineering (MBSE) and Product Lifecycle Management (PLM) currently merge towards holistic, software-supported System Lifecycle Management approaches [3], PLM also becomes connected with Internet-of-Things (IoT) applications and digital twin implementations [4, 5]. The

T. Dickopf (✉) · C. Apostolov
CONTACT Software GmbH, Martin-Luther-Straße 8, 67657 Kaiserslautern, Germany
e-mail: thomas.dickopf@contact-software.com

C. Apostolov
e-mail: christo.apostolov@contact-software.com

S. Forte · J. C. Göbel
Institute of Virtual Product Engineering—VPE, University of Kaiserslautern, Gottlieb-Daimler-Straße 44, 67663 Kaiserslautern, Germany
e-mail: forte@mv.uni-kl.de

J. C. Göbel
e-mail: goebel@mv.uni-kl.de

© The Author(s), under exclusive license to Springer Nature Switzerland AG 2021
D. Krob et al. (eds.), *Complex Systems Design & Management*,
https://doi.org/10.1007/978-3-030-73539-5_16

virtualization of product validation processes [6] describes a fundamental capability of successful engineering enterprises [2], and Closed-loop Systems Engineering becomes a reality by the integration of virtual validation techniques and field connectivity.

One central research question is how field data of current product instances and families can be used reliably and in a standardized way for system validation and improvement purposes using MBSE concepts. A system design validation approach [7] connecting SysML-based (OMG Systems Modeling Language) systems validation with aggregated field data is presented in the following sections. This contribution focuses on the management and the analysis of operational data (e.g., aggregation of usage information or error rates) and its use in the system design to improve the current product version and new product generations. The implementation and validation of the presented concepts took place on a SoS research testbed at the University Kaiserslautern using the PLM and IoT technology CONTACT Elements (CE) by CONTACT Software.

2 State of the Art

2.1 Smart Products and Their Lifecycle Management

Smart Products are discussed as cyber-physical-based products with a certain degree of autonomy and the capability to communicate and interact with different actors [1, 8–10]. In order to maximize customer value, these products are integrated into different smart environments [10], which consists of various smart products, services, and connecting infrastructures. However, this integration leads to an immense increase in complexity and will change business disruptively [9]. Therefore, new collaboration and cooperation models should be defined so that the autonomous system of systems vision can become a reality. Furthermore, companies have to optimize their operational processes, methods, and tools regarding the complexity to make them manageable in the future.

Supporting the management of such System of Systems, Systems Lifecycle Management (SysLM) was established as a concept to manage system-related data along the lifecycle of contemporary smart products and the systems they are integrated into [11, 12] (Fig. 1). Therefore, on the one hand, Product Lifecycle Management (PLM) has been extended with a general information management concept for the early interdisciplinary product development phase, where model-based approaches are used (cf. MBSE). On the other hand, through the collection and analysis of data of field devices and assets (cf. IoT), the product usage and support phase has gained significantly in importance [11, 12], which enables new possibilities such as product reconfiguration [13].

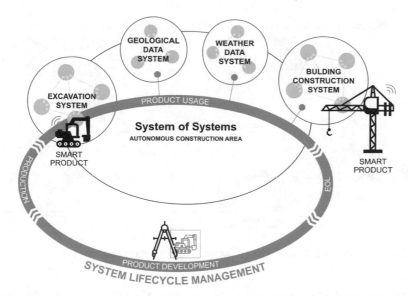

Fig. 1 An exemplary representation of a smart product as part of a system of systems during its life cycle [20]

2.2 Digital Twin

A central element of the digital transformation is the digital twin [14]. As a virtual image of a field device, it reflects its physical state in a current and historicized form. The digital twin forms the link between field data (system environment, construction status data, technical condition, operating data), service information, analysis models, and development data (e.g., product data, bills of material, documents, process data). Digital twins define access from the application level to information logistics and become part of IoT solutions' building architecture [15].

There is not one universal digital twin. The concrete instantiation depends on use cases and business models of its various application areas. Consequently, a large number of definitions have been introduced and refined over the last few years. An overview of definitions found in literature can be found in [16–18].

Figure 2 represents the data model for the instantiation and use of digital twins. If the digital twin is placed at the center of information logistics, then the data processing steps can be traced from the sensor level via the gateways to central data management. A field device with sensors sends telemetry data to an IoT platform in a typical application, such as its location and other operating parameters. To simplify the creation of new digital twins, a digital master is available as a template. It represents a blueprint of a virtual product from which twins can be created or copied, which then—enriched with serial numbers, classification features, and data from the product creation process—represent a real field device, a machine, or even a complete production environment. Digital twins can be grouped in two ways.

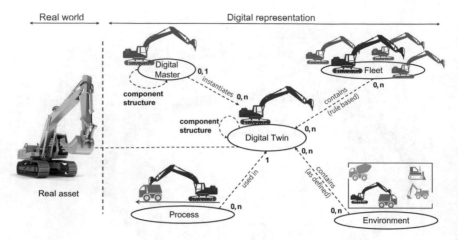

Fig. 2 A data model for the instantiation and use of digital twins (based on [15])

Based on defined rules, digital twins can be grouped into fleets to analyze the same categories' assets. Environments, on the other hand, allow different digital twins to be encapsulated, allowing entire plants or machine parks to be modeled and analyzed. Machine processes can be logged as usage processes in the digital twin so that data is available for the individual processes. A classification of the processes with subsequent analysis helps users detect anomalies in individual machines, products, or even batches. A feedback of the results into product development in the sense of closed-loop engineering would support fast product improvements and strategic product planning [15].

2.3 Closed-Loop Systems Engineering

To achieve shorter development cycles and a high level of quality assurance with a minimized number of iteration [19], Closed-Loop Systems Engineering (CLE) describes an approach for successfully handling product complexity. Fast digital information flows and fast access processes for product components supported by agile methods at the operational level are just two core CLE ideas. In this context, the following strategical concepts of how best-in-class companies handle issues associated with complexity successfully has been identified [2, 20]:

- Virtual prototypes (models): use of virtual prototypes (starting with formal, executable system models) for the verification and validation of preliminary designs (drafts) before physical prototypes are built
- Simulation: use of simulation as an essential enabler and its application as early as possible in development processes (system analysis and design phase)

Fig. 3 Expandable system improvement concept for complex systems (based on [20, 21])

- Improvement loops: completing the continuous improvement loop by recording the actual simulation results and feeding this information back in to calibrate and improve future simulations.

By applying the CLE characteristics mentioned above in combination with the benefits of system lifecycle management and IoT, a fully model-based approach for optimizing today's complex, multidisciplinary smart systems have been developed. The approach focuses on three specific improvement loops, which lay their improvement emphasis in the lifecycle phases of interdisciplinary system development, system verification and validation, and system usage (cf. Fig. 3) [20, 21].

Model-in-the-loop
Model-in-the-loop [20, 21] means executing a system model in the corresponding authoring modeling tool or combination with additional simulation tools [19, 22] to verify the system's design and the early validation of partial solutions or system components. The model execution aims to assess how a system, particular object, or phenomenon will behave over time. Either because the system is not yet fully defined or available, or because it cannot be implemented directly due to cost, time, resources, or risk constraints, model execution enables us to gain system understanding by utilizing simulation. Application scenarios for model execution are [23]:

- Evaluate design alternatives
- Select the best set of parameters
- Verify system with its constraints
- Perform requirement compliance analysis
- Perform what-if analysis.

Twin-in-the-loop

Twin-in-the-loop [20, 21] describes an approach for the design and validation of the further digital twin of a system, its management platform, as well as concomitant business and service processes. By using the possibilities of the model-in-the-loop concept, this step can take place early in the system development phase, when the system does not yet exist. It enables a first configuration of the digital twins and their associated business and service models by simulating field devices by connecting a simulation-ready system model to an IoT operational system. The twins will be implemented later on. This approach brings the following advantages to the system development process:

- Increased dependability of the system design
- Faster implementation of physical and digital subsystems dependent on not yet existent elements of the system
- Faster handover of the system to the customer.

System-in-the-loop

System-in-the-loop [20, 21] addresses the concept of system improvement based on the seamless integration of the design and operational usage of smart products. Individual products or entire product populations are optimized through planning strategies and retrofitting concepts [24, 25], closing the loop between system architectures and digital twins. For this purpose, the field information has to be referenced precisely and broken down appropriately to permit the targeted use of the results to develop products and processes further. The return of field data into development requires close integration between PLM systems (the core of engineering data management) and IoT operational systems (the core of operational data management), which can lead to the following advantages for enterprises and employees:

- Targeted system improvement through analyzed and evaluated field data
- Involvement of customers in the ongoing optimization processes
- Improvement of strategic product planning based on usage-based market observations.

3 Closing the Loop Between Digital Twins and System Architectures

This section presents a methodical procedure and its application to represent one possibility of implementing the system-in-the-loop concept. The focus is on the interaction and information exchange between MBSE approaches in the early design phase and IoT application in the system's operational use.

For a better understanding, the procedure is illustrated using a selected application scenario. After a brief description of the scenario, both the methodical concept (five-step approach) and its direct implementation at the application scenario using specific modules of the CONTACT Elements platform are presented in the following. The concept is divided into two main parts. The first part considers the generation, management, and operational data analysis (steps 1–4). The second part considers integrating and providing this information to optimize the system, respectively, its architecture (step 5). The Essential steps are:

1. Create or instantiate digital twins
2. Classify the digital twins
3. Define the fleets
4. Analyze the data
5. Feedback to development.

For the illustration of the developed concept, a hydraulic excavator was chosen as an example system. In previous publications [11, 12, 20, 21], the excavator considers a cybertronic element in various cybertronic systems. For example, as part of an autonomous construction site or an automated forestry work or dyke work SoS. Here, the focus lies on the aggregation of operational data of different variants of one product generation and the markets in which the excavators are used. That means that on different markets (e.g., continents, countries, or even regions), different system variants of a product generation (e.g., chain or wheel drive; adjustable or monoblock boom; electric or internal combustion engine) are used differently.

For the evaluation of the concept and especially for the generation of semi-realistic field data, a research testbed of an autonomous construction area was established in cooperation between the Institute of Virtual Product Engineering (VPE) at the University of Kaiserslautern and CONTACT Software GmbH. In general, the testbed focuses on research questions in the field of model-based development and the use and optimization of complex smart product systems from a methodological and information-technological perspective. The testbed software implementation is based on the open platform technology CONTACT Elements (CE) and its numerous specialized modules to support digital transformation.

3.1 Management and Analysis of Operational Data

For a data-driven system improvement, it is necessary that each device's field data is accessible and managed in a corresponding infrastructure. By creating a virtual image of these devices, operational data becomes more usable and enables various benefits (cf. Sect. 2.2). The first and most essential step is to create a digital twin or instantiate it from a digital master (*step 1—using the CE module "Digital Twin"*). The field device's generated data will be sent via a standardized communication protocol to the digital twin managed by the IoT operating system and provided to various users (e.g., users, maintenance or service technicians, device manager) for various purposes [20]. In order to merge the correct data for the analysis case under consideration, digital twins can be bundled in fleets (*step 3—using the CE module "Device Management"*). As mentioned above, a fleet describes a group of twins band together based on specific rules. These rules could be based on the system's properties and features or its sub-components or based on other characteristics, like the market where the system is used. Therefore, digital twins can be classified according to their characteristics (*step 2—using the CE module "Universal Classification"*).

In general, classification approaches are used to support the organization of data objects based on similar or associated characteristic-based object groups within a hierarchically structured characteristics catalog to reduce redundancies, cut the time required for data maintenance, and improves the efficiency of characteristic-based queries. Consequently, these approaches offer an optimal and sophisticated approach to define fleets in a targeted manner. Based on the previously chosen methodical procedures, the digital twin classification occurs after its creation or has already been transferred from the digital master by instantiation and inheritance. Once the fleet has been created concerning the desired analysis context, the individual twins' data can be merged and used for the analysis (*step 4—using the CE module "Monitoring and Analytics"*). The analysis itself, depending on its goal, can also be done in several ways. For example, the aggregated data analysis could be performed directly in the IoT platform by defining and applying key performance indicators (KPIs), which could then automatically trigger new processes if necessary. On the other hand, analysis or machine learning tools can also be integrated into the IT infrastructure for more complex analyses.

Figure 4 illustrates the prototypical realization using the excavator fleet scenario in *CONTACT Elements for IoT*. While 4.1 shows how the excavator is classified according to its specific characteristics (e.g., type of drive, type of boom, or the market for which the product was manufactured), 4.2–4.4 illustrates the excavator fleet dashboard of the European market. 4.2–4.3 shows a widget at the dashboard body, which lists all twins belonging to the fleet, including static attributes as well as current states and geographical positions. 4.4 points to the dashboard header, where attribute values can be displayed. This case represents KPIs according to the usage time and its proposal distribution to the different working modes.

3.2 Provision and Utilization of Operational Data

This approach's final step is to provide and re-integrate the operational data—analyzed, evaluated, and prepared—as additional information back into the development process to utilize it to improve the current system and the following product generations. These could be, e.g., usage profiles, system load factors, utilization contexts, system failures, or engineering changes. According to close the loop between operational usage and engineering development, seamless integration between the IoT operational Systems (the core of operational data management) and the PLM systems (the core of engineering data management) is required (*step 5—using the CE module "Closed Loop Engineering"*). The data exchange between the PLM and IoT operating systems is described in Fig. 5.

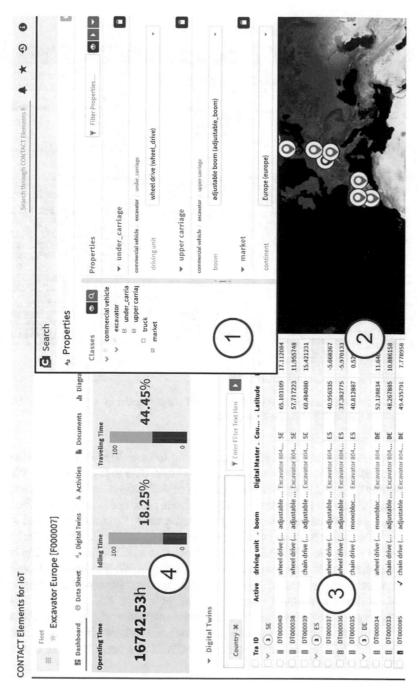

Fig. 4 Twin classification and fleet management in CONTACT elements for IoT

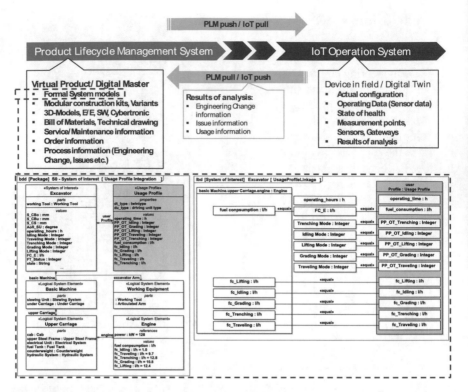

Fig. 5 Push and Pull concept between PLM and IoT supplemented by a possible integration of operating data into SysML-based system models

Looking back at the initial idea of combining IoT and MBSE to support system improvements, it is necessary to link the operational date to formal system models, representing the architecture of a smart product or SoS. Depending on the information content and the context of the system improvement, these could be new requirements for the system, derived from measures generated in the IoT platform, based on issues or usage information that can be integrated into the system model through its extension by new elements. Additionally, Fig. 5 illustrates an exemplary extension of the excavator model by adding the new element *Usage Profile*. The analyzed values of the excavator fleet analysis (cf. Fig. 4) could be used to optimize the system in the sense of the introduced model-in-the-loop concept in Sect. 2.3.

For the data exchange between the PLM system and the modeling tool to create the system architecture, both corresponding methods and software interfaces are necessary. One prototypical solution based on CONTACT Elements has been developed in the German research project mecPro^2 [3].

4 Conclusion

Combining the concepts of MBSE and IoT, this paper introduced a Closed-loop systems engineering approach to the systematic improvement of smart product systems. Whether from the simulation in the development phase or generated by the real system during product usage, the resulting information's re-integration into the system development through improvement loops achieves enormous benefits. The presented concept for improvement of smart product systems includes three loops: (I) model-in-the-loop; (II) twin-in-the-loop; and (III) system-in-the-loop, whereby the last loop has been discussed in this paper in more detail by using a 5-step approach to improve a field device or a further product generation by re-integrating the operational data into the development process.

References

1. Tomiyama, T., Lutters, E., Stark, R., Abramovici, M.: Development capabilities for smart products. CIRP Ann. **68**(2), 727–750 (2019)
2. Paquin, R.: The Systems Engineering Closed Loop Process: The Key for Validation. Aberdeen Group (2014)
3. Eigner, M., Koch, W., Muggeo, C. (eds.): Modellbasierter Entwicklungsprozess cybertronischer Systeme—Der PLM-unterstützte Referenzentwicklungsprozess für Produkte und Produktionssysteme. Springer, Berlin (2017)
4. Abramovici, M., Göbel, J.C., Savarino, P.: Virtual twins as integrative components of smart products. In: IFIP Advances in Information and Communication Technology Product Lifecycle Management for Digital Transformation of Industries, pp. 217–226 (2016)
5. Abramovici, M., Dang, H.B., Göbel, J.C., Savarino, P.: Systematization of IPS2 diversification potentials using product lifecycle data. Procedia CIRP 288–293 (2016)
6. Stark, R., Krause, F.L., Kind, C., et al.: Competing in engineering design—the role of virtual product creation. CIRP J. Manuf. Sci. Technol. **3**, 175–184 (2010)
7. Gregorzik, S.: Smart business blueprints—how digital business gets going in the Internet of Things. Contact Software White Paper, 2017. https://www.contact-software.com/en/white-paper/smart-business-blueprints/
8. Abramovici, M.: Smart products. In: Laperrière, L., Reinhart, G. (eds.) CIRP Encyclopedia of Production Engineering. Springer, Berlin (2015)
9. Porter, M.E., Heppelmann, J.: How smart, connected products are transforming companies technology & operations. Harvard Bus. Rev. (2015)
10. Mühlhäuser, M.: Smart products: an introduction. In: Communications in Computer and Information Science Constructing Ambient Intelligence, AmI 2007 Workshops Darmstadt, Germany, pp. 158–164 (2007)
11. Eigner, M., Dickopf, T., Apostolov, H., Schäfer, P., Faißt, K.G., Keßler, A.: System lifecycle management—initial approach for a sustainable product development process based on methods of model based systems engineering. In: Fukuda, et al. (eds.) 11th IFIP WG 5.1 International Conference—PLM 2014, pp. 287–300. Springer, Heidelberg (2014)
12. Eigner, M., Dickopf, T., Apostolov, H.: System lifecycle management—an approach for developing cybertronic systems in consideration of sustainability aspects. In: Takata, et al. (eds.). Procedia CIRP—The 24th CIRP Conference on Life Cycle Engineering, Kamakura, Japan, 8–10 Mar 2017. Elsevier Procedia, pp. 128–133 (2017)

13. Abramovici, M., Göbel, J.C., Hoang, B.D.: Semantic data management for the development and continuous reconfiguration of smart products and systems. CIRP Ann. **65**(1), 185–188 (2016)
14. Gartner. Gartner Hype Cycle for Emerging Technologies. https://www.gartner.com/en/new sroom/press-releases
15. Göckel, N., Müller, P.: Entwicklung und Betrieb Digitaler Zwillinge. In: Eigner, M. (ed.) Zeitung für wirtschaftlichen Fabrikbetrieb—Digitaler Zwilling. Hanser, Band 115 (2020)
16. Negri, E., Fumagalli, L., Macchi, M.: A review of the roles of digital twin in CPS-based production systems. Procedia Manuf. **11**, 939–948 (2017)
17. Stark, R., Damerau, T.: Digital twin. In: Chatti, S., Laperrière, L., Reinhart, G., Tolio, T. (eds.) The International Academy for Production Engineering, CIRP Encyclopedia of Production Engineering. Springer, Berlin (2019)
18. Göbel, J.C., Eickhoff, T.: Konzeption von Digitalen Zwillingen smarter Produkte. In: Eigner, M. (ed.) Zeitung für wirtschaftlichen Fabrikbetrieb—Digitaler Zwilling. Hanser, Band 115 (2020)
19. Di Maio, M., Kapos, G.D., Klusmann, N., Atorf, L., Dahmen, U., Schluse, M., Rossmann, J.: Closed-loop systems engineering (CLOSE): integrating experimentable digital twins with the model-driven engineering process. In: 4th IEEE International Symposium on Systems Engineering, ISSE (2018)
20. Dickopf, T., Apostolov, H., Müller, P., Göbel, J.C., Forte, S.: A holistic system lifecycle engineering approach—closing the loop between system architecture and digital twins. In: Putnik, G. (ed.) Procedia CIRP—29th CIRP Design Conference 2019, Póvoa de Varzim, Portugal, 8–10 May 2019. Elsevier Procedia, pp. 538–544 (2019)
21. Dickopf, T.: A holistic methodology for the development of cybertronic systems in the context of the Internet of Things. Shaker (2020)
22. Schluse, M., Atorf, L., Rossmann, J.: Experimentable digital twins for model-based systems engineering and simulation-based development. In: 2017 Annual IEEE International Systems Conference (SysCon 2017)
23. Strolia, Z., Pavalkis, S.: Building executable SysML model—automatic transmission system (Part 1). 2017. https://blog.nomagic.com/building-executable-sysml-model-automatic-transm ission-system-part-1/
24. Albers, A., Reiss, N., Bursac, N., Richter, T.: iPeM—Integrated Product Engineering Model in Context of Product Generation Engineering. Elsevier (2016)
25. Massmann, M., Meyer, M., Dumitrescu, R., von Enzberg, S., Frank, M., Koldewey, C., Kühn, A., Reinhold, J.: Significance and challenges of data-driven product generation and retrofit planning. In: Putnik, G. (ed.) Procedia CIRP—29th CIRP Design Conference 2019, Póvoa de Varzim, Portugal, 8–10 May 2019. Elsevier Procedia, pp. 992–997 (2019)

Collaborative Engineering Method for More Electric Aircraft (MEA)—Tradeoffs and Informed Decision-Making Process

Jean-Rémy Imbert and Nathan Marguet

1 Introduction

It is a fact: industrial products are complex!

They use a multitude of different components and technologies. They are associated with technical systems that pose real integration difficulties.

In an organization, an individual is no longer in a position to master alone the design, manufacturing, industrialization, maintenance, evolution and recycling aspects of the product.

Collaboration within teams is obviously more than necessary:

- access to information in a secure way;
- share it effortlessly;
- distribute it in real time;

is essential!

In response to these problems, a new type of tool—and the methods that go with them—is beginning to emerge among manufacturers. These collaborative "platforms" can be classified in two categories:

- "monopolistic platforms", which solve communication difficulties between tools and disciplines,

    - by standardizing the supported formats and unifying the Design environment;

J.-R. Imbert (✉)
Digital Product Simulation K.K., Kioi-cho 3-16, Kioi-cho Kaneda Blddg. 6F, Chiyoda-ku, Tokyo 102-0094, Japan
e-mail: jean-remy.imbert@dps-fr.com

N. Marguet
Digital Product Simulation, 108 av. Jean Moulin, 78170 La Celle-Saint-Cloud, France
e-mail: nathan.marguet@dps-fr.com

© The Author(s), under exclusive license to Springer Nature Switzerland AG 2021
D. Krob et al. (eds.), *Complex Systems Design & Management*,
https://doi.org/10.1007/978-3-030-73539-5_17

– by simply replacing the business tools, external to the platform, with the
  platform's tools.

These "monopolistic platforms" are meant to be the central point where everything
is supposed to happen.

• "agnostic platforms", which,

    – do not standardize the Design environment and supported formats;
    – but take an aggregation approach to content with advanced connection
      capabilities.

Such platforms also provide a "notification and control center" (dashboard) where
engineering decisions can be made.

However, the scope of their control over the Design process is intended to be
limited, allowing organizations to retain all of their tools. Above all, it allows companies to define new ways of working, independent of the platform used. The strength
of this type of platform therefore lies in their ability to connect to existing engineering
applications.

We present in this paper the approach we have implemented for the study of
electromechanical alternatives to an initially mechanical actuator.

2   More Electrical Aircraft

Our use case is part of a fundamental technological change. It results from the race
towards the all-electric aircraft. The first responses to this challenge are leading to
the emergence of a hybrid electric or "More Electric Aircraft" (MEA) [1, 2].

If we forget for a moment the economic motivations of aircraft manufacturers,
this change stems from regulations concerning the reduction of CO_2 gas emissions.
The battle will therefore be fought on several fronts and in several stages. Indeed, the
difficult revolution that is pushing for the decarbonization of propulsion is underway.
It will lead to the replacement of turbojet engines by electric motors. But for a long
time now, the industry has been mobilized in a quest to reduce the mass of aircraft.

Historically, a mix of hydraulic and pneumatic energy drives non-propelling
systems. Advances in power electronics, fault-tolerant architecture and flight
control systems make it possible to envisage their replacement by lighter equipment. The technological maturity of Electro-Hydrostatic Actuator (EHA) and
Electro-Mechanical Actuator (EMA) is accelerating the electrification of aircraft
(Fig. 1).

This trend is fueled by the collaboration of aircraft makers with their equipment
manufacturers and suppliers, who are imagining new architectures and designing
systems with high energy density and electrical intensity.

On the scale of the entire aerospace industry, the More Electric Aircraft (MEA)
concept holds the promise of significant improvement. It is an essential element
of value creation: aircraft weight and fuel consumption can be reduced. The MEA

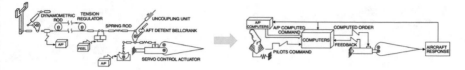

Fig. 1 From mechanical to fly-by-wire flight controls

concept opens up new ways to optimize aircraft performance. It increases flight reliability and should reduce operating and maintenance costs, and ultimately the total lifecycle cost of the aircraft.

3 Electro-mechanical Actuator

To be able to turn, an aircraft is equipped with ailerons. These aerodynamic control surfaces allow the plane to rotate around its roll axis. They are located on the trailing edge of the wing and lowered or raised thanks to the actuator, object of our study (Fig. 2).

Our case study focuses on the implementation of an Electro-Mechanical Actuator (EMA) to replace mechanical flight controls.

Engineering teams are reviewing concurrently three Design architectures:

- screw-nut system (1);
- connecting rod-crank system (2);
- direct-drive system (3) (Fig. 3).

The Electro-Mechanical Actuator is equipped with different sub-components, such as motor, reduction gear, etc. The activity related to the definition and integration of this mechatronic system into the aircraft wing is becoming a challenge for the teams who will be dealing with many aspects.

This project requires skills in several technologies and several disciplines. It is therefore quite naturally that the teams are organized in silos. Silos has developed over time following the logic of clustering specific expertise to favor efficiency.

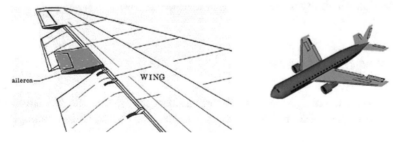

Fig. 2 Location of an aileron on an aircraft

Fig. 3 Three design
architectures

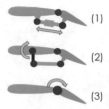

However, at crucial moments throughout the project the quality and speed of communication take precedence over the technical aspects.

4 Organizations

4.1 Current Organization

Poor communication between teams—whether organized in silos or not—is linked to various factors. More and more teams are distributed, their members are scattered over several locations or even several continents. Differences in time zones and language make teamwork more complicated.

Furthermore, these "logistical difficulties" can be compounded by infighting between these different teams or their members. Human factors make the mechanisms of collaboration very subtle…[3, 4]

Despite or because of the existing procedures and structure imposed by the PLM, PDM, SPDM tools, there is not always an effective process that coordinates disciplines and articulates tasks. Sequential work is then naturally favored: Team #1 waits for Team #2 to send its results before being able to work.

Some data is missing, some must be reprocessed and some is out of date or have become useless.

Thus, communication difficulties and divergences in internal objectives are the cause of a lot of wasted time, oversights and even errors.

Changes and excessive redesign phases extend the development time as they result in the need to solve unforeseen problems. Project costs are impacted by the organization's inability to detect inconsistencies early on.

4.2 New Organization

It is necessary to change the way of working. For the EMA implementation study we organized the collaboration from a new angle. The solution we encourage:

Bridging silos to improve Designs

Fig. 4 "Bridging silos"

To explain this approach in a few words, we can compare disciplines and teams to pieces of puzzles that do not fit together perfectly. Our methodology is an answer to many of the problems mentioned above.

It consists in providing the missing bridge that connects all these disciplines and teams together (Fig. 4).

This central piece—an agnostic platform—is a facilitator. It checks the consistency of the information available to the teams and distributes this information. It verifies the conformity of the technical hypotheses with the company's business rules.

Right from the start of the Design process, this platform highlights technical issues, inconsistencies and even errors.

The connection between the different software applications involved in the product Design is possible at any time. Teams push their most recent data to the platform and get the necessary information from other disciplines in return. All participants benefit from the pooling of information.

This makes it easier for them to find up-to-date data: with no extra effort, they have access—on demand—to the key information they need for their joint activity. The Design process is more agile because it constantly adapts to changes in each silo.

This connected—and collaborative—engineering approach activates a very fluid communication between the company's silos.

By enabling project teams to resolve conflicts earlier, by offering mechanisms for detecting inconsistencies and errors, the agnostic platform reduces the overall product Design time, and therefore its cost.

The emulation generated between teams plus the time saved, pushes engineers to redeploy their energy on innovation. They explore a larger Design space. They study a larger number of configurations. They optimize their Design.

The value of their product is thus increased.

5   Execution Using the Karren Platform

The Electro-Mechanical Actuator implementation project—in all its complexity—will be managed on several macro levels, using three conceptual objects.

5.1   The Requirements: "Requirements View"

The EMA requirements define the project issues. They are made available to stakeholders in karren. They can be completed manually. But karren's connection capabilities also allow to aggregate a selection of project-specific requirements that would be stored and managed in a specialized software, (DOORS, Rectify, Excel, etc.). The connector principle offers the possibility of using multiple data sources [5, 6].

The karren platform thus becomes the "Single source of truth" (SSOT) shared by the project team (Fig. 5).

The requirements used are of a technical nature. They directly concern the actuator or are closely related to the constraints of its environment:

- "actuator will be supplied with 28 Vdc";
- "maximum current intensity will be 50 A".

The requirements may also be of a different nature and not directly mathematical:

- "purchased components must be purchased within the EU";
- "EMA must be beautiful"... (why not?)

Finally, product requirements can define performance objectives:

- maximization: "reaction time should be 20% better than on the # model";
- minimization: "mass of the system must be less than 12.5 kg".

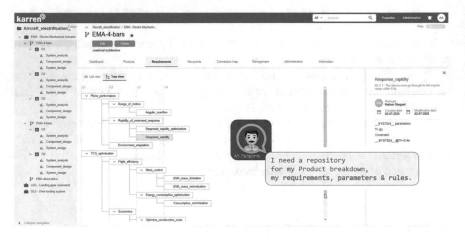

Fig. 5 Karren repository for requirements

5.2 The Business Knowledge: "Product View"

The strength of agnostic platforms lies in the fact that they manage only the product parameters necessary and sufficient for collaboration. These are key parameters, in the sense that they allow stakeholders from neighboring silos or subcontractors to be able to work.

It is very important to tell that the platform manages the parameters as objects. Each parameter can be valued or not. Each parameter can have one or more instances. Each one being linked to different candidate configurations of the product.

These platforms do not compete with PLM, PDM or SLM tools. They connect to them. karren only needs to manages a short list of parameters and values. Those that are necessary for the collaboration community and the progress of the business.

In the case of the actuator, this could be:

- gear ratio;
- overall dimensions (height, width, depth);
- mass;
- cost.

The variety of disciplines covered is unlimited as long as they can be described with input and output parameters.

Business knowledge is also managed in the platform through "rules". These rules are the mathematical materialization of lower-level requirements. They can also express relationships between parameters:

- "EMA_Beauty = true, if Mean_Subjective_Aesthetic_Evaluations > 5";
- "Reaction = Force × Distance_D".

Because the "Product view" object stores all the parameter objects, possibly valued, "Product view" is the object through which each of the Design proposals—i.e. configurations—is managed (Fig. 6).

It is necessary that during the course of the project different alternatives can be imagined. It is interesting that each proposed configuration is challenged in relation to the functional requirements of the product. The "Product view" object allows all this.

5.3 The "Viewpoint"

The EMA implementation project involves several disciplines, several teams. Each one has technical interests to defend.

For example, if we consider the reduction ratio the EMA gearing, from the mechanical strength point of view it should be small. From the point of view of the transition time between "Aileron up" and "Aileron down", the value of this same parameter should be high.

Fig. 6 A candidate configuration is challenged on requirements and business rules

Obviously, there is no question of producing two actuators, one "fast", the other "resistant". But it is quite judicious, (necessary) to find the best compromise, compatible with the data, the constraints, the estimates of each discipline.

Since Viewpoints are related to disciplines, the connectors of the agnostic platform retrieve—on demand—data generated by other software applications and value the corresponding parameters. The parameters can be either input or output parameters.

The platform aggregates this information to its database and then distributes it to other software tools. The exchange is then done via Application connectors following the same mechanism.

The platform maps the dependencies between Requirements, Viewpoints and Products (their parameters, their business rules) (Fig. 7).

The karren platform can connect to applications such as:

- Excel, Matlab, Dymola, GT Power, modeFRONTIER, Isight, CATIA, 3DEXPE-RIENCE, DOORS, and many others.

5.4 Manage the Complexity

At any time,
   Contributors of a Viewpoint, (e.g. mechanical strength) can:

- verify that their candidate configuration complies with business rules and meets product requirements.

Contributors from another Viewpoint, (e.g. System control) can:

- also check their candidate configuration, (business rules, requirements);

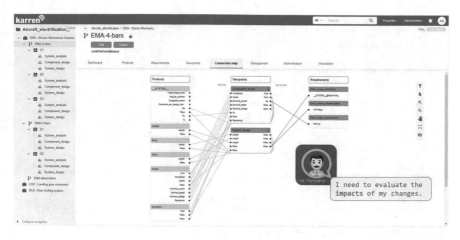

Fig. 7 Mapping of products, viewpoints and requirements

- propose a configuration with parameter values that are different from the values of the other Viewpoints;
- know the sets of values chosen by the other teams that are part of the collaboration perimeter.

Product Owner or Project Manager can:

- check the compliance status of each proposed configuration with respect to product requirements and business rules;
- highlight incompatibilities between disciplines (e.g. if a common value is not yet found for a common parameter);
- initiate discussions in order to reach a compromise;
- arbitrate and decide, in order to resolve a technical conflict.

The karren platform thus frees up the Design process by allowing teams to work in parallel while being regularly synchronized (Fig. 8).

Because the platform aggregates the essential information that enables the level of maturity and convergence of the project to be known, it also quite naturally offers an excellent monitoring tool.

The alternatives studied that fail to meet product requirements are detected and discarded. Conflicts are identified earlier. They can be arbitrated at a time when changes and evolutions are still possible and less costly.

All disciplines have the same level of information and can therefore better understand the issues and constraints of the Departments connected to them.

What-if scenarios are easily challenged. On this point, the example of the three EMA architectures illustrates the capabilities and interest of a collaborative engineering methodology.

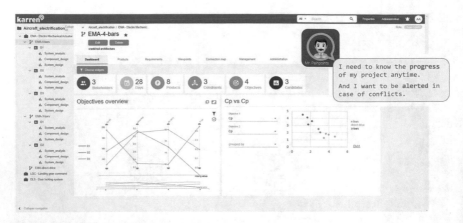

Fig. 8 Dashboard and various KPI for steering a project

These platforms do not yet integrate Artificial Intelligence technologies (as such) into the technical decision-making process. Nevertheless, the level and quality of information made available to engineers enables them to make informed decisions.

Once an optimal solution is found, karren stores it as a reference. Likewise, the decision steps (the configuration history) are kept and could be reused in future or similar projects. karren keeps track of the "when, who and why" to justify the decisions.

6 Conclusion

The innovation process can be a particular form of "*Stratégie tâtonante*" (trial and error strategy). Under this term arise:

- "intentionality", (e.g. the intention to create a product);
- "uncertainty", (e.g. to achieve objectives, to meet requirements);
- "unpredictability", (e.g. which of the architectures is the most efficient?).

Innovation is a collective progression led by organized entities that need the right balance between "order", (science, routines, procedures) and "disorder", (brainstorming, chance).

The silo organizations resulted from a logic of clustering specific expertise to favor efficiency. Whether we like it or not, they have proven their effectiveness. It is illusory—even counterproductive—to pretend to "break down" these silos. However, it is relevant to want to bridge them.

We have deployed a collaborative Design methodology equipped with the karren "agnostic" platform. As a result, karren overcomes three obstacles:

- without loss of time, the distributed stakeholders find the central information they need for their work;
- convergence cycles are quick because they are agile enough;
- difficulties arising from the heterogeneity of software applications, languages, cultures and divergent interests are avoided.

The engineering teams studied simultaneously several configurations of each of the three candidate architectures. karren allowed reducing the number of Design iterations. This was achieved by making exchanges between disciplines fluid. And by challenging each technical proposal with respect to product specifications and requirements (verification of conformity), from the upstream definition phases and throughout the project.

The multidisciplinary Design context of this "complex" product presented a large number of antagonistic singularities. karren facilitated the convergence (convergence verification) of the teams towards feasible solutions by constantly identifying technical conflicts.

Project management has greatly benefited from the platform's information and notification capabilities. Deadlocked configurations and non-performing proposals were eliminated earlier. Many tradeoffs between disciplines were facilitated by karren. This has led to optimal solutions. Finally, karren has shown the relevance of our approach in the fact that this methodology successfully executes:

- a collaborative design process with tradeoffs and informed decisions.

References

1. Qiao, G., Liu G., Shi Z., Wang Y., Ma S., Lim T.C.: A review of electromechanical actuators for more/all electric aircraft systems. SAGE J. Proc. IMechE Part C: J. Mech. Eng. Sci. **232**(22) (2018)
2. Qi, H., Fu, Y., Qi, X., Lang, Y.: Architecture optimization of more electric aircraft actuation system. Chin. J. Aeronaut. **24**(4), 506–513 (2011)
3. Temri, L.: Les processus d'innovation: une approche par la complexité. IXème Conférence Internationale de Management Stratégique AIMS (2000)
4. Signoretti, L.: How to overcome the challenges from the organizational silos. In: Conference in HEC Paris (2020)
5. Aeroval Inc.: Website https://moreelectricaircraft.com/
6. Airbus Commercial Aircraft: E-fan X—A giant leap towards zero-emission flight

Design of Test Flight Mission Planning and Playback Verification System Based on STK

Jieshi Shen, Bingfei Li, and Cong Chen

1 Introduction

As a costly and risky project, flight test should take into full consideration the factors such as test site, meteorological conditions and critical envelope. Therefore, it puts forward strict requirements on the planning of the flight mission. But due to the coupling factor, as a systematic project, it is very difficult to carry out the planning of the test flight mission with mathematical calculation method and the workload is too big. Therefore, at present, the index request is put forward to certain phases of the test flight according to the test purpose in the flight test. Most of the time, it still relies on the pilot's subjective experience to control the aircraft. On the other hand, the flight data obtained from the flight test is very valuable. Through the playback of the flight process by the flight data, on the one hand, it can help do the auxiliary calculation analysis on the flight. On the other hand, it can form a flight test database to quickly reproduce and review a certain flight process with a simulation method to achieve the purpose of demonstration and mission reuse. Satellite Tool Kit (STK) software can conduct convenient and rapid analysis of land, sea, air, and space missions, and can display the analysis results in the form of graphics and text. Its three-dimensional visualization function can import the real terrain, images and 3D solid models, and the degree of the scene simulation is high. It is the authoritative software for visual simulation in the field of aeronautics and astronautics. However, for flight test missions only, STK software is too large in size and relatively complex in use, and there is no management database support. Aiming at the conventional mission mode existing in flight test, this paper developed a flight mission planning and data playback system based on STK to achieve the purpose of rapid flight test task auxiliary design and pre-test flight simulation. It also supports real test flight data playback, visually restores the flight test process, and uses airborne flight data for relevant calculations to assist in the analysis of flight test results [1–3].

J. Shen (✉) · B. Li · C. Chen
Chinese Aeronautical Radio and Electronics Research Institute, Shanghai, China

© The Author(s), under exclusive license to Springer Nature Switzerland AG 2021
D. Krob et al. (eds.), *Complex Systems Design & Management*,
https://doi.org/10.1007/978-3-030-73539-5_18

2   Key Technology of System

2.1   STKX Component Technology

STKX components allow developers to seamlessly integrate the two-dimensional and three-dimensional visual interfaces and various data analysis capabilities in STK into their applications. STKX supports the forms of COM components and ActiveX controls to be embedded in applications that support OLE (Object Linking and Embedding). It supports C++, .Net and Java development environment, and provides a series of classes, interfaces and events that support STKX components to interact with third-party applications. At the same time, it can also receive and execute Connect commands sent by third-party connection control software through the API functions provided by STKX [4, 5].

STKX components mainly contains four kinds of controls, which are:

1.   AGI Map Control

AGI Map Control is a two-dimensional graphical display control that displays a two-dimensional map of the globe. Developers can seamlessly integrate it into their own applications. By writing the corresponding event response code, functions such as partial zoom in and zoom out of the map in the two-dimensional map can be realized.

2.   AGI Globe Control

AGI Globe Control is a three-dimensional graphical display control, which is the core of the three-dimensional display module in STK. Various operations on the user interface can be controlled by establishing mechanisms such as messages and events.

3.   Analysis Engine Interface Components

The analysis engine interface component interface provides a channel for the application to connect to the STK analysis engine, which can be directly invoked in applications without or with a GUI interface, and the calculation and simulation of the established tasks can be completed by sending control commands.

4.   Graphics Analysis

Graphics Analysis control is a spatial simulation analysis tool, with functions such as azimuth and altitude analysis, area analysis, and shadow analysis.

The simulation pattern of the STKX components is shown in Fig. 1. STKX components can be seamlessly integrated into user's secondary developed applications as separate functional modules. Each of these processing units interacts with the various integrated STKX components through the application interface. The latter transmits the corresponding processing results back to the former through local calls, and the processing unit then proceeds with the returned information. STKX internal event listener can realize STKX internal event monitoring, and users can also write event handlers to respond, so as to control the simulation scene more accurately.

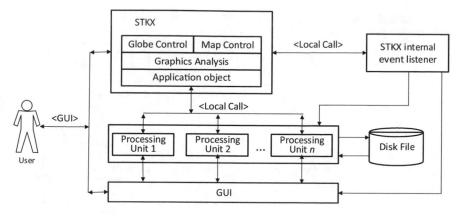

Fig. 1 Simulation mode for STKX components

2.2 Complex Flight Test Data Mapping Transformation

The amount of real data generated during a test flight is enormous. On the other hand, each type of test flight data has a complex format and many items. In order to load the data efficiently and restore the test flight process effectively, it is necessary to sample and process the huge data before importing the test flight data into the system, to reduce the data pressure. After sampling the data, it is necessary to filter the list of data and select the simulation scene data that can be supported by the system to be added to the dynamic properties of the entity.

2.2.1   Test Flight Data Sampling

Data sampling can effectively reduce the pressure on the database during system loading. Take the GPS data recorded in flight as an example. If a set of data is sampled at the interval of 0.01 s during the flight sampling process, the recorded data of the aircraft will reach 60,000 sets after only 10 min of flight. However, such high precision data is not needed in the simulation system. Therefore, the system designed the sampling process before importing the real test flight data to reduce the data pressure. The sampling process is shown in Fig. 2.

The amount of file data processed by sampling is reduced to 2^{-N} groups, and the time points of data are evenly distributed, with little impact on data distortion. The contents of the header file refer to the file description and data list title in the flight data file, as shown in Fig. 3. In practice, each type of test flight data file has inconsistent header contents and inconsistent number of lines, but usually no more than 20 lines. Therefore, the system cut the first 20 lines of the file during the sampling process to avoid the loss of the header (Table 1).

Fig. 2 Test flight data
interlaced sampling process

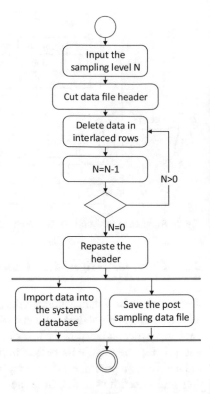

Fig. 3 Data support
relationship

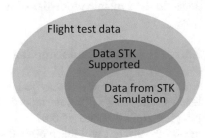

2.2.2 Flight Data Conversion

After the test flight data is sampled, the complex test flight data is converted to the data supported by the simulation scene to realize the playback drive of the aircraft. The way to convert the flight data is to map the data types in the flight data file to the data supported by STK. Generally, the input data supported by STK is the inherent parameters and kinematics information of the entity, while many data types generated by real test flight cannot be supported by STK.

Take Table 2 as an example. The table shows the data recorded by the aircraft

Table 1 Header contents of GPS data file

Project:		********-gps			
Program		******************************			
Profile		*********************************			
Source		GPS *********************			
Datum		WGS84, (processing datum)			
UTC Offset:		18 s			
Local time:		+8.0 h, CCT [***************]			
LocalUTC	Latitude	Longitude	H-Ell	VEast	VNorth	VUp
HMS	Deg	Deg	m	(m/s)	(m/s)	(m/s)

Table 2 GPS data mapping relation

File data	STK Data
UTC	Absolute time in STK
ID	STK Object ID
Longitude	The longitude of the aircraft at the certain time
Latitude	The latitude of the aircraft at the certain time
Height	The height of the aircraft at the certain time
East speed *	The east speed of the aircraft at the certain time *
North speed *	The north speed of the aircraft at the certain time *
Vertical speed *	The vertical speed of the aircraft at the certain time *

GPS during the test flight. The data required by STK can be extracted and imported into the database by means of list mapping.

The data listed in Table 2 can be mapped to STK supported data types, but in fact, After loading the model with absolute time and latitude and longitude position information, the eastward speed, northward speed and vertical speed can all be obtained through STK simulation calculations (as shown in Fig. 3). And as long as time precision is high enough, the value calculated by the simulation is very close to the actual value. Therefore, when data is imported into the aircraft model, the system only considers the injection time and location information in the design. The speed information contained in the file (indicated by * in the table) is calculated by STK simulation itself, which greatly reduces the amount of imported data and improves the efficiency of data injection.

2.3 Model Library Support for Model Equipment

STK defines a rich basic object model library, but there is no model-oriented equipment database support. In the course of use, there will be cases of repeated tedious load model assembly and parameter setting of the equipment. This system uses the user-defined plug-in interface provided by STK software to design the model library support of model equipment. Through the association between the tables, the database manages the unified file management of combat entities (including aircraft, ships, ground facilities, vehicles and radars, etc.) and payloads (such as sensors, radars, receivers, transmitters, etc.). The parameters are templated to achieve the purpose of rapid construction and reuse of equipment models. Figure 4 shows the relationship between a certain type of aircraft and the load in the database.

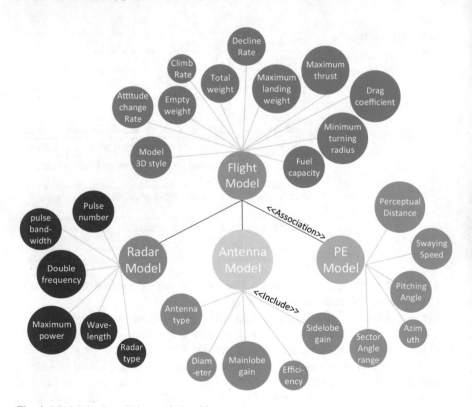

Fig. 4 Model database linkage relationships

3   System Prototype Implementation

3.1   System Structure Frame Design

The logic architecture of the flight mission planning and playback verification system is divided into four layers from bottom to top: data layer, engine layer, business layer, and interface layer, as shown in Fig. 5. The main functions of each layer structure are:

1. data layer: providing storage management of scene data, model entity data, geographic information data, simulation result data and flight test data, etc., implemented by means of relational database + files;
2. engine layer: STKX-based simulation engine providing various display and simulation computing services for upper layer businesses, including two-dimensional/three-dimensional digital battlefield display, system simulation engine and related simulation models;
3. business layer: including the functional modules for carrying out mission planning, simulation deduction, data playback and other business development. Mainly providing services such as battlefield environment configuration, model entity deployment, simulation deduction control, simulation deduction event scheduling, simulation deduction evaluation, flight unit data playback, flight test mission playback and simulation recording.

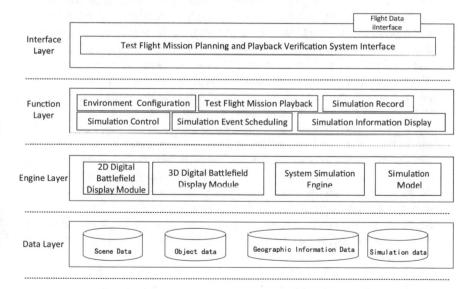

Fig. 5 System logical structure hierarchy

4.   interface layer: graphical operation interface, function button controls invoking the business layer services in a bundled manner, and windows displaying two-dimensional/three-dimensional scenes, simulation results and database data. The real test flight data is loaded into the entity database through the flight data access interface.

3.2   System Function Framework

According to the requirements of flight test mission, the system function was designed in a process of research. The system has the characteristics of two application scenarios and four-step design flow in use. The two application scenarios are forward test flight mission planning and test flight playback verification. The difference between the two is that in the flight-mission planning, the aircraft/payload parameters and trajectories are artificially created as expected values, while in the playback verification process, the data files obtained from real flight tests are loaded (Fig. 6).

The design flow is embedded in the main interface of the system to achieve the purpose of guiding the designers to standardize use. The system is mainly divided into the following steps in use:

1.   Scene environment design: scene creation, management, and construction of flight test basic environment, including terrain/texture data loading, hydrology, electromagnetic and atmospheric environment parameter settings;
2.   Flight test entity deployment: set the parameters of the aircraft and onboard loads (photoelectric, radar, antenna, transmitter, receiver and other sensors) participating in the flight test; input the flight path trajectory of the aircraft, there are different input modes in the two cases of test flight mission planning and test flight playback verification: (1) mission planning: there are two forms of waypoint input and mission mode. Waypoint input is to input the aircraft's latitude and longitude position and speed information at the critical moment to determine the aircraft's trajectory; mission mode is to establish common flight mission styles, such as takeoff, hover, flight at the trajectory as 8, etc., and to determine the flight trajectory based on the center position and trajectory geometric parameters. (2) Playback verification: the airborne GPS, inertial navigation and in-network status file data obtained during the real test flight are added to the aircraft trajectory information, and the real flight trajectory of the object aircraft is generated in the three-dimensional space of the system.
3.   Scene simulation control of test flight mission: scene control is provided in the simulation process, including the modification of simulation step size and the switching of simulation perspective;
4.   Simulation/playback results output: output simulation analysis results, such as aircraft fuel consumption, flight distance, inter-aircraft sensor visibility, communication link interference analysis, etc.

The database stores the scene environment template and aircraft equipment model, from which the users can add the scene environment or aircraft model with complete

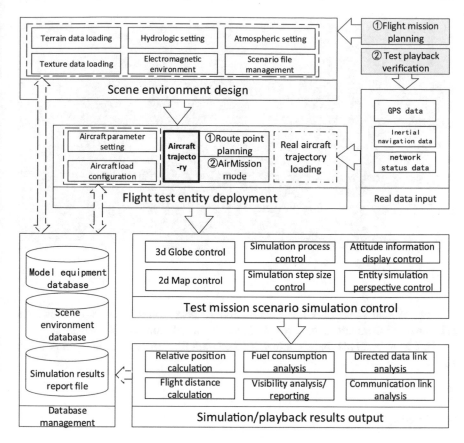

Fig. 6 System function framework

parameter settings. At the same time, it also supports the selection of scenes or aircraft entities on the interface and stores them in the data model library to achieve the purpose of reuse.

3.3 System Prototype Implementation

The system is implemented in C# language. Figure 7 shows the GUI interface of the system. The interface is mainly divided into six areas:

Area ① is the main display area, including three-dimensional and two-dimensional map display and imported data view;

Area ② is the toolbar, mainly for scene, entity parameter settings, simulation control and simulation analysis selection etc.;

Area ③ is the entity browser, mainly for existing entity display;

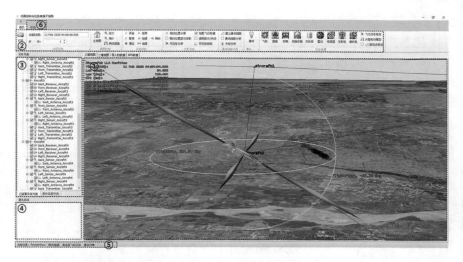

Fig. 7 System GUI

Area ④ and area ⑤ is the information display window. Area ④ shows the relationship information, and area ⑤ shows the simulation data information;

Area ⑥ for data playback control, the real data file can be selected to import into it.

4 Application Cases

4.1 Flight Mission Planning

In a test flight mission, the system is used to visually plan the test flight mission. The four aircraft involved in the test flight (with directional antennas installed in the front, rear, left and right of the fuselage) flew to the designated airspace in formation according to the prescribed plan. After arriving at the designated airspace, No. 1 aircraft performed roll action, No. 2 aircraft performed large radius barrel roll and hover maneuver action, No. 3 aircraft performed horizontal S flight and combat turn action, and No. 4 aircraft performed climb, dive, backflip, combat turn and other actions. The visual aircraft motion design is shown in Figs. 8, 9, 10 and 11.

A simulation is performed before the test flight to check the stability of the communication chain network under maneuvering conditions (the left wing of No. 2 shows the radiation effect of the directional antenna). Figure 10 shows the link situation of the communication chain under ideal conditions through system simulation analysis. After flying for 1570 s, No. 1 aircraft tilted due to the roll action, and the antenna on the right side of the fuselage was disconnected from the No. 2 aircraft in the right airspace. At the same time, it entered the communication area with the right and rear

Fig. 8 Aircraft 1 performing a roll

Fig. 9 Aircraft 2 performing a barrel roll

Fig. 10 Aircraft 3 performing a combat turn

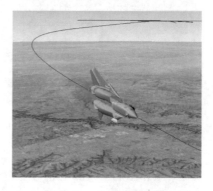

antennas of No. 3 aircraft and No. 4 aircraft in the front left airspace. The simulation results show that the stability of directional data link may be affected by the large angle change of fuselage (Fig. 12).

Fig. 11 Aircraft 4
performing a backflip

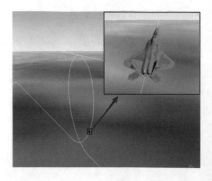

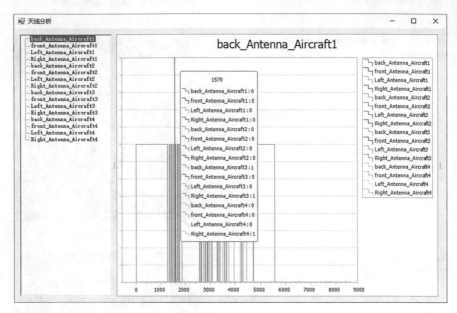

Fig. 12 Connectivity curve between the right antenna of aircraft 1 and the antenna of other aircrafts

4.2 Test Flight Four-Unit Network Playback

During a test flight, the aircraft recorded GPS data and network data. By importing
GPS data files and networking data files into the system, the test flight process of the
four-unit network was played back. The flight test mission planning and playback
verification system directly displays the flight trajectory and situation information
of the tested aircraft in three-dimensional space, and visualizes the playback of the
entire flight test process (Figs. 13 and 14).

Fig. 13 Playback of
four-unit network

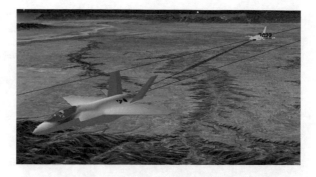

Fig. 14 Online status data
display

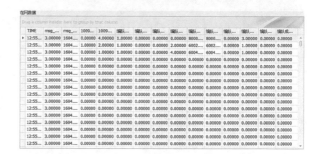

5   Conclusion

Visual mission planning can help identify risk items and test points for flight tests. Through the design of flight test mission planning and replay verification system, this paper provides a flight test planning method, carries out situational mission planning and visual action design for the aircraft to be tested, and quantifies the flight test time and task arrangement. During the simulation process, analysis methods such as geomorphological observation, execution of fuel analysis, and flight distance test were used to ensure the safe development of flight test missions. At the same time, the developed real test flight data playback function can effectively utilize the data, which is of great significance in flight test engineering [6].

References

1. Huang, J., Dang, T., Zhao, Y.: An approach for integrating VC with STK and its application in simulation. Comput. Simul. 1 (2007)
2. Raghu, N., Tejaswini, G.V., Aparna Rao, S.L.: Simulation of 13 panels phased array antenna by using STK tool. Int. J. Electron. Commun. Eng. **2**(4), 19–24p (2013)
3. Zhang, Z., Xu, Y., Zeng, G.: Analysis on the STK-based space mission simulation. J. Acad. Equip. Command Technol. **1**(011) (2006)

4. Beering, D.R., Tseng, S., Hayden, J.L., et al.: RF communication data model for satellite networks. MILCOM 2009-2009 IEEE Military Communications Conference. IEEE, pp. 1–7 (2009)
5. Caijuan, Z.: STK and its application in satellite sys-tem simulation. Radio Commun. Technol. **33**(4), 45–46 (2007)
6. Shiyi, W.: Simulation application of STK in avionics sensor equipment test environment. J. Comput. Appl. Softw. **36**, 1 (2019)

IMA Dynamic Reconfiguration Modelling and Reliability Analysis of Tasks Based on Petri Net

Zhiao Ye, Shuo Wang, and Jiapan Fu

1 Introduction

The avionics systems tend to be more and more modular and integrated, which is playing a more and more important role in the complicated systems [1]. Integrated Modular Avionics (IMA) is becoming the most advanced stage of avionics development today. Also, all the subsystems are deployed on the Common Functional Modules (CFM). It achieves the high physical synthesis of the systems and functional synthesis.

IMA systems have been bringing a number of advantages because of the flexible architecture till now, which is in great need of high reliability. What's more, the popularity of automation make it possible that the system could have the ability to heal itself. When the system fails, it can recover to ensure the reliability, such as degradation and dynamic reconfiguration, which is an efficiency way to keep the system safe. Compared to the static reconfiguration, the dynamic reconfiguration does not affect the normal operation of the aircraft when resources are reconfigured. Dynamic reconfiguration process means that when the fault occurs, the system configuration will change. The application running on the failed module will stop and reconfigure, and restart on a new module. In the event of failure, the process of dynamic reconfiguration can guarantee the reliability of the system.

Dynamic reconfiguration improves the systems' flexibility while reducing systems' hardware redundancy. Moreover, the reliability of dynamic reconfiguration straightly affects the reliability of the systems. Especially for mission critical systems, the reliability of dynamic reconfiguration needs to be evaluated and ensured that system's safety is met. However, each reconfiguration process consists of several

Z. Ye (✉) · S. Wang · J. Fu
Key Laboratory of Synthetic Technology of Avionics, China National Aeronautical Radio Electronics Research Institute, Shanghai 200241, China

S. Wang
e-mail: wang_shuo@careri.com

tasks, and the reliability of the computational reconfiguration must be derived from the reliability of each task. This is also the goal of this paper.

Too help us analyze and verify the system, several approaches are taken, such as Model Driven Development (MDD), which avoids a waste of time and money and is invested to the system modification after the system development finishes. Also analysis at early stage before actions can ensure the system's quality [2]. MDD based on the architecture becomes an important research domain for complicated embedded systems. Then the architecture and structure languages become important.

We have carried out a study on the reliability of each mode of the dynamic reconfiguration process, and Quan Zhang has proposed IMA reconfiguration modeling and reliability analysis based on AADL after the study [3]. But all the analysis is not about the tasks of specific dynamic process. What we need to do now is to decompose the dynamic reconfiguration process into a few of steps of the conversion process in the system modal, and to complete the analysis of the reliability of the system for each task, which is, task reliability.

This paper proposes a model based on reliability analyzing method for the process of dynamic reconfiguration. In this method, we translate the system modeled by AADL into Petri net and set the reliability property parameters for simulation to verify if the dynamic reconfiguration process meets reliability. The rest of the paper is organized as follows. In Sect. 2, the development and mechanism of the dynamic reconfiguration of IMA is discussed. In Sect. 3, Petri net are introduced. In Sect. 4, a reliability evaluation method is proposed to simulate the dynamic reconfiguration and calculate the reliability of each task in the Petri net. Section 5 presents a case study. The last part is the conclusion.

2 IMA and Dynamic Reconfiguration

The IMA concept proposes an integrated architecture with application software portable across an assembly of common hardware modules. An IMA architecture imposes multiple requirements on the underlying operating system. It realizes all kinds of the functions of avionics with CFMs. The ASAAC standards define the module functions and interfaces of CFM.

The use of common modules to IMA system has the ability to reconfigure hardware and software resources, while the software system in the blue print system makes the system in the software structure can also be changed according to demand.

The blue print system is a set of management software running in the application software by which the information on the system can be centrally managed. The system only needs to modify the blueprint on the configuration when it changes. Blue print also provides a database that can be accessed offline during the loading application phase, which can also be located in the mass storage module for on-the-fly access for online troubleshooting, configuration or reconfiguration and communication management.

Fig. 1 Process of dynamic
reconfiguration

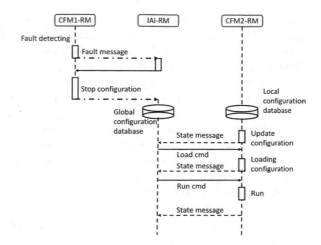

IMA system's general function module and the blueprint disposition system enable the system to change the target system by changing the blueprint system configuration at any time, this change ability is the IMA system flexibility direct embodiment.

Dynamic reconfiguration, which means system configuration changes between the processes. The trigger can be reconstructed by the system mode conversion, system failure or test and maintenance instructions to complete, the schematic diagram shown in Fig. 1. Dynamic reconfiguration utilizes the excessive resources shared among the subsystems to improve the performance. In consideration of the complexity of dynamic reconfiguration, the reliability of the process is in badly demand. But there are few evaluation methods for the process of dynamic reconfiguration.

3 Tools for Modeling and Analyzing

As we know, AADL is the basic tool that used to model the system which can describe the architecture of the embedded real-time system and its functional or nonfunctional properties in a standard way. The AADL model can accomplish the process of design, analysis and verification. Furthermore, SAE has published a series of annex, such as graph annex, error model annex [4], data annex, ARINC653 annex and behavioral annex [5]. Among them, the behavioral annex can describe components' behavior by state and transition which could define trigger conditions and transition actions. The transition contains data receiving or sending, execution time etc.

AADL model includes software components and software platforms. The software components is for software architecture modeling, including processes, threads, thread groups, subroutines, data. The execution platform is used for modeling hardware and operating systems, including the processing unit, virtual processing unit, a memory, a bus, a virtual bus, and peripherals. An AADL contains at least one system component, which can be hierarchically partitioned by subsystem components. The components are connected, bound, accessed and invoked to describe the hierarchical system structure. AADL is generally applicable to qualitative and quantitative analysis of the software architecture [6]. The schema in AADL can represent the configuration state of a system or component [7]. The corresponding thread and connection configuration process during modal conversion is showed in Fig. 2.

In this paper, we use a modal to represent the configuration state of the system. The association between the two modes can be represented by a behavioral attachment. In other words, behavior can be described by behavioral attachments. The process of dynamic reconfiguration could be modeled in AADL. In general, the main modeling correspondence is shown in Table 1.

Then the AADL-based model is transformed into a Petri net to simulate the dynamic reconfiguration process. Petri net is a kind of information flow model in

Fig. 2 Mode transition in AADL

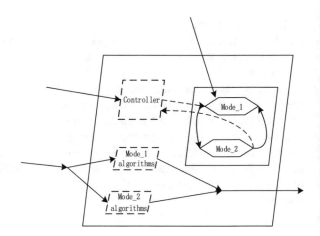

Table 1 Corresponding rules in ARINC653 annex

ARINC653 entity	Module	Virtual		Process
AADL entity	Processor	Virtual Processor	Process	Thread
Inter-partition communication	Queuing ports		Sampling ports	
Inter-process communication	Event data port connection		Data port connection	
Intra-partition	Buffers	Blackboards	Semaphore	Events
Inter-thread communication	Event data port connection	Shared data component or Data port connection	Shared data	Event port connection

the network structure with the ability of parallelism, uncertainty, and synchronization and description distribution. It provides a formal modeling method based on graphics and mathematics for our analysis of complex systems. And it is widely and effectively used in many other respects. As a graphical tool, a Petri net can be a data stream and a network. Although as a mathematical tool, Petri nets can be used to set state equations, algebraic formulas and other mathematical models.

The basic Petri nets are composed of places, transitions, and arcs that connect them. In the basic Petri net, one place can describe a state of the system. A transformation represents a process or event that changes the system. The arc contains the arc from the location to the transition point and from the transition to the location [8]. There can be several tokens in each place. The token in the system means the state of the system at that time. Classical Petri nets can simulate states, events, conditions, and so on. But in fact there are some shortcomings. For example, it has no concept of time and can easily cause state explosion. In addition, it does not describe the transition probabilities. Thus, in this approach, the classical Petri net adds reliability probabilities to it [9].

4 Task Reliability Evaluation

The traditional reliability evaluation methods which is applied to the system structure mainly include the reliability block diagram, Fault Tree analysis [10], Markov models [11], Petri net and so on. We have done a lot of analysis of the reliability of dynamic reconstruction, but for the dynamic reconfiguration, the reliability of each task analysis has not done yet. As a matter of fact, in order to analyze the reliability of the whole dynamic reconfiguration process, we would first decompose to the system modal and every step of the conversion process. The reliability analysis of each task of the system is carried out, that is, the task reliability and task reliability are defined as the ability to complete a defined function within a mission profile. When a task reliability value is determined, only the association failure affecting the task is counted.

The dynamic reconfiguration process can be decomposed into several tasks as shown in Fig. 3. First of all, the reliability of each task is analyzed, and then the

Fig. 3 Data backup involves the components

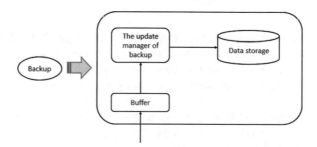

reliability of the reconstruction process is calculated in series, if the parallel case, the same applies.

Second, when the link was destroyed, which involves thread interrupt procedures and transmission link interrupt program, the calculations would be based on the reliability of the two software. Similarly, when establishing a new link, we should consider the establishment of transmission links and the establishment of thread-link procedures in the virtual channel to calculate the reliability. Because the new partition involves the memory resources of the module, building the program in the partition, so the reliability of the memory component is a consideration for the calculation of the new partition's program reliability.

The reliability of each state transition is calculated by the reliability of each software component, and then the reliability of the whole process can be calculated by modeling.

4.1 AADL Modeling

The dynamic reconfiguration can be modeled by AADL. A mode represents a configuration of active subcomponents and connections in the static architecture. The structure and composition are described by the elements in AADL. Initial mode just represents the initial state of the system. When reconfiguration process happens triggered by errors, the state of system represented by mode can change, which is called the mode transition. Behavior and Error model annex can be used to represent the system's different behavior and error transitions.

The behavior annex is used here to describe the details of the system's process, such as the actions, time cost and memory occupation. After the model transition given below, and then we analyze the properties of the behavior annex, including the reliability of the components, as well as the reliabilities of the links between components and so on. Then the reliability of the model is analyzed and the reliability of the whole model is obtained. More details about the modeling method are shown in Fig. 4.

4.2 Transit into Petri Net

In order to make it more convenient to analyze the reliability of dynamic reconfiguration, we transit AADL model to Petri net and make the corresponding rules for the transition. The transition relations are listed in Table 2.

For the calculation of reliability problems, and for the conversion of the model, the AADL would be translated into a stochastic Petri nets. The failure rate of the components is represented by the transition probabilities in the Petri net, and the failure recovery probability of each component is also expressed by the transition probabilities of Petri nets.

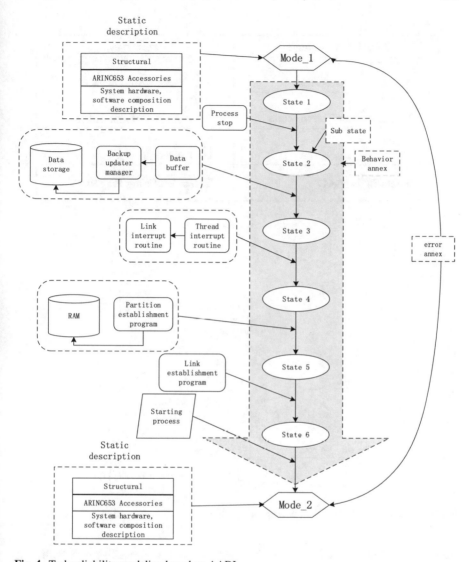

Fig. 4 Task reliability modeling based on AADL

5 Case Study

In this section, to calculate the reliabilities, we would first do the model conversion, to translate the AADL model into a stochastic Petri nets.

The states of the components involved to complete the tasks in AADL model are translated into the Petri net. The failure rate of the components is represented by the transition probabilities in the Petri net, and the failure recovery probability of each component is also expressed by the transition probabilities of the Petri net. The data

Table 2 The transition rules

AADL Components	Petri Net
⬡ Active Mode	⊙ Place With Token
⬡ Non-Active Mode	◯ Place Without Token
⤳ Transition	⬚ Transition
Reliability Property	Transition Firing Property

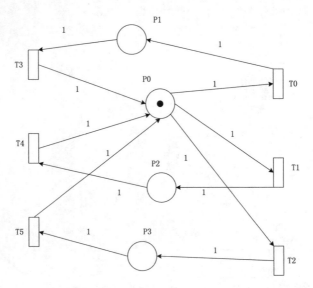

Fig. 5 The petri net model of data backup task components

backup, for example, involves three components, three components of the fault state with p_1, p_2, p_3 that the state of failure with p_0 said the formation of the model are as Fig. 5.

5.1 Reliability Analysis Based on Stochastic Petri Net

As is known for all, the stochastic Petri nets with a finite number of transitions are isomorphic to a Markov chain. The Markov process is a stochastic process, and in this paper, homogeneous Markov processes would be used to analyze the task reliability.

First of all, Markov process is a theory that the next state of the system depends on the current state of the system, and for any natural number n, the time point $0 \leq t_1 \leq t_2 \leq \cdots \leq t_n$ have the relationship below:

$$P\left[X_n(t_n) = i_n | X(t_1) = i_1, \ldots, X(t_n) = i_{n-1}\right]$$
$$= P\left[X(t_n) = i_n | X(t_{n-1}) = i_{n-1}\right]$$
$$i_1, i_2, i_3, \ldots, i_n \in S$$

Among them, the probability of the system transitioning to the next state is independent of the previous evolution of the system, i.e., the system state is only relevant to the current state and not to the previous evolution. The corresponding Markov chain can be constructed by solving the reachable set of stochastic Petri nets.

In the stochastic Petri net, the transition delay can be regarded as a continuous random variable xi, which should obey the exponential distribution.

According to the correlation theorem of the stationary distribution of the Markov chain and the Chapman-Kolmogorov equation:

$$\begin{cases} P^* \otimes Q = 0 \\ \sum_{i=1}^{k} P^*(M_i) = 1 \end{cases}$$

The matrix Q is the transfer rate matrix of the Markov process. q_{ij} $(i \neq j)$ depends on the state diagram of the Markov chain. q_{ij} is the rate value on the arc when there is a directed arc between the marker M_i and the marker M_j in the graph; q_{ij} equals zero when there is no arc. The elements on the diagonal of matrix Q have the relation that $q_{ij} = \sum_{i \neq j} q_{ij}$.

In order to calculate the reliability of each sub-state task in IMA dynamic reconfiguration, we taking data backup as an example, it is assumed that data backup involves three components: data update management, buffer pool and data memory, and there are four states of these three components, the state M1 means that the three components are working properly, one of M2 M3 M4 means one of the components is in a state of failure. The transfer rate between the various states represents the migration reliability, as shown in the Markov chain in Fig. 6.

$$\begin{pmatrix} -0.6 & 0.1 & 0.3 & 0.2 \\ 0.2 & -0.2 & 0 & 0 \\ 0.2 & 0 & -0.2 & 0 \\ 0.1 & 0 & 0 & -1 \end{pmatrix}$$

Equations can be obtained according to the formula:

$$YQ = 0 \& y_1 + y_2 + \cdots + y_n = 1$$

Like below:

Fig. 6 The Markov chain of
the data backup

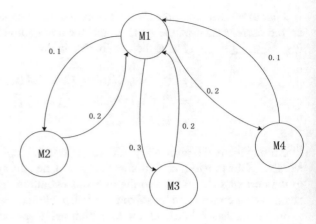

$$\begin{cases} -6y_1 + 2y_2 + 2y_3 + y_4 = 0 \\ y_1 - y_2 = 0 \\ 3y_1 - 2y_3 = 0 \\ 2y_1 - y_4 = 0 \\ y_1 + y_2 + y_3 + y_4 = 1 \end{cases}$$

Then, we could get the steady state transition probability of each M_i by solving the linear equations:

$$\begin{cases} y_1 = P[m_1] = 0.2 \\ y_2 = P[m_2] = 0.1 \\ y_3 = P[m_3] = 0.3 \\ y_4 = P[m_4] = 0.4 \end{cases}$$

The task reliability in this case is:

$$R = P[m_1] = 0.2$$

5.2 Related Works

Similarly, we could find all the tasks happening in the dynamic reconfiguration process as is shown in Fig. 4, and the task completion reliability can be calculated for each sub state.

Since the dynamic reconfiguration process is a series system, the task reliability of the whole process can be obtained by multiplying the task completion reliability of each sub state.

6 Conclusion

In this paper, based on Petri net, the model of IMA dynamic reconfiguration process tasks' reliability calculation is established. Adding the required components with their reliability attributes of each sub-state task to the model, by which the original dynamic reconfiguration process model would be enriched. It will be a great usage in analyzing the safety of dynamic reconfiguration in IMA system.

In the future work, there would be still much work to do to improve and perfect the Petri net model and the probability calculation method. Currently, the calculation still needs to be done in the AADL model made by Java-developed tools. And we are exploring new ways to build models and calculate the reliability [12].

References

1. Watkins, C.B., Walter, R.: Transitioning from federated avionics architectures to integrated modular avionics. Clerk Maxwell, J. (ed.) A Treatise on Electricity and Magnetism, vol. 2, 3rd edn., pp. 68–73. Clarendon, Oxford (2007)
2. Feiler, P.H., Lewis, B.A., Vestal, S.: The SAE architecture analysis & design language (AADL) a standard for engineering performance critical systems. In: Computer Aided Control System Design, 2006 IEEE International Conference on Control Applications, 2006 IEEE Inter-national Symposium on Intelligent Control, 2006 IEEE. IEEE (2006)
3. Zhang, Q., Wang, S., Liu, B.: IMA reconfiguration modeling and reliability analysis based on AADL. In: IEEE 4th Annual International Conference on. Cyber Technology in Automation, Control, and Intelligent Systems (CYBER). IEEE (2014)
4. SAE: Architecture Analysis and Design Language (AADL) Annex Volume 1. SAE standards AS5506/1 (2006)
5. SAE: Architecture Analysis and Design Language (AADL) Annex Volume 2. SAE standards AS5506/2 (2011)
6. Hecht, M., Alexander, L., Chris, V.: A tool set for integrated software and hardware depend-ability analysis using the architecture analysis and design language (AADL) and error model annex. In: 2011 16th IEEE International Conference on Engineering of Complex Computer Systems (ICECCS). IEEE (2011)
7. Rugina, A.E., Kanoun, K., Kaaniche, M.: An architecture-based dependability modeling framework using AADL. In: Iasted International Conference on Software Engineering & Applications, pp. 222–227 (2007)
8. Teruel, E., Franceschinis, G., Pierro, M.D.: Well-defined generalized stochastic petri nets: a net-level method to specify priorities. IEEE Trans. Softw. Eng. **29**(11), 962–973 (2003)
9. Chiola, G., Marsan, M.A., Balbo, G., et al.: Generalized stochastic Petri nets: A definition at the net level and its implications. IEEE Trans. Softw. Eng. **19**(2), 89–107 (1993)
10. Han, X.T., Yin, X.G., Zhe, Z.: Application of fault tree analysis method in reliability analysis of substation communication system. Power Syst. Technol. **28**(1), 56–59 (2004)
11. Dugan, J.B., Randy, V.B.: Reliability evaluation of fly-by-wire computer systems. J. Syst. Softw. **25**(1), 109–120 (1994)
12. SAE: Architecture analysis & design language (AADL). AS5506, SAE International, 2004

Integrated Digital MEMS Design for Movement-Control System

Chenglong Duan, Yongsheng Hu, and Tao Tao

1 Introduction

Requirements of inertial navigation and movement control for future tactical and strategic machines or weapons are: short start-up time, low power consumption, large dynamic range, strong shock and vibration resistance, medium and high accuracy, small size and weight, easily used and maintenance, and long performance guarantee period and so on [1–4]. Improving the integration of the inertial system based on MEMS, selecting MEMS instruments that have advantages in volume and cost, and using highly integrated embedded computer technology to achieve the miniaturization of the strapdown inertial system, fully absorbing the international advanced experiences of combination, integration and modularization can make the product more suitable for the needs of movement-control system.

Using integrated circuits to replace discrete components and modularizing complex hardware and software design technologies will help improving system reliability, testability, maintainability, accuracy, and expanding applications, increasing production, and reducing costs [5–9]. In accordance with the above ideas, one MEMS device was used to construct a micro-mechanical inertial measurement device, and the real-time and in-depth comprehensive compensation for the system error presented by the MEMS sensitive device were performed, and the formal model was verified.

This article mainly explains the following:

1. Analyzing the basic character of MEMS system and the key technology of miniaturization design, and elaborate the miniaturization design technology of the navigation computer;

C. Duan (✉) · Y. Hu · T. Tao
Avic Beijing Keeven Aviation Instrument Company, Beijing, China

Y. Hu
e-mail: mengxiangkaixuan@163.com

2. In order to make the system meet the requirements of environmental test conditions such as vibration, shock, high and low temperature (−55–70 °C), the environmental adaptive design of the MEMS system was carried out, including static measurement accuracy in severe high and low temperature environments, and dynamic character, noise in the vibration environment, etc. The signal has the ability to resist high-frequency noise interference [4], that is to say, to ensure that the movement control system has sufficient phase reserve and amplitude reserve in the oscillating mode, meanwhile to ensure that the system does not appear jitter and other problems;

3. In response to the requirements of movement control system feedback control, dynamic character tests were carried out, modelling component signals as objects, optimizing signal measurement accuracy, improving real-time performance, and reducing delay coefficients in order to meet high maneuverability and improve maneuvering characteristics, that is, "movement is smart, static is stable" [5].

2 MEMS Miniaturization Design

The MEMS component includes interface boards, solution boards, power boards, MEMS gyroscopes, accelerometers and other major modules. The length, width, and height are 130 mm * 75 mm * 84 mm, and the weight is 0.75 kg. The embedded navigation computer (DSP + FPGA) completes the simulation, digital signal processing, providing external digital (serial 422, 429 optional) pitch, roll, heading triaxial angular rate, forward, lateral, normal triaxial acceleration, pitch, roll attitude information, circuit design uses modular technology realizes power supply and cross-linking of internal and external signals. The design of the miniaturized structure was designed to facilitate heat dissipation and meet mechanical strength requirements. MEMS gyroscopes and meters are installed orthogonally without interference in size for easy disassembly, as Fig. 1.

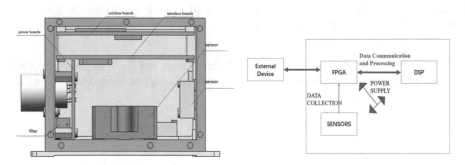

Fig. 1 The Design of MEMS

3 Digital Signal Processing

3.1 Architecture

The embedded navigation computer uses DSP and FPGA architecture to realize signal processing and interaction. The system includes completely self-test, data synchronization and communication, solution, program storage and other functions. It integrates FPGA interface processing and DSP computing capabilities. FPGA performs logic, timing, interrupt control, digital signal synchronization and transmission, which is multi-threaded execution; DSP performs inertial navigation algorithm and inertial device depth compensation processing, which is single-threaded, determines the appropriate DSP interrupt period based on the actual solution cycle, FPGA performs digital communication optimization which is beneficial to reduce digital signal delay and improve dynamic response and data transmission accuracy.

3.2 Digital Signal Optimization

The HB6096 signal transmission rate is set to 100 K. The actual transmission time is determined according to the number of semaphores (such as 12) and the number of bits (such as 32 bits). The delay can be accurately measured. The sending time occupied by the data (label number), each sending a number takes 0.3 ms, a total of 4 ms, see the last signal in Table 1. The solution period is 0.35 ms, the external communication period is 5 ms, and the interrupt period is selected as 1 ms or 5 ms.

Table 1 Measured time

Label	Original code	Bus time	Intervals (ms)
1	EFC2F259	1:27 s.144.505.340	0
2	E40A0019	1:27 s.145.655.340	1.1
3	E4080099	1:27 s.146.005.340	1.5
4	6440A0D9	1:27 s.146.355.340	1.8
5	E0000039	1:27 s.146.705.340	2.2
6	600000B9	1:27 s.147.055.340	2.5
7	E00160CB	1:27 s.147.405.340	2.9
8	E000A02B	1:27 s.147.755.340	3.2
9	7FFFE0AB	1:27 s.148.105.340	3.6
10	E009C04B	1:27 s.148.455.340	3.9
11	600200 EB	1:27 s.148.805.340	4.3
12	E0FF406B	1:27 s.149.155.340	4.6
13	EFC2F259	1:27 s.149.505.340	5

Actual measurement results using a bus analyser (the format of time is Minute/second/millisecond/microsecond/nanosecond):

For signals with high requirements of low delay characteristics, taking angular rate as an example, it should be sent first to reduce the phase lag of data with more information.

4 Modeling

The angular rate signal required by the movement-control system is analyzed as an object, and the system identification is equivalent to the corresponding transfer function model, which is helpful for the design of complex control rates.

For the movement control system, the angular rate signal's amplitude frequency response in the operating frequency band has no large overshoot, slow attenuation, and low phase frequency response.

4.1 Time Domain Characteristics

The system outputs discrete digital signals in the time domain. The filter is set to reduce the noise level of the output signal. When analyzed under the vibration spectrum, the signal shows good stability, as Fig. 2.

The frequency domain characteristics reflect the dynamic character of the signal, which means the response characteristics of the signal when the input quantity changes with time.

Through frequency correction (amplitude, phase compensation), the frequency characteristics of the object can be changed, damping characteristics can be changed, overshoot can be reduced, and phase delay can be reduced as following Tables 2 and 3:

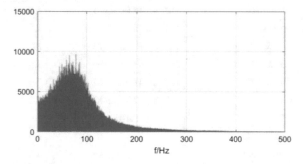

Fig. 2 High-frequency noise resistance to high frequence (>100 Hz)

Table 2 Filter information with s and z domain

Model	Link
	Filter
Correction function	$\dfrac{2.75s^3+1752s^2+0.0004405s+0.00000522}{s^3+1733s^2+0.0006953s+0.00000522}$
	Transfer function (z domain):
	$\dfrac{1.28z^3-2.07z^2+0.6565z+0.2454}{z^3-1.138z^2-0.2604+0.5095}$

Table 3 Object model

Model	Link
	Object
function	$\dfrac{4555.7}{s^2+39.216s+4581.7}$

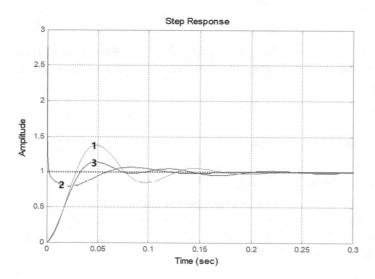

Fig. 3 Filter design result of step response

The filter's step response is line 2 and the line 3 is the step response of the final object as Fig. 3 seen.

4.2 Bandwidth Configuration

Designing digital filters to achieve the goal of MEMS gyroscope output's bandwidth configuration is as follows: the four curves indicate the amplitude-frequency and phase-frequency characteristics when the bandwidth is configured at 16, 33, 66, and

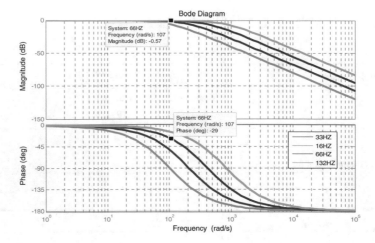

Fig. 4 Bandwidth configuration

132 Hz. In the corresponding low frequency band (within 107 rad/s), the amplitude has no overshoot, and the phase changes linearly throughout the frequency band as Fig. 4 seen.

4.3 Time Delay Analysis

(1) Theoretical analysis:

For the communication of the all-digital system, taking 429 as an example, the communication loop and communication period are positively related to the delay of the signal. By analyzing the communication path of the signal, the approximate delay can be determined as Figs. 5 and 6 seen.

(2) Test method:

Through the test method, the delay and amplitude-frequency characteristics of the signal can be accurately measured, and the filter settings can be adjusted to meet the application requirements of specific scenarios (higher real-time performance and smaller steady-state error).

4.4 Dynamic Character Test

Required equipment as Fig. 7 seen: angular vibration table (TDC-4, 0.03–60 degrees per second adjustable range, frequency range 0–Hz as simulation source), frequency response meter, D/A conversion equipment, acquisition equipment.

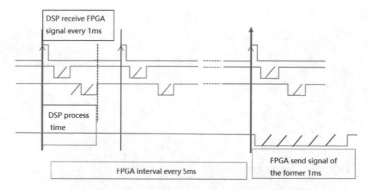

Fig. 5 Communication path

Fig. 6 Signal timing test

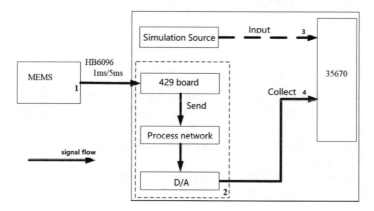

Fig. 7 Test method

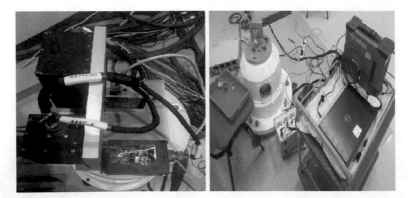

Fig. 8 Test instruments

Frequency sweep test method:

The 35,670 frequency response instrument provides a frequency sweeping command signal, which drives the angular vibration table to generate a variable frequency angular rate as signal (1) The MEMS component responds to the normal harmonic as signal (2) The signal is converted into analog signals 3 and 4 by a signal conversion board. OR36 multi-channel dynamic signal acquisition instrument or use 35,670 to simultaneously acquire signals 3 and 4, compare the two signals, and use FFT to analyze the dynamic characteristics of the angular rate of the MEMS component.

Note: The dynamic characteristics of the angular vibration table must be accurately known. After calibration, it can be used for the dynamic characteristics analysis of specific related signals (Fig. 8).

4.5 Results

Angular rate accuracy test as example: a rate point test every 10°/s in the range of −300 to 300°/s (Symmetric interval), Comparing the accuracy between the given signal and the acquired signal.

After analysis, the error is controlled within 0.02°/s, and the linearity is good as Figs. 9 and 10 seen.

After analysis, the dynamic characteristics of MEMS are equivalent to a typical second-order overdamping link. The damping ratio is greater than 1. It has good damping characteristics.

1. The amplitude-frequency characteristics are naturally attenuated. The configuration is different and the bandwidth is different. Slow attenuation within 20 Hz as Fig. 10 seen;

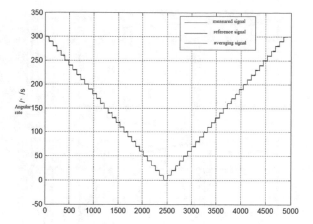

Fig. 9 The result of accuracy test (−55–70 °C)

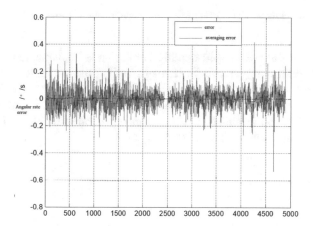

Fig. 10 The result of accuracy

2. Phase-frequency characteristics are related to delay. In order to obtain a system with a small delay, an optimal digital system can be designed (1 ms interruption, 429 data is updated in time, 5 ms communication), signal communication is optimized, phase within 20 Hz Hysteresis (90° prevails) meets the needs of the control system as Fig. 11 seen.

5 Conclusion

Highly integrated, and miniaturized MEMS system used in movement control systems which can provide angular rate signals and other inertial information more than 6 axis that meet the needs of high precision and high dynamic character is

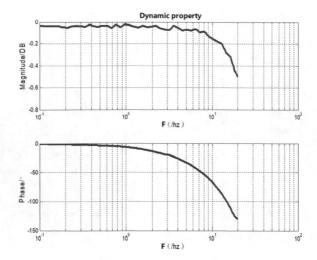

Fig. 11 The result of dynamic character (Slow attenuation within 20 Hz)

designed with method of MBSE; digital filter design is implemented to achieve the goal of bandwidth configuration in terms of setting the amplitude frequency and phase frequency according to the requirements of using; reducing the time delay by optimizing the digital signal. The dynamic character test method used in this article which is effective can establish an object transfer function model to determine the amplitude and phase characteristics of the system.

Reference

1. Nikolic, V., Motamedi, S., et al.: Extreme learning machine approach for sensorless wind speed estimation. Mechatronics **34**, 78–83 (2016)
2. Johansen, T.A., Cristofaro, A., Sorensen, K., Hansen, J.M.: On estimation of wind velocity, angle-of-attack and sideslip angle of small UAVs using standard sensors. In: International Conference on Unmanned Aircraft Systems 510–519 (2015)
3. Neumann, P.P., Bartholmai, M.: Real-time wind estimation on a micro unmanned aerial vehicle using its inertial measurement unit. J. Sens. Actuators A Phys. 235:300–310 (2015)
4. Rhudy, M.B. Larrabe, T., Chao, H., Gu, Y., Napolitano, M.: UAV attitude, heading, and wind estimation using GPS/INS and an air data system. In: Aiaa Guidance Navigation & Control Conference, pp. 1–10 (2013)
5. Fravolini, M.L. Pastorelli, M., Pagnottelli, S., Valigi, P., Gururajan, S.: Model-based approaches for the airspeed estimation and fault monitoring of an Unmanned Aerial Vehicle. Environ. Energy Struct. Monit. Syst. 18–23 (2012)
6. Cho, A., Kim, J., Lee, S., Kee, C.: Wind estimation and airspeed calibration using a UAV with a single-antenna GPS receiver and pitot tube. IEEE Trans. Aeros. Electron. Syst. 109–117 (2011)
7. Fu, Q., Liu, Y., Liu, Z., Li, S., Guan, B.: Autonomous in-motion alignment for land vehicle strapdown inertial navigation system without the aid of external sensors. J. Navigate. 1312–1328 (2018)

8. Li, J., Wang, Y., Li, Y., Fang, J.: Anti-disturbance initial alignment method based on quadratic integral for airborne distributed POS. IEEE Sens. J. 4536–4543 (2018)
9. Scherzinger, B.M. (1996) Inertial navigator error models for large heading uncertainty. In: Proceeding IEEE PLANS, pp. 477–484 (1996)

MBSE Adoption for Cabin Temperature Control System Design of Civil Aircraft

Junjie Ye, Wei Guo, Junjie Xue, Qi He, Yong Zhao, and Liangyu Zhao

1 Introduction

Aircraft environment control system is one of the support systems for normal aircraft operation, which is of great significance to comfort and safety of aircraft. As an important part of aircraft environment control system, cabin temperature control system (CTCS) is becoming more complicated with the increase of comfort and safety requirements. Therefore, the challenge of the design of CTCS is increasing. The traditional system design method, Document-Based System Engineering (DBSE), has been unable to meet the challenges fundamentally, which has led to a series of problems: low understanding consistency, poor communication, high cost of design and so on. It is necessary that Model-Based Systems Engineering (MBSE) method

J. Ye · L. Zhao (✉)
School of Aerospace Engineering, Beijing Institute of Technology, No. 5, Zhongguancun South Street, Haidian District, Beijing 100081, China
e-mail: zhaoly@bit.edu.cn

J. Ye
e-mail: 3220180063@bit.edu.cn

W. Guo
Beijing Aeronautical Science and Technology Research Institute of COMAC, Beijing 102211, China
e-mail: guowei2@comac.cc

J. Xue
Beijing Institute of Electronic System Engineering, Beijing 100854, China
e-mail: bitxue@foxmail.com

Q. He
Shaanxi Aero Electric Co., Ltd., Xi'an 710077, China

Y. Zhao
AVIC Aerospace System Co., Ltd, Beijing 100028, China
e-mail: zhaoyong4429@sina.com

D. Krob et al. (eds.), *Complex Systems Design & Management*,
https://doi.org/10.1007/978-3-030-73539-5_21

265

is applied to solve these problems in the design of CTCS. The application of MBSE makes the design of CTCS more agile and the requirements is satisfied wholly. The research subject in this article, CTCS, belongs to the air conditioning elements under the aircraft environmental control system. Figure 1 shows a typical aircraft system architecture hierarchy [1].

The essence of MBSE is systems engineering. MBSE is a methodology of systems engineering coming into being, which is developing rapidly and recognized as an effective solution of designing more and more complex systems [2]. For this reason, MBSE promotes the communication between different domains and understanding consistency of system in design. Seen from Fig. 2, compared with the traditional DBSE method, MBSE transforms design information into models and transfers between various fields and it follows a so-called V cycle [3, 4].

A move away from traditional Document-Based Systems Engineering (DBSE) to Model-Based Systems Engineering is a growing trend around the world. Kaslow researches on CubeSat Model-Based System Engineering (MBSE) reference model, it's claimed that MBSE holds the promise of reducing the burden of team [5]. Hai

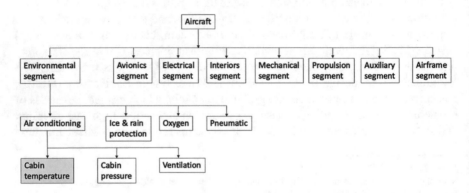

Fig. 1 Typical aircraft system architecture

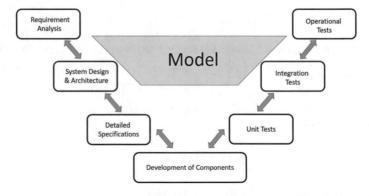

Fig. 2 V cycle for MBSE

applies MBSE method in the design and simulation of civil aircraft landing gear brake system [6]. Motamedian concludes that MBSE helps engineers and managers to complete the design of system engineering project much easier and at lower cost [7]. MBSE consists of three important aspects: modeling language, modeling tool and modeling method. In this paper, a widely used modeling language, SysML, is employed, which is defined by Object Management Group (OMG) [8, 9].

MBSE method is applied to the design of civil aircraft CTCS in this paper, and the logic of CTCS is analyzed and translated into SysML model. Based on the logic analyzed by SysML model, a Simulink model is created. At the same time, the integration of SysML model and Simulink model is completed to implement co-simulation of the overall design. The verification of the model is achieved through co-simulation. The modeler who is about to apply MBSE also starts from needs and then completes requirements analysis, function analysis and system architecture analysis. The difference is that MBSE implements these processes through a model to maintain understanding consistency and promote ease of reuse [9].

2 Models of CTCS

The role of CTCS is to allow the crew to control the temperature of control cabin (CONT CAB), forward cabin (FWD CAB) and afterward cabin (AFT CAB) through the temperature control panel to maintain the desired temperature.

There are many limitations to create a model of the entire system of interest in one modeling language: validated libraries, a language fitting the problem domain perfectly, and so on [10]. That's in the case, the SysML language and Simulink is used in this paper to build discrete model and continuum model of CTCS respectively.

2.1 Discrete Model of CTCS

The establishment of CTCS discrete model is divided into problem domain and solution domain. The purpose of the problem domain is to analyze the stakeholder needs and further refine it with the SysML model elements, which is able to implement the operational analysis and functional analysis [11]. Black box and white box are two phases in the problem domain. The former focuses on how system of interest, CTCS, interacts with the environment, and the latter clarifies the problem of how CTCS operates. The analysis process from black box to white box is a refinement process, step by step.

In common with DBSE, stakeholder requirement is also the initial input for MBSE. The relevant regulatory requirements and internal requirements of CTCS are not focused on in this paper, but only takes user needs as stakeholder requirements and makes them as inputs for demand analysis. According to investigation and analysis, the user needs of CTCS are as follows:

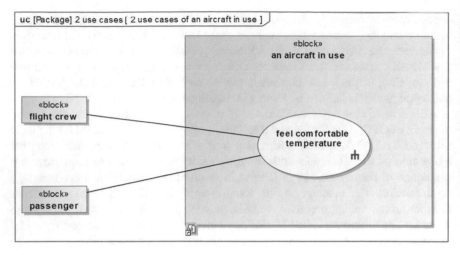

Fig. 3 Use case diagram of the black box phase

- Set temperature: it must be possible to set and maintain desired temperature in the CONT CAB, FWD CAB and AFT CAB.
- Display temperature: the CTCS shall be display the temperature of cabin.
- Heat and cool air: the CTCS shall be able to heat and cool air of cabin appropriately.

During the black box phase of problem domain definition, the collected user needs of CTCS are translated into SysML model elements to prepare for subsequent demand allocation and creation of traceability relationships between stakeholder needs and other SysML model elements. Use case is a SysML model element which refines functional stakeholder needs [11]. Compared with stakeholder needs of CTCS described by natural language, use case diagram can expose what system of interest, CTCS, wants to achieve and the external environment more accurately. The use case diagram of the black box phase for CTCS is shown in Fig. 3.

Based on the identified use case, the primary scenario of the use case is captured. As seen from Fig. 4, an activity diagram is created, in which CTCS and supposed user of the system (flight crew and passengers) are captured in a form of swim lane partitions. Obviously, the activity diagram is created with the premise that modeler observe CTCS from a black box perspective.

The black box analysis is completed, an activity diagram from white box perspective is established in order to further analyze the system architecture. It provides a basis for analyzing the system architecture of CTCS. Block definition diagram and internal block diagram are created to define the external interfaces and internal logic architecture which is shown in Fig. 5. There are three types of interfaces in CTCS, which are the interface for transmitting energy (mainly electric power), the interface for transmitting information, and the interface for transmitting materials (mainly heat air entraining). Based on the definition of the interfaces, the internal logic architecture

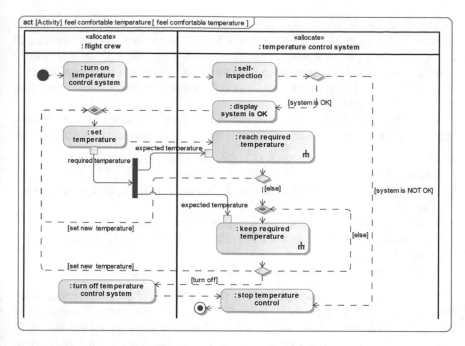

Fig. 4 Activity diagram of CTCS

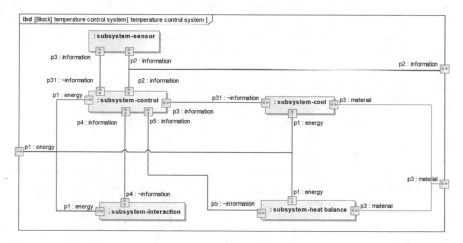

Fig. 5 Internal block diagram of internal logic architecture

of CTCS is captured: control subsystem, sensor subsystem, UI subsystem, cooling subsystem and heating subsystem.

The final task of the problem domain is to review the process of the problem domain definition to create a requirements traceability matrix. This process associates previously activities with stakeholder needs that were originally translated into SysML model elements, ensuring that all stakeholder needs have been assigned and that all activities can be traced back to at least one stakeholder needs rather than being out of thin air [11].

The stakeholder needs in the problem domain are converted into the system requirement during solution domain, where the system requirements are as follows:

- Display set of temperature: the CTCS shall be able to display the set of expected temperature.
- Display temperature of cabin: the CTCS shall be able to display the temperature of cabin.
- Power requirement: the CTCS shall be supplied with electric.
- Air entraining requirement: the CTCS shall be supplied with air entraining for heating and cooling.
- Manual control: the CTCS can adjust the temperature of cabin manually.

Solution domain implies a change of perspective: it changes from system of interest to system under design. It defines a more precise model of the design for the system [11]. During this phase, we can refine to any expected hierarchy, or in other words can iterate to any expected hierarchy. Solution domain can also implement trade-off analysis while providing different design solutions by multiple teams, this paper only analyzes to the subsystem level. The corresponding system logical architecture (see in Fig. 6) and state machine diagram (see in Fig. 7) are created.

Lastly, the key point must be pointed out is a traceability relationship is established to confirm that all requirements are assigned clearly. Figure 8 is a requirements

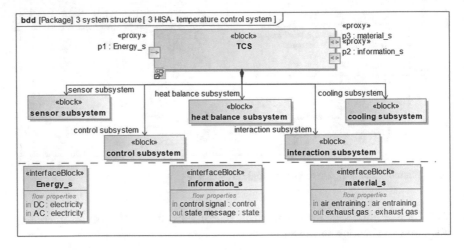

Fig. 6 Block definition diagram of system architecture

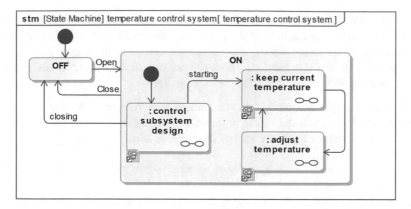

Fig. 7 State machine diagram of system

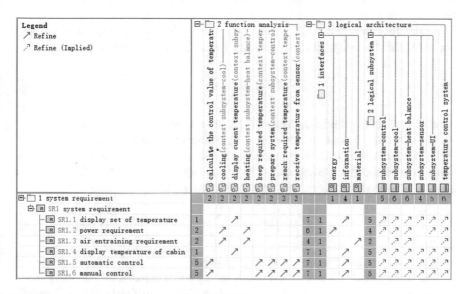

Fig. 8 Requirements traceability matrix of CTCS

traceability matrix, which shows the traceability relationship between requirements and functions or physical architecture.

2.2 Continuum Model of CTCS

In this section, a continuum model of CTCS will be created with Simulink. The theory and design of current CTCS of Simulink have been quite mature. This article is based on the methods and data provided in the literature to complete the creation of related Simulink models [12]. Meanwhile, the logic architecture of Simulink is based on the IBD diagram in SysML. The continuum model of CTCS can be divided into three parts, the cabin model, the controller model, and the sensor model.

Firstly, cabin model is created. Under the premise of the following assumptions, the thermal equilibrium equation of the cabin is derived.

- The cabin and the ambient air do not conduct heat transfer because of the thermal insulation layer enclosed by the outer wall of the cabin;
- The radiant heat of the outer wall of the cabin is zero.

$$T_c(t_c s + 1) = X_c g_h + Y_c \theta_{cg} \tag{1}$$

where t_c is equivalent time constant; T_c is the temperature of cabin; X_c is amplification of the effect of air supply on cabin temperature; g_h is relative change of air supply of mixed air; Y_c is amplification factor of effect of temperature of air supply on cabin temperature; θ_{cg} is the temperature of air supplied to cabin.

Next comes the sensor model. Pt film resistor is widely used in aircraft temperature control system as temperature sensors, the following is transfer function,

$$G(s) = \frac{K_{Pt}}{T_{Pt}s + 1} \tag{2}$$

where K_{se} is amplification factor; T_{se} is time constant of Pt film resistor.

The controller model includes common temperature electricity bridge, amplifier and related actuators including valves and electric motors. The former three can be simply expressed as a scaling factor and the transfer function of the latter (electric motors) is following as typical control logic is adopted.

$$G(s) = \frac{K_{em}}{s} \tag{3}$$

where K_{em} is amplification of electric motor.

The factors of CTCS mentioned above refer to related literature, which is shown as follow in Table 1 [12]. It should be noted that the factors of temperature electricity bridge, amplifier and valves are expressed as, K_{eb}, K_a and α respectively. And K_c satisfies $K_c = X_c g_h + Y_c \theta_{cg}$.

The Simulink model of CTCS is shown in Fig. 9.

This continuous model also needs to be converted to an FMU using Function Mock-up Interface (FMI) plug-in component to implement Co-simulation.

Table 1 The factors of CTCS

K_c	T_c	K_{Pt}	T_{Pt}	K_{em}	K_{eb}	K_a	α
0.3827	30	0.4994	0.5	7.9449e−05	1	9.2350e+05	0.0084

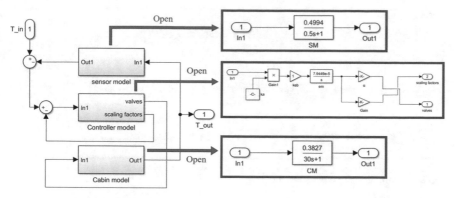

Fig. 9 Simulink model of CTCS

3 Integration of Discrete Model and Continuum Model

Due to the complexity of system under design and the limitations of the existing digital technology, the implement of the overall simulation of CTCS needs to be built into a discrete model (SysML model) and a continuum model (Simulink model). Because of that, integration of discrete model and continuum model as a problem arises. Functional Mock-up Interface (FMI) is an interface plug-in component that is independent of modeling software which is popular in various domain [13]. It can translate kinds of models into FMU for Co-simulation. The implementation process is shown in Fig. 10.

First of all, the Simulink model of the CTCS shown in Fig. 9 is converted into an FMU through the FMI Mock-up Interface plug-in component. It should be noted that the input of expected temperature and the output of cabin temperature must be expressed by the input port and output port in Simulink, respectively, so that the input ports or output ports in the FMU will form. Otherwise, no ports will appear when

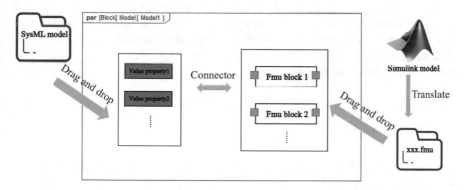

Fig. 10 The process of co-simulation

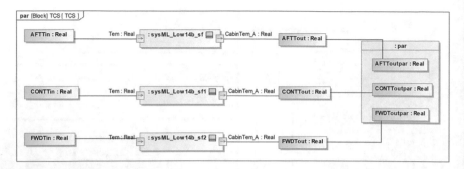

Fig. 11 Parametric diagram of system

the FMU is dragged and dropped into the parameter diagram. The correctness of the FMU of CTCS is validated through FMUChecker.

The FMU is then dragged and dropped into a parametric diagram built in the SysML model and associated with the value properties of the corresponding block (see in Fig. 11). Finally, the discrete model and continuum model of CTCS are fused together through the link of the FMU to form one model.

4 Simulation and Verification

This section simulates the complete CTCS model created above to verify the rationality of the architecture logic of system under designed and the interfaces between the subsystems. In order to facilitate the simulation and observation of the output results a corresponding GUI is established in the SysML model. By observing the simulation results, as shown in Fig. 12, it can be verified whether the system achieves the expected effect or not. And the result of Simulink model is shown in Fig. 13 as reference.

The simulation result implies that the system can adjust the temperature of each cabin to the corresponding expected temperature. The alarm light turns red, as shown in Fig. 12, when the afterward cabin reaches limited temperature. It can be inferred that the logic of system for giving an alarm is correct. Obviously, the logical of the designed system is reasonable.

5 Conclusion

A SysML model that describes the system architecture and dynamic behavior of the CTCS is created in this paper. Then, a Simulink model describing the heating balance subsystem and the cooling subsystem is established and transformed into an

Fig. 12 Result of co-simulation

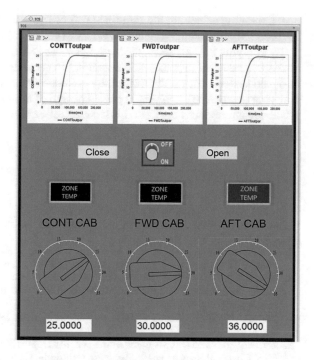

Fig. 13 Result of Simulink model

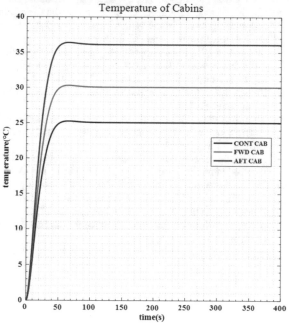

FMU, thereby, achieving the establishment of the overall model of CTCS. Finally, the model is simulated to obtain the effect to verify the correctness of the model.

By applying the MBSE method, design reuse is improved. Secondly, the transfer of models between disciplines is beneficial to an understanding consistency of the designed system. The correctness of the previous design has also been improved (via verifying in advance), avoiding repeated design of the system. All in all, it improves efficiency and reduces costs, and it is an effective way to help implement complex systems.

References

1. Jackson, S.: Systems Engineering for Commercial Aircraft. Taylor Francis Ltd., United Kingdom (2015)
2. Gregory, J., Berthoud, L., et al.: The long and winding road: MBSE adoption for functional avionics of spacecraft. J. Syst. Softw. **160**, 1–11 (2019)
3. Hutchinson, J., Whittle, J., Rouncefield, M.: Model-driven engineering practices in industry: social, organizational and managerial factors that lead to success or failure. Sci. Comput. Program. **89**, 144–161 (2014)
4. Casse, O.: SysML in Action with Cameo Systems Modeler. ISTE Press Ltd. and Elsevier Ltd., Great Britain and the United States (2017)
5. Kaslow, D.: CubeSat Model-Based System Engineering (MBSE) Reference Model-Application in the Concept Lifecycle Phase. Pasadena, California, USA, 31 August-2 September, 2015
6. Hai, X., Zhang, S., Xu, X.: Civil aircraft landing gear brake system development and evaluation using model based system engineering. In: Proceedings of the 36th Chinese Control Conference. Dalian, China, 26–28 July, 2017
7. Motamedian, B.: MBSE applicability analysis. Int.-J. Sci. Eng. Res. **4**, 1–7 (2013)
8. Colombo, P., Khendek, F., Lavazza, L.: Bridging the gap between requirements and design: an approach based on problem frames and SysML. J. Syst. Softw. **85**, 717–745 (2012)
9. Delligatti, L.: SysML Distilled: A Brief Guide to the Systems Modeling Language. Addison Wesle, USA (2013)
10. Widl, E., Judex, F., Eder, K., et al.: FMI-based co-simulation of hybrid closed-loop control system models. In: International Conference on Complex Systems Engineering. Storrs, CT, USA, 9–11 September, 2015.
11. Aleksandraviciene, A., Morkevicius, A.: MagicGrid Book of Knowledge: A practical guide to Systems Modeling using MagicGrid. No Magic, Inc, Allen Texas, USA.
12. He, C.: Research on Integrated Control Technology for Civil Aircraft Electromechanical Systems. Nanjing University of Aeronautics and Astronautics (2010)
13. Blochwitz, T, Otter, M., et al.: The functional mock-up interface for tool independent exchange of simulation models. In: Proceedings of the 8th International Modelica Conference. Dresden, Germany. 20–22 March, 2011

Model-Based Airplane Energy Management

Chao Zhan, Yanxia Mao, Liang Yan, and Thomas J. Benzie

1 Introduction

With the increasing electrification of civil aircraft system, the trend that traditional aircraft electrical power, pneumatic and hydraulic sources are independent of each other has changed [1]. In the latest generation of civil aircraft, the Boeing B787 adopts a multi electric system architecture. The traditional engine bleed air is replaced by compressed air from an electrically driven system. And the hydraulic source is also driven by Engine Driven Pumps (EDP's) and Electro Mechanical Pumps (EMP's), so the requirements for power load are greatly increased. The Airbus A350 series still adopts the system architecture of traditional engine bleed and dual EDPs. Two different architectures show that there is no optimal architecture, but the most balanced architecture. Then, there are many aircraft load systems, especially tens of thousands of electrical equipment. Therefore, how to ensure the energy matching, especially in the preliminary design stage, when it is necessary to determine the structure of the three major energy sources? Some research show it shall be consideredin the aircraft level [2]. The traditional way is either based on human experience or based on Excel for statistical analysis, which is inefficient and error prone. There are some papers was released to show the system level modeling and simulation based on some modelling tools [3–6].

C. Zhan (✉) · T. J. Benzie
Shanghai Aircraft Design and Research Institute, COMAC, Shanghai, China
e-mail: zhanchao@comac.cc

T. J. Benzie
e-mail: tombenzie@comac.cc

Y. Mao · L. Yan
Beijing Runke Tongyong Technologies Co,. Ltd, Beijing, China
e-mail: myxfighting@126.com

L. Yan
e-mail: iowins@sina.com

Based on the function model of MBSE (Model-Based Systems Engineering), the architecture simulation is carried out by building the system model, which can effectively analyze and confirm the architecture in the early stage and reduce the risks for the later design.

This paper mainly introduces a method of building physical architecture based on Modelica to simulate and analyze the system architecture, so as to realize the refined energy management of the whole aircraft, ensure the reasonable decomposition of model library, and confirm the purpose of the system architecture scheme at the early stage of design. Through this method, the accuracy of the whole aircraft energy management index is improved, the redundancy of energy demand is reduced, and the design efficiency is improved.

2 Airplane Energy Management in COMAC

In the previous models, the three major energy sources are relatively independent, belonging to three ATA chapters. The structure and load analysis of the three major energy sources (pneumatic, hydraulic and electrical) were often carried out independently. However, with the deepening of system electrification, especially the trend of hydraulic system electrification, it is impossible to reflect the overall state of the whole aircraft simply by the trade-off of a single system architecture load. Therefore, it is necessary to make overall coordination at the aircraft level to ensure the balance between various energy sources.

At present, COMAC's energy management is still in the exploratory stage in the overall aspect. In the previous models, the load of each energy system is still based on experience, and statistical analysis is carried out through Excel. With the introduction of MBSE concept into COMAC, model-based simulation analysis is also used for reference by various disciplines. For example, based on thermal field analysis, environmental control system improves the accuracy of air source load demand; landing gear brake system also carries out system simulation analysis of hydraulic load demand based on Simulink. This paper introduces the modeling and simulation of the whole aircraft energy system based on Modelica, and makes a comprehensive analysis and trade-off of the various working conditions that may exist in the system, which is also a step forward for COMAC in energy management.

3 Model-Based Airplane Energy Management

3.1 Overview

Modelica Language is an object-oriented, non-causal multi-domain language. This object-oriented modeling approach allows users to build specialistic models in a familiar and intuitive way, with a feature of strong flexibility and compatibility. Furthermore, almost all Modelica tools fully support Functional Mock-up Interface (FMI) standard, which allows model integration from specialistic tools through Functional Mock-up Units (FMUs) and coupling of multi-physics.

In the process of MBSE, Modelica is good at solving logical aspects of. To make up for the shortages and take advantages of both languages, we adopt the SysML-Modelica transformation method of OMG SysPhs and realize the mapping between SysML architecture and Modelica architecture.

Considering the flexibility and multi-physics of an airplane, building the system architecture of an airplane based on Modelica can meet the overall design needs.

3.2 Model-Based Multi-physics Architecture Design

According to the compositions of an airplane, we build hardware architectures such as Electrical Power subsystem, Environmental Control subsystem, Hydraulic subsystem and Engine subsystem. Also including avionics subsystem and cockpit subsystem that related to software and HMI for airplane system integration.

3.2.1 Hierarchical Architecture Design

Just like any general object-oriented languages, Modelica has the same three basic features that is inheritance, polymorphism and encapsulation. Based on these features, we can define hierarchical layers of airplane architectures. Higher level architecture models define the connectors, relations and compositions of lower level architecture models.

For example, in the top level layer of an airplane, composition relations and connectors of all subsystems (i.e., Electrical Power subsystem, Environmental Control subsystem, Hydraulic subsystem and Engine subsystem) are defined. Meanwhile, those models that describes the external interfaces are defined as parent class. In the subsystems level layer, taking Electrical Power subsystem as example which usually consists of main power supply, emergency power supply, secondary power supply, auxiliary power supply and other electrical equipment, composition relations and connectors are defined and different Electrical Power architectures are formed with different relations and connections. All these architectures are inherited from the same parent class and constitute different blue prints of Electrical Power subsystem. Benefiting from the same parent class, all these subsystem architectures can be replaced quickly without extra effort in the top level and we can get different architectures of an airplane. Similarly, we can decompose the architecture continually until satisfaction. The advantages is apparent, it can bring more reusability and simplicity for system design, and improve design efficiency. The hierarchical architectures of an airplane are shown in Fig. 1.

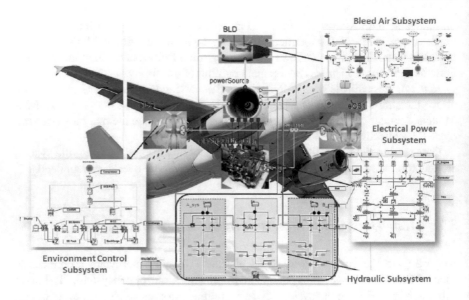

Fig. 1 Hierarchical architecture model of airplane

With the architecture definition, the model library hierarchical structure (see Fig. 2) has been built, which will be beneficial to model extension and management.

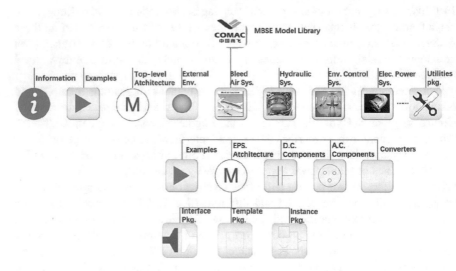

Fig. 2 Model library hierarchical architecture

3.2.2 Domain-Specific Models Definition

In this section, it will show that how we built domain-specific model, taking electrical power subsystem as example. Aircraft electrical power system is the general name of aircraft power supply system and various electrical equipment. The power supply system includes the aircraft power supply system and the aircraft power transmission and distribution system. The latter is used to distribute and manage energy. Electrical equipment includes aircraft flight control system, powerplant, avionics, environment control system (ECS), lighting, anti-icing and cabin services systems etc. The function of aircraft power supply system is to ensure reliable supply of electric power to electrical equipment, especially important electrical equipment directly related to safety.

In the aircraft level, the purpose of analyzing of electrical power supply system is to collect the energy consumption, put forward power requirement for engines and reasonable index for bus-bar and other key components, considering load demands under different flight conditions. For this purpose, the real physical properties of the components are simplified. For example, we use combinations of resistance, capacitance and inductance to replace various types of electrical loads. The typical models and model library are shown as Fig. 3.

The model library can be used to build various components of electrical power system, such as three-phase generator, ac bus, dc bus, transformer rectifier, dc inverter, battery, switch and so on. The architecture is shown as Fig. 4, whose branches can automatically calculate the total voltage and current according to the load connected downstream, in order to calculate the total power.

Other subsystems models are built in the same way. Shown as Fig. 5.

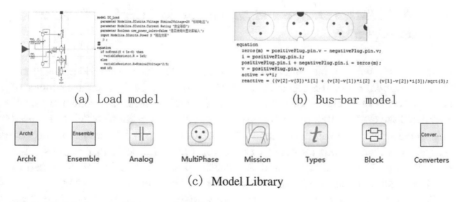

(a) Load model   (b) Bus-bar model

(c) Model Library

Fig. 3 Electrical power subsystems models

Fig. 4 Electrical power
subsystem architecture
model

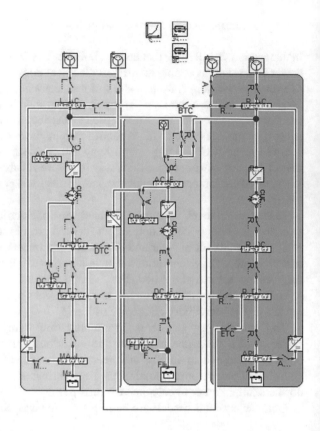

3.3   Model Based Design and Simulation for Airplane Energy Management

In this section, we simulate and evaluate two cases. The first example demonstrates the real application of the electrical power system models. Performance index of key equipment in electrical power system are defined through statistical analysis of electrical loads with dynamics features in different flight conditions. Therefore, based on the electrical system architecture model, the electrical power system performance analysis was carried out by calling Excel and finish the design of various flight sequence, flight load management, power distribution logic and 975 loads properties for various situations (see Fig. 6). The results are shown in Fig. 7 and Fig. 8.

Compared with the traditional method for statistical calculations based on Excel spreadsheets, based on Modelica model, it has apparent advantages. On the one hand, one is able to get electrical power system intuitive and dynamic change of state and is beneficial for the designer to determine power distribution error, on the other hand, one can get more accuracy envelope of system index, can effectively reduce the redundancy design and improve design accuracy and efficiency.

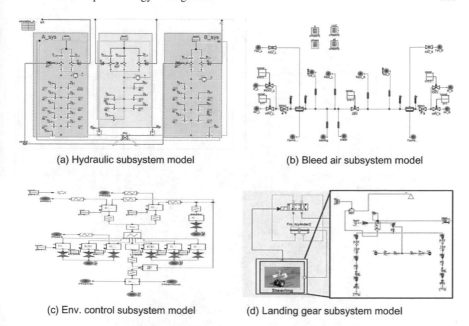

(a) Hydraulic subsystem model

(b) Bleed air subsystem model

(c) Env. control subsystem model

(d) Landing gear subsystem model

Fig. 5 Some other subsystem architectures

Fig. 6 A typical load configuration spreadsheet

The second example is the system simulation with the whole aircraft mission phase. In this example, the whole flight mission is defined as pre-takeoff phase, climbing phase, cruising phase, landing phase and taxiing phase. Results are shown in Fig. 9.

During the taxiing phase, the aircraft starts to power up with the approximate power of 58 KW. At the same time, the APU is turned on, and the aircraft begins to bleed, with the approximate bleed demand of 12171LBS/h. All this demands a certain power requirement for the engine, so there will be fuel consumption and the fuel compensation of 1600LBS. In the results, the variation trend of fuel compensation at different phases is consistent with the total demand of the power used engine bleed, the power used in the electrical system and the shaft power of the engine.

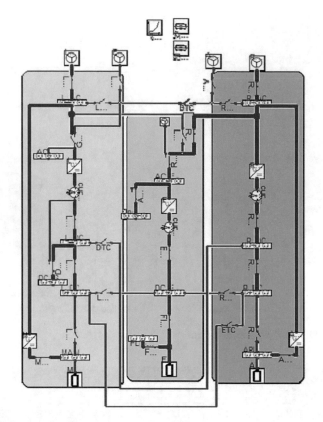

Fig. 7 Simulation result under a certain work condition

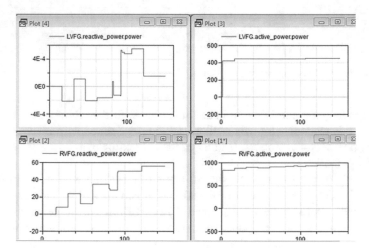

Fig. 8 Reactive power results of typical components during a certain flight sequence

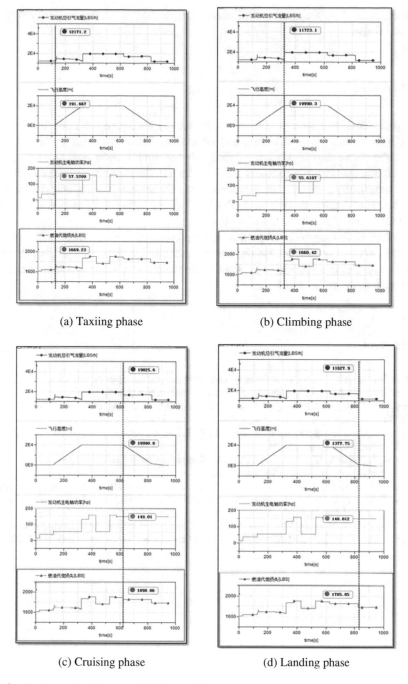

(a) Taxiing phase   (b) Climbing phase

(c) Cruising phase   (d) Landing phase

Fig. 9 Airplane system energy simulation results

The results show that the system energy architecture is reasonable and ready for additional analysis.

4 Conclusion

This paper discuss a new way to carry out energy load analysis and management through Modelica-based modeling and simulation. Secondly, through the standard model library, we can quickly build different system architecture and carry out energy trade-off analysis of different schemes. And through the standardized excel table, we can automatically convert it into load model to help improve the statistical efficiency and accuracy. Finally, the whole model is driven by state machine to carry out the simulation analysis of the whole flight phase.

At present, the method and path have been confirmed. Later work will focus on the scenario model based on SysML driven the Modelica system model, break down the barriers of two modeling languages, and open the way from problem domain to solution domain model in MBSE.

References

1. AbdElhafez, A.A., Forsyth, A.J.: A review of more-electric aircraft. In: 13th International Conference on Aerospace Sciences and Aviation Technology (2009)
2. Wang, D., Zheng, Y., Zhang, Q.: Energy management in aircraft design. Aeronaut. Sci. Technol. (2013)
3. Schallert, C.: A novel tool for the conceptual design of aircraft on-board power systems, SAE 2007 Aerotech Congress (2007)
4. Martin, K., Martin, O., Loic, R.: A multi level approach for aircraft electrical systems design 6th international modelica conference (2008)
5. Martin, K., Antonio, G., Jiabin, W., Johann, B.: A components library for simulation and analysis of aircraft electrical power systems using Modelica. In: 13th European Conference on Power Electronics and Applications (2009)
6. O'Connell, T., Russell, G., McCarthy, K., Lucus, E.: Energy management of an aircraft electrical system. In: 46th AIAA/ASME/SAE/ASEE Joint Propulsion Conference & Exhibit (2010)

Model-Based Product Line Engineering with Genetic Algorithms for Automated Component Selection

Habibi Husain Arifin, Ho Kit Robert Ong, Jie Dai, Wu Daphne, and Nasis Chimplee

1 Introduction

In today's systems engineering, Product Line Engineering (PLE) is necessary to develop a complex system at a lower cost [1–4] and shorter time-to-market [1–3]. The need for improving business profitability by maximizing the benefits of reuse has kept growing constantly [1]. In addition, PLE is not only useful for optimizing the cost and time of acquisition and ownership, but also helpful for system engineers to optimize the product or service quality [5].

H. H. Arifin (✉) · H. K. Robert Ong · W. Daphne · N. Chimplee
Dassault Systèmes, Vélizy-Villacoublay, France
e-mail: habibihusain.arifin@3ds.com

H. K. Robert Ong
e-mail: hokitrobert.ong@3ds.com

W. Daphne
e-mail: daphne.wu@3ds.com

N. Chimplee
e-mail: nasis.chimplee@3ds.com

J. Dai
Vision MC, Shanghai, China
e-mail: jie.dai@visionmc.com

© The Author(s), under exclusive license to Springer Nature Switzerland AG 2021
D. Krob et al. (eds.), *Complex Systems Design & Management*,
https://doi.org/10.1007/978-3-030-73539-5_23

287

One of the frequently asked questions in applying Model-Based Systems Engineering (MBSE) methodologies after its fundamental has been covered is how to adopt MBSE into PLE [6] (also known as Model-Based Product Line Engineering (MBPLE) [4]). The question arises because there are high and increasing needs of the variety of product lines, customized products, or different designs for trade study analysis [6]. The benefits of using models to support the PLE activities are as follows.

- Bringing MBPLE (MBSE and PLE) together to allow users to model product lines in industry standard or de facto standard formats [4].
- Optimizing system architectures, performing trade-off studies, performing verification and validation, and promoting the development of cohesive operational, functional and physical architectures of the system [1].
- Helping establish meta-model consistency and traceability of the different engineering artefacts [1]. For example, by linking *150% model* to requirements, the traceability and compliance to the requirements can be maintained. Product line feature selection will not only result in a *100%-product model* but also the *100% model* of the product line's *150%* of requirements [7]. (*The terms 150% model and 100% model will be discussed in the next section*).

MBPLE alone may be insufficient to overcome the problems of a *Rolling Stock* product line (*the term Rolling Stock will be discussed in the next section*). A large automotive company can produce tens of thousands of variants in a product line [8]. Complete benefits of PLE in a "Rolling Stock" full product line can be achieved with a deeper analysis at the business level [1]. For example, *"How much assembling cost for the selected components for a particular product variant?"* To answer this question, the synthetization of candidate physical architectures is required. It allows system engineers to address the concerns about performance, cost, reliability, availability, and security [5] in an earlier phase because a physical architecture defines relationship among physical system elements allocated from a *100%-logical-architecture model* [5]. However, when the component selection is conducted, the dimension of architecture models (i.e., feature model, *150% model* with its variation points, and *100% model*) and the number of their possible permutations/combinations [6] are growing exponentially. To address this issue, automation is needed because it can provide greater efficiencies in searching within a large solution space [9].

Heuristic algorithms, e.g., Genetic Algorithms (GA) [10] have been applied successfully in optimization and evaluation of problems in many engineering domains [11, 12]. Some related studies have proven that algorithms are suitable to fulfill the need for automation in the component selection of design synthesis in MBSE [13, 14].

The objective of this paper is to extend MBPLE with genetic algorithms to automate component selection of design synthesis for *100%- physical-architecture model*. The paper shows how to leverage genetic algorithms in an existing MBPLE to perform a deeper trade study analysis at the component level to address the concerns about performance (*i.e., output power and speed*) and physical (*i.e., mass*) requirements. The contribution is expected to help systems engineers reduce the human efforts to perform traditional (brute-force) trade studies which are time-consuming [15] and error-prone [16].

Figure 1 shows the *150%-logical architecture model* used in this paper, which is illustrated in a block definition diagram (BDD). The "Car" block is the target system (also known as system of interest (SOI)). The block owns 7 part properties (*i.e., engine, spoiler, sunRoof, door, wheel, chassis, and lamp*) and 2 sub-part properties (*i.e., filamentBurnoutDetector and autolevelingMotor*) under lamp's part property. (*For more details, see* Fig. 11).

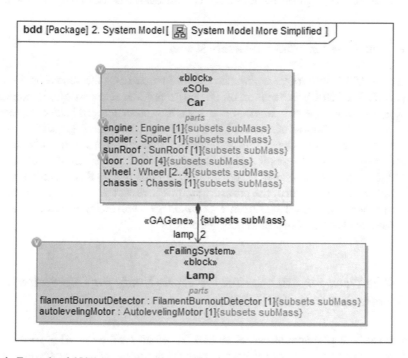

Fig. 1 Example of *150%-Logical-Architecture Model*

2 Literature Reviews

2.1 MBSE + PLE = MBPLE

Product-Line Engineering (PLE) is a system engineering process that helps managing the underlying architectures of the product platforms or portfolio of a company to maximize the benefits of reuse [1]. The reuse concept is usually achieved by classifying the families of similar products [17], which can be adapted and customized to a specific target market or customer's needs [5]. The entire family of similar products is also known as a product line [17]. A family of similar products can share a set of assets and common production facilities [1] [8]. A product line consists of multiple different concrete products called product variants [17]. Bringing MBSE and PLE together can maximize the benefits of using modeling techniques to support the system engineering process activities [1] as discussed in the introduction section.

2.1.1 150%- and 100%-Architecture Model

The terms *150%* and *100%-architecture model* come from the second generation PLE (2GPLE). 2GPLE is centered on the entire product line portfolio to engineer, manage, and deliver it as a single, feature-based, automated production system. It starts with the definition of stakeholder's visible product characteristics (external variants) and internal product characteristics (internal variants) before going into the definition of all possible detailed product performances and different solutions [1]. In 2GPLE, *150% model* (superset model) is a common design that encompasses the entire variability spectrum of the products [17]. Where, *100% model* (subset model) is a particular design produced from *150% model* by narrowing it down according to the feature choices of that particular product [17].

2.2 The Need for Automation for Trade Study for Component Selection from 100% Model

A trade study is an essential part of an MBSE design synthesis [18]. It means "*the decision-making actions that select from various requirements and alternative solutions on the basis of net benefit to the stakeholders*" [18]. Trade study analysis or trade-off study helps select the fittest configuration to the specified criteria among many possible solutions [19]. Component-selection is a process of design synthesis to select a set of feasible components from a component library (also known as Bill of Material), which satisfies the requirements of the *100%-logical-architecture model*. Therefore, the trade study is usually needed to help system engineers select the right design and components.

As discussed in the introduction section, automotive manufacturing represents the most challenging environment for systems PLE today [7, 20]. A large automotive company can produce tens of thousands of variants in a product line [8]. This is why the products of automotive industry are also known as *Rolling Stock* products. A *Rolling Stock* product is a complex system where the complexity is exponentially increasing (more complex functionalities, more integration with other systems) [1]. An example of a product line is a *Car* system with 9 components shown in Figure 1. The engine has 2 variants: gasoline and diesel. For each engine variant, there are 2 sub-variants: atmo type and turbo type. Once the product configurator has decided which specific variant the car will use, the trade study needs to be performed in the physical architecture. According to the *Bill of Materials* in Table 7, each sub-variant of engine has 3 possible components to select. Therefore, the number of configurations for the engine alone will be *12 possibilities*. This example has not been applied to the whole components of the car. Based on the example, it is clear that both variant and component configurations give the decision maker a large solution space to explore and evaluate. Therefore, an automation for the trade study is really needed since a linear-approach is more efficient when only a few variations are involved [14].

2.3   *Genetic Algorithms for Component Selection*

Genetic Algorithm (GA) is an optimization method based on Charles Darwin's evolution theory. The theory states that only the strongest individual will survive [21]. It is considered as one of the best search tools for finding an optimal solution in a large searching space [22]. GA explains the mating selection process in biological terms where the representation is known as deoxyribose nucleic acid (DNA). The process starts when parent chromosomes in a population mate and exchange their genes, which resulted in their offspring inherited the parents' characteristics. A gene mutation that causes the offspring to be different from the parents can occur. During the evaluation process, surviving offsprings are measured by their fitness value [23]. (*The important terms and concepts of genetic algorithms are shown in Tables 3 and* 4). Two advantages of using GA over the other searching and optimizing algorithms:

- GA can be used in a wide range of applications and to perform searching in a complex solutions space [24].
- GA are less likely to be led astray by the local optima because it has an entire set of solutions spread throughout the solution space [25].

2.4   Focus of the Paper

To conduct a trade study, system engineers usually need to design a number of alternatives *100%-logical-architecture model* of a system and manually analyze them to find the best design for component level [23, 26]. This process is often time and cost consuming [15] and error-prone [16]. The complexity of systems created two issues in component selection [18, 27]: (*1*) *The problem in performing an extensive search to discover the component combination(s) through many possibilities; and* (*2*) *The problem that occurs when there is no feasible combination exists to satisfy the requirements.* This paper focuses mainly to solve the first problem by designing a decision support system to assist the systems engineers and stakeholders performing the component-selection process with fewer efforts and resources. GA is used to automate the capability to find the best solution among a huge number of alternatives [28]. Some related studies have been conducted by [23, 26, 27, 29, 30].

3   Methodology

3.1   Building 150% Model with MBPLE and Producing 100% Model

Building Feature Model. A Feature Model is a hierarchical decision tree that identifies and defines possible product variations and distinguishes product characteristics in a product line [2, 31]. Features in the feature model can be placed into Root Feature Group directly or into several subgroups (Feature Groups) with multiple grouping levels.

- Root Feature Group, located only at the tree root, is the root of a feature model. In this paper, it is represented by a class with the stereotype «RootFeatureGroup» applied.
- Feature Group is located only at the tree leaf. Feature Group is the subgroup of features that are connected to the Root Feature Group. In this paper, it is represented by a class with the stereotype «FeatureGroup» applied.

Building *150% Model* with Variation Point and Feature Impact. A Variation Point is a particular point (variable) in the *150% model* that is connected and varies according to the feature choices. In other words, Variation Points are constraints of a systems model. In this paper, when a variation point is applied to a model element, a little (v) icon will appear on the element. This paper used several kinds of variation points as follows:

- Existence Variation Point: If a model element can exist or not exist in some variants. To apply the existence variation point, the constraints of the model element is applied with the stereotype «ExistenceVariationPoint».

- Primitive Property Variation Point: If a property of a model element can be set as a Boolean/String/Numeric, e.g. the number of doors of a car can be set as a numeral (2 or 4 doors). To apply the existence variation point, the constraints of the model element is applied with the stereotype «PrimitivePropertyVariationPoint».
- Element Property Variation Point: If a property of a model element can be set as another type of the model element, e.g. a part/reference property has a type of a block. To apply the existence variation point, the constraints of the model element is applied with the stereotype «ElementPropertyVariationPoint».

Feature Impact is a dependency relationship between a feature and a variation point. It is used to specify which features from the feature model that impact a variation point. The condition of impact is constructed with an expression that can be built with some scripting languages such as JavaScript, Groovy, Python, etc.

Define Variant Configuration. Configuration is an instance model of a Feature Model. In this paper, a configuration is represented by an instance. The value of each instance slot represents the value of a feature instance of the configuration. The values can be modeled as:

- Boolean (true/false): If the feature can be applied or not (Yes-No feature).
- Enumeration: If the feature has multiple alternative choices. The alternatives are modeled as enumeration literals.

Producing *100% Model* with Variant Realization. Variant Realization is a model transformation to produce *100% model* from *150% model* according to the selected variant configuration and its constraints (variation points).

3.2   Selecting Components for 100% Model with Genetic Algorithms

Figure 2 shows the methodology used to illustrate the steps of using GA for component selection for *100% model*. The processes are divided into 3 main sections: the inputs set, the outputs set, and the configuration. The inputs set requires a SysML 1.5 model and an instance catalog (*Bill of Materials*). The *SysML 100%-model* must be exported to a UML XMI 2.5 format. Both XMI format and instance catalog are transformed respectively to a structure model and a well-structured array of instances so that GA machine can recognize them. Finally, they are ready for the evaluation process with GA machine. GA parameters are a set of parameters to define the number of populations, maximum generations/evolutions, and crossover/mutation method and probability factor. A configuration represents a solution of the selected components. The outputs set wraps the best solution(s) that fits the target requirements and weighting factors.

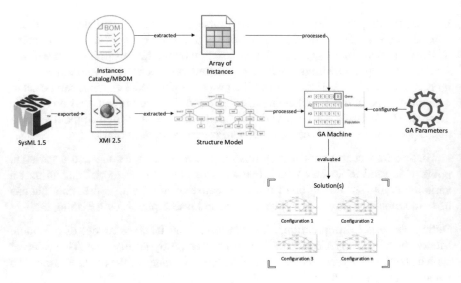

Fig. 2 Component selection with genetic algorithms

Apply GAProfile to *100% Model*. GAProfile describes the custom stereotypes used to represent the GA component selection concepts. Figure 3 shows 3 stereotypes that extend SysML to model the concepts of variants with genetic algorithms. The «FitnessValue» stereotype defines the fitness-value property of a configuration. The «GAGene» marks down the property included in the trade study. The «SOI» stereotype can be used to define the target system of a model. The system of interest (SOI) is the system whose life cycle is under consideration [5]. Any component that will be included in trade study analysis must be applied with the «GAGene» stereotype.

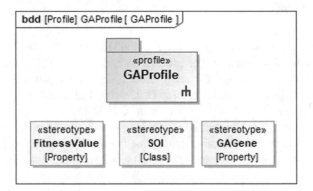

Fig. 3 GAProfile

Derive FitnessValue from Requirements. The trade study is based on a systematic comparison of *100% model* of a car measured with respect to performance requirements (*i.e., output power and speed*) and physical requirements (*i.e., mass*) (See Fig. 4). Based on the given requirements, three MOEs can be derived to satisfy those requirements [5] (See Fig. 5): *maxMass, maxSpeed*, and *maxOutputPower*. The optimization and evaluation of alternatives will select the best selected components with the highest fitness value. To calculate the fitness value, the MOEs measurements and the weighting factors (ratio) [5] are needed.

Use Constraint Block for Fitness Function. The use of fitness function is quite broad in the application of GA. A fitness function can specify a constraint to evaluate an alternative solution. To get the best results, it is necessary to involve domain experts for each measurement to define the best design of fitness constraints. Figure 6 shows the parametric models to evaluate the solutions and all MOEs are bound to a constraint parameter owned by a constraint property that has a type of constraint block. The following equations are used to calculate the fitness value of the solution:

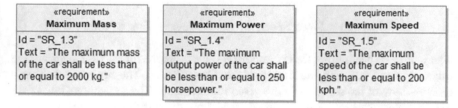

Fig. 4 Example of requirements

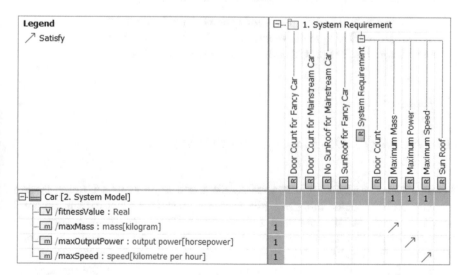

Fig. 5 Satisfy MOEs to requirements

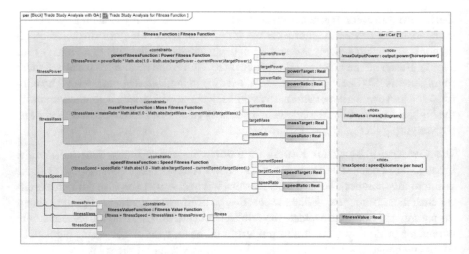

Fig. 6 Parametric diagram of fitness function

$$\text{fitness} = \text{fitnessSpeed} + \text{fitnessMass} + \text{fitnessPower} \tag{1}$$

where, fitnessSpeed, fitnessMass, and fitnessPower are the fitness of *maxSpeed*, the fitness of *maxMass*, and the fitness of *maxOutputPower*. The fitness is measured to know how fit the solution to the given requirements. The parameters below are the examples of target requirements derived from the stakeholder needs. The ratio defines the weighting factor for each target requirement. The higher the ratio is, the higher the target requirement is prioritized in the trade study.

speedTarget = 200 kph, with *speedRatio* = 0.5 (50% of total fitness).
massTarget = 2000 kg, with *massRatio* = 0.2 (20% of total fitness)
powerTarget = 250 hp, with *powerRatio* = 0.3 (30% of total fitness)

4   Results

4.1   Result of 100% Model from 150% Model

Feature Model and Configurations. As discussed in previous section, a feature model shall be modeled independently from any specific product variant. Figure 7 shows the example of a complex feature model. The « RootFeatureGroup » is "ComplexFM". To manage the complexity, the features are grouped into several «FeatureGroup» i.e., "Exterior", "Interior", "Safety", "Lamps", and "SunRoof".

Figure 8 shows the example of configurations, represented by the instances of

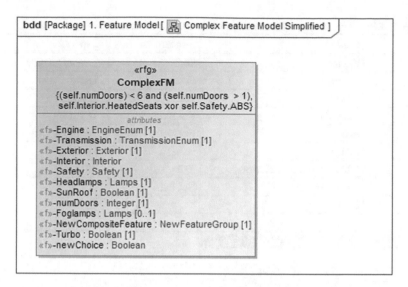

Fig. 7 Example of feature model

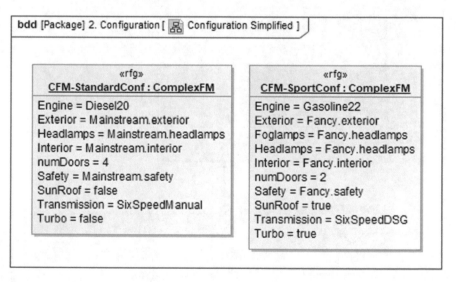

Fig. 8 Example of configurations

the feature model, StandardCar and SportCar variant configurations. The differences between these configurations are shown in Table 1.

100% Models of SportCar Variant. Figure 9 shows a *100% model* example as the result of variant realization for the SportCar variant configuration in Fig. 8. In this paper, the component selection is applied only to the SportCar variant. Another example of the StandardCar variant can be seen in Fig. 12.

Table 1 Differences between StandardCar and SportCar configurations

No.	Feature	StandardCar	SportCar
1	Engine	TurboDiesel	AtmoGasoline
2	Chassis	Exist	Exist
3	Wheel	Exist	Exist
4	SunRoof	Not Exist	Exist
5	Spoiler	Not Exist	Exist
6	Lamp	Exist	Exist
7	FilementBurnoutDetector	Exist	Exist
8	AutolevelingMotor	Not Exist	Exist
9	Door	4 Doors	2 Doors

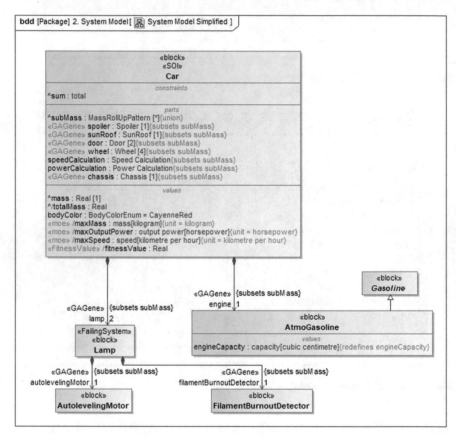

Fig. 9 Example of *100% Model* of SportCar variant

4.2 Results of Component Selection of 100% Model

Structure Model and RNL Tree. The structure model extends *100%-logical –architecture model*. A structure model is not necessary a copycat of its SysML model. It takes only the required elements to perform the genetic algorithms. In this preliminary study, the structure model includes part property, value property, SOI block, constraint block, and constraint parameter from SysML 1.5 model. Table 6 shows the example elements that already extracted from SysML 1.5 model. The *Stereotype(s)* shows the stereotype applied to an element. The *Level* shows the depth of an element in the structure model. The *NodeId* and *ParentId* are auto-generated by the system to define the owner/parent of an element. The *Name* is a given name of an element. The *Value* is used to represent the value of an element (the default value if it has value at the beginning). The Relationship is the relation connector to its owner/parent or binding element. The *Formula* is used to represent the constraint of a constraint block. A way to visualize a structure model is by plotting them into a Root-Node-Leaf (RNL) tree. Figure 10 shows an example of RNL tree based on the elements in Table 6. The elements are extracted from the structure model to illustrate the hierarchical relationship between elements [32]. The root refers to the 0th level's node of a tree or the entry point to a structure tree. A node is an element in the middle level of the structure tree. The root is also a node without a parent. A leaf or also referred to as a terminal node, is simply a node without children.

Instances (Bill of Materials). Table 7 contains the instances set for each component extracted from a *Bill of Materials*. Each component has a number of parameters called *Instance Parameters*, e.g., an "Engine" has *name, mass,* and *engineCapacity*. The *GAGene* value is an additional parameter auto generated sequentially with an encoding/representation algorithm. In this paper, ASCII upper-case characters are used to represent the chromosomes. The details of variables to set up the *GAGene*

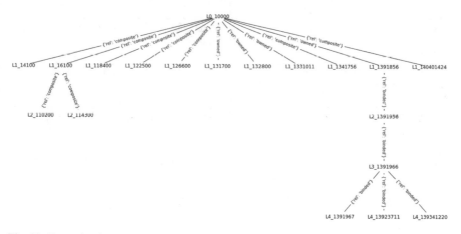

Fig. 10 Example of root-node-leaf tree

values are as follows:

$$\text{maxCandidate} = \text{distance}^{\text{sizeOfGene}} = 26^{\text{sizeOfGene}} \tag{2}$$

where,

- maxCandidate >= maxInstance and sizeOfGene must be the minimum value
- maxInstance = maximum number of instances in all instance types
- ASCII upper case:

  - lowerBound = "A"
  - upperBound = "Z"

- distance (A-Z) = 26 characters

4.2.1 Solution(S)

Solution(s). Table 2 shows the example of top-five best solutions from component selection. Each solution is represented with a chromosome sequence ['Atmo-Gasoline', 'Lamp', 'FilamentBurnoutDetector', 'AutolevelingMotor', 'Spoiler', 'SunRoof', 'Door', 'Wheel', 'Chassis'] and its MOEs. For example, the 1st best solution has chromosomes ['B', 'A', 'B', 'B', 'C', 'B', 'D', 'C', 'C']. According to the chromosome sequence, this solution is assembled with ['AtmoGasoline': atmo-Gasoline3.5, 'Lamp': lampA, 'FilamentBurnoutDetector': filamentB, 'Autoleveling-Motor': autoB, 'Spoiler': spoilerC, 'SunRoof': sunRoofB, 'Door': doorD, 'Wheel':

Table 2 Examples of solution(s)

Chromosome	MOE			Fitness value
	maxMass (kg)	maxOutputPower (hp)	maxSpeed (km/h)	
['B','A', 'B', 'B', 'C', 'B', 'D', 'C', 'C']	1755.62	250	151.0973307695852	0.853305326923963
['B', 'C', 'C', 'B', 'B', 'D', 'A', 'D', 'C']	1749.38	250	151.2767712040366	0.8531299280100915
['B', 'A', 'A', 'A', 'C', 'B', 'A', 'A', 'C']	1724.14	250	152.01138747111003	0.852442468677775
['B', 'A', 'B', 'B', 'A', 'B', 'A', 'A', 'C']	1722.82	250	152.05020054249442	0.852407501356236
['B', 'C', 'A', 'C', 'C', 'C', 'B', 'B', 'A']	1246.9	250	169.3517766866317	0.8480694417165793

wheel19inc, 'Chassis': chassisC]. The fitness value of the solution is approximately 0.85 with the upper bound is 1.0, meaning that the solution is 85% fits the target requirements.

5 Conclusions and Future Works

This paper demonstrates the use of genetic algorithms to perform an automated component selection for *100% model* in the design synthesis of MBSE. A *100% model* is a logical architecture resulted from variant realization of MBPLE. The application presented here is for component selection with trade study analysis for a product variant of a car (SportCar variant). The optimization and evaluation are applied with fitness value that is measured with a set of MOEs and weighting factors derived from the given requirements. For future work, genetic algorithms are not only applicable for *100% model*, but are also feasible for automated configuration selection for *150% model*.

Appendices

See appendix Tables 3, 4, 5, 6, and 7; Figs. 11 and 12.

Table 3 GA terms and concepts [13, 21, 22, 25 ,33–36]

GA terms and concepts	Description
Population	Population is the highest-class hierarchy of the domain. It is a set of individuals which forming the start and the end of an evolution step. A population consists of a list of organisms with different phenotypes
Phenotype	Phenotype is the actual representation of an individual
Genotype	Genotype is a collection of chromosomes. It is the structural and immutable representative of an individual
Chromosome	Chromosome is a collection of genes which is constructed from multiple genes
Gene	The basic building block of DNA. It represents distinct aspects of the solution as a whole, just as human genes represent distinct aspects of individuals, e.g., eye color
Allele	The value of a gene. It contains the actual information of the encoded solution, e.g., blue and brown for the possible colors of eyes
Representations	Representation/encoding/phenotype-genotype mapping can be described as a list of solutions which allows the computers to understand the process. The results of encoding are the chromosomes
Crossover	Crossover/recombination is the same as simulating the mating process to pass the genetic traits of two parents' chromosomes to their successors by exchanging and compounding their genes
Mutation	Mutation is unexpected change(s) on the allele in a chromosome. Mutation is useful to keep the variety of the individuals in the whole population

Table 4 GA component selection terms and concepts

GA component selection terms and concepts	Description and implementation
Structure model and its elements	The **structure model** is implemented in a Root-Node-Leaf (RNL) tree. Each **element** represents an XMI25Element (SysML element). See Table 6 and Fig. 10
GAGene	A **GAGene** can be applied to a property of structure model as a gene on chromosome. The "Chassis" in the "Car" structure model is an example of a **«GAGene»**. See Fig. 1
Instance	An **instance** is a possible element of a component. For example, "chassisA" is an instance of instance type "Chassis". See Table 7
Instances set	The **instances set** consists of all candidate instances for each GAGene. See Table 7

(continued)

Table 4 (continued)

GA component selection terms and concepts	Description and implementation
Instance parameters	The **instance parameters** distinguish each element of instances set. For example, "Chassis" has 5 candidate instances. Each of them has different instance parameters (*name* and *mass*). See Table 7
Configuration	A **configuration** is a valid set of variants and the core. A solution of genetic algorithm is represented in a **configuration**. See Table 2
Fitness Functions	A **constraint** specifies rules for a valid set of configurations. It is implemented with the **fitness functions**, derived from the constraint block. The constraint blocks are derived from the requirements. See Fig. 6

Table 5 Table of components of used example (car)

No.	Component	Is VariationPoint	Is GAGene	Attribute	Unit

1	Engine has subtypes:
	 • Gasoline has subtypes:
	   – AtmoGasoline
	   – TurboGasoline
	 • Diesel has subtypes:
	   – AtmoDiesel
	   – TurboDiesel
		Yes	Yes	Enginecapacity	Cubic centimetre

				Mass	Kilogram
2	Chassis	No	Yes	Mass	Kilogram
3	Wheel	No	Yes	Diameter	Centimeter
				mass	Kilogram
4	SunRoof	Yes	Yes	Mass	Kilogram
5	Spoiler	Yes	Yes	Mass	Kilogram
6	Lamp	Yes	Yes	Mass	Kilogram
7	FilementBurnoutDetector	Yes	Yes	Mass	Kilogram
8	AutolevelingMotor	Yes	Yes	Mass	Kilogram
9	Door	Yes	Yes	Mass	Kilogram

Table 6 Table of structure model from *100% Model*

Stereotype (s)	Level	NodeID	ParentID	Name	Value	Relationship	Formula
['Block', 'SOI']	0	L0_10000	None	Car	None	None	None
['PartProperty', 'GAGene']	1	L1_14100	L0_10000	engine	None	Composite	None
['PartProperty', 'GAGene']	1	L1_16100	L0_10000	lamp	None	Composite	None
['PartProperty', 'GAGene']	2	L2_110200	L1_16100	filamentBurnoutDetector	None	Composite	None
['PartProperty', 'GAGene']	2	L2_114300	L1_16100	autolevelingMotor	None	Composite	None
['PartProperty', 'GAGene']	1	L1_118400	L0_10000	spoiler	None	Composite	None
['PartProperty', 'GAGene']	1	L1_122500	L0_10000	sunRoof	None	Composite	None
['PartProperty', 'GAGene']	1	L1_126600	L0_10000	door	None	Composite	None
['PartProperty', 'GAGene']	1	L1_1341756	L0_10000	wheel	None	Composite	None
['PartProperty', 'GAGene']	1	L1_140401424	L0_10000	chassis	None	Composite	None
['ValueProperty', 'MOE']	1	L1_131700	L0_10000	maxMass	None	Owned	None
['ValueProperty', 'MOE']	1	L1_132800	L0_10000	maxOutputPower	None	Owned	None

(continued)

Table 6 (continued)

Stereotype (s)	Level	NodeID	ParentID	Name	Value	Relationship	Formula
['ValueProperty', 'MOE']	1	L1_1331011	L0_10000	maxSpeed	None	Owned	None
['ValueProperty', 'FitnessValue']	1	L1_1391856	L0_10000	fitnessValue	None	Owned	None
['ConstraintParameter']	2	L2_1391956	L1_1391856	fitness	None	Binded	None
['ConstraintBlock']	3	L3_1391966	L2_1391956	Fitness Value Function	None	Binded	Fitness = fitnessSpeed + fitnessMass + fitnessPower
['ConstraintParameter']	4	L4_1391967	L3_1391966	fitnessMass	None	Binded	None
['ConstraintParameter']	4	L4_13923711	L3_1391966	fitnessSpeed	None	Binded	None
['ConstraintParameter']	4	L4_139341220	L3_1391966	fitnessPower	None	Binded	None

Table 7 Table of instances (bill of materials)

No.	Instance type and subtype	GAGene	Instance parameters		
1	Engine		*name*	*mass*	*engineCapacity*
	• AtmoGasoline	A	atmoGasoline3.2	369	3200
		B	atmoGasoline3.5	403	3500
		C	atmoGasoline4.0	461	4000
	• Turbo Gasoline	A	turboGasoline3.2	369	3200
		B	turboGasoline3.5	403	3500
		C	turboGasoline4.0	461	4000
	• AtmoDiesel	A	atmoDiesel2.4	272	2362
		B	atmoDiesel2.8	327	2835
		C	atmoDiesel3.2	369	3200
	• TurboDiesel	A	turboDiesel2.4	272	2362
		B	turboDiesel2.8	327	2835
		C	turboDiesel3.2	369	3200
2	Chassis		*name*	*mass*	
		A	chassisA	1000	
		B	chassisB	1250	
		C	chassisC	1500	
3	Wheel		*name*	*mass*	*Diameter*
		A	wheel17inc	8	43.18
		B	wheel18inc	8.47	45.72
		C	wheel19inc	8.94	48.26
		D	wheel20inc	9.41	50.8
		E	wheel21inc	9.88	53.34
		F	wheel22inc	10.4	55.88
		G	wheel23inc	10.8	58.42
		H	wheel24inc	11.3	60.96
4	SunRoof		*name*	*mass*	
		A	sunRoofA	54.4	
		B	sunRoofB	63.5	
		C	sunRoofC	72.6	

(continued)

Table 7 (continued)

No.	Instance type and subtype	GAGene	Instance parameters	
		D	sunRoofD	81.7
		E	sunRoofE	90.7
5	Spoiler		*name*	*mass*
		A	spoilerA	5.44
		B	spoilerB	6.35
		C	spoilerC	7.26
		D	spoilerD	8.16
		E	spoilerE	9.07
6	Lamp		*name*	*mass*
		A	lampA	4.99
		B	lampB	5.44
		C	lampC	5.9
		D	lampD	6.35
		E	lampE	6.8
7	FilementBurnoutDetector		*name*	*mass*
		A	filamentA	0.27
		B	filamentB	0.32
		C	filamentC	0.36
		D	filamentD	0.41
		E	filamentE	0.45
8	AutolevelingMotor		*name*	*mass*
		A	autoA	2.27
		B	autoB	2.72
		C	autoC	3.18
		D	autoD	3.63
		E	autoE	4.08
9	Door		*name*	*mass*
		A	doorA	54.4
		B	doorB	59
		C	doorC	63.5
		D	doorD	68
		E	doorE	72.6

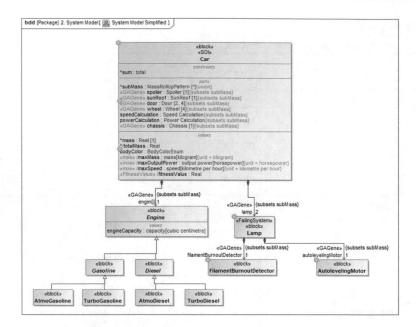

Fig. 11 Completed *150% model*

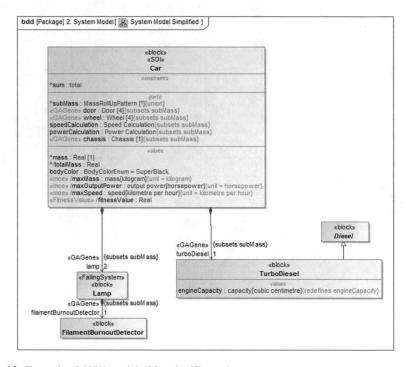

Fig. 12 Example of *100% model* of StandardCar variant

References

1. Chalé Góngora, H.G., Ferrogalini, M., Moreau, C.: How to boost product line engineering with MBSE—a case study of a rolling stock product line. In: Complex Systems Design & Management, Springer International Publishing Switzerland, pp. 239–256 (2015)
2. Young, B., Clements, P.: Model based engineering and product line engineering: combining two powerful approaches at raytheon. In: 27th Annual INCOSE International Symposium (IS 2017), Adelaide, Australia (2017)
3. Krueger, C., Clements, P.: Systems and software product line engineering. In: Software Product Lines: Going Beyond, Springer, Berlin, Heidelberg, pp. 511–512 (2013)
4. Hause, M., Hummell, J.: Model-based product line engineering—enabling product families with variants. In: 2015 IEEE Aerospace Conference, Big Sky, MT, USA (2015).
5. INCOSE: Systems Engineering Handbook: A Guide for System Life Cycle Processes and Activities, Wiley (2015)
6. Weilkiens, T.: Variant Modeling with SysML, MBSE4U (2014)
7. Wozniak, L., Clements, P.: How automotive engineering is taking product line engineering to the extreme. In: 19th International Conference on Software Product Line, Nashville Tennessee (2015)
8. Bolander, W.J., Clements, P.C., Krueger, C.: It takes a village: why PLE technology solutions require ecosystems of PLE technology providers. In: 26th Annual INCOSE International Symposium (IS 2016), Edinburgh, Scotland, UK (2016)
9. Tong, C., Sriram, D.: Artificial Intelligence in Engineering Design: Volume III: Knowledge Acquisition, Commercial Systems, And Integrated Environments, Academic Press (1992)
10. Goldberg, D.E.: Genetic Algorithms in Search, Optimization and Machine Learning. Addison-Wesley Longman Publishing Co., Inc, Boston, MA USA (1989)
11. Winter, G., Periaux, J., Galan, M., Cuesta, P.: Genetic Algorithms in Engineering and Computer Science, New York, NY. Wiley, USA (1996)
12. Harman, M., Jones, B.: Software engineering using metaheuristic innovative algorithms. In: International Conference on Software Engineering (ICSE), Toronto, Ontario, Canada, Canada (2001)
13. Cagan, J., Campbell, M.I., Finger, S., Tomiyama, T.: A framework for computational design synthesis: model and applications. J. Comput. Inf. Sci. Eng. 5(3), 171–181 (2005)
14. Arifin, H.H., Ong, H.K.R., Daengdej, J., Chimplee, N., Sortrakul, T.: Automated component-selection of design synthesis for physical architecture with model-based systems engineering using evolutionary trade-off. In: INCOSE International Symposium, vol. 28, no. 1, pp. 1296–1310 (2018)
15. Dinger, R.H.: Engineering design optimization with genetic algorithms. In: Northcon/98 Conference Proceedings, Seattle, WA, USA, USA (1998)
16. J.M. Branscomb, C.J. Paredis, J. Che, M.J. Jennings, Supporting multidisciplinary vehicle analysis using a vehicle reference architecture model in SysML. In: Conference on Systems Engineering Research (CSER'13), Atlanta, GA (2013)
17. No Magic, Inc., "Product Line Engineering," 2020. [Online]. Available: https://docs.nomagic.com/display/PLE190/Product+Line+Engineering. Accessed 7 Feb 2020
18. Spyropoulos, D., Baras, J.S.: Extending design capabilities of SysML with trade-off analysis: electrical microgrid case study. In: Conference on System Engineering Research, Atlanta, GA (2013)
19. No Magic: No Magic Documentation. No Magic, Inc., 11 July 2017. [Online]. Available: https://docs.nomagic.com/. Accessed 19 July 2017
20. Flores, R., Clements, P., Krueger, C.: Mega-scale product line engineering at general motors. In: Software Product Line Conference (2012)
21. Kramer, O.: Genetic Algorithm Essentials. Springer Nature, Gewerbestrasse, Cham (2017)
22. Chakraborty, R.: Genetic algorithms and modeling. 10 August 2010. [Online]. Available: http://www.myreaders.info/html/soft_computing.html. Accessed 16 Nov. 2017

23. Arifin, H.H., Chimplee, N., Ong, H.K.R., Daengdej, J., Sortrakul, T.: Automated component-selection of design synthesis for physical architecture with model-based systems engineering using evolutionary trade-off. In: INCOSE International Symposium, vol. 28, no. 1, pp. 1296–1310 (2018)
24. Lazko, O.: Genetic algorithms application for components parametric synthesis optimization. In: Modern Problems of Radio Engineering, Telecommunications, and Computer Science, Lviv-Slavsko, Ukraine (2006)
25. Meffert, K.: JGAP Documentation (2017). [Online]. Available: http://jgap.sourceforge.net/doc/jgap-doc-from-site-20071210.pdf. Accessed 19 July 2017
26. Robert Ong, H.K., Sortrakul, T.: Comparison of selection methods of genetic algorithms for automated component-selection of design synthesis with model-based systems engineering. In: 9th International Science, Social Science, Engineering and Energy Conference, Bangkok, Thailand (2018)
27. Nassar, N., Austin, M.: Model-based systems engineering design and trade-off analysis with RDF graphs. In: Conference on Systems Engineering Research, Atlanta, GA (2013)
28. Albarello, N., Welcomme, J.-B., Reyterou, C.: A formal design synthesis and optimization method for systems architectures. In: 9th International Conference on Modeling, Optimization & Simulation MOSIM'12, Bordeaux, France (2012)
29. Nassar, N.N.: Systems engineering design and tradeoff analysis with RDF graph models. University of Maryland (2012)
30. Arifin, H.H., Robert Ong, H.K., Daengdej, J., Novita, D.: Encoding technique of genetic algorithms for block definition diagram using OMG SysML™ notations. In: INCOSE International Symposium, Orlande, Florida (2019)
31. Krueger, C.W.: Multistage configuration trees for managing product family trees. In: 17th International Software Product Line Conference, Tokyo, Japan (2013)
32. Lau, K.: Swift algorithm club: swift tree data structure. 11 July 2016. [Online]. Available: https://www.raywenderlich.com/1053-swift-algorithm-club-swift-tree-data-structure
33. Wilhelmstotter, F.: JENETICS: Library User's Manual. 2017. [Online]. Available: http://jenetics.io/manual/manual-3.8.0.pdf. Accessed 19 July 2017
34. Frye, A.: Genetic algorithms and pareto-frontiers. California Polytechnic: Aero Department, 20 March 2017. [Online]. Available: https://www.youtube.com/watch?v=k4AxbXSy76U&t=200s. Accessed 2 Oct 2017
35. Obitko, M.: Introduction to genetic algorithms. 1998. [Online]. Available: http://www.obitko.com/tutorials/genetic-algorithms/. Accessed 16 Nov 2017
36. Frey, S., Fittkau, F., Hasselbr, W.: Search-based genetic optimization for deployment and reconfiguration of software in the cloud. In: ICSE, San Francisco, CA, USA (2013)

Model-Based System Engineering Adoption for Trade-Off Analysis of Civil Helicopter Fuel Supply System Solutions

Liu Weihao, Guo Yuqiang, Chen Qipeng, and Zhao Hui

1 Introduction

For complex system development, it is required that the requirements management, architecture development and production teams should effectively collaborate to provide timely, useful, and cost-effective products. Traditional Document-Based System Engineering (DBSE) method stores and transfers project information in documents, which is labor-intensive and needs plenty of manual analysis, review and inspection. MBSE is the formalized application of modeling to support system requirements, design, analysis, optimization, verification and validation [1, 2]. By using system modeling tools, an integrated, consistent and clear system model can be generated to give a unified representation, allowing for good understanding of the relationships between stakeholders, organizations and their impacts. Moreover, the system model provides the opportunity to link various domain-specific tools together to produce a model-based framework for a systems engineering project, and more complex verification and validation of the system can be conducted before a real system is produced.

In recent years, exploratory work on how to use MBSE method to realize integrated product design, simulation, optimization and verification in the digital virtualization environment has been carried out by many famous enterprises worldwide including NASA, Boeing and Airbus. It has been demonstrated that using MBSE method can significantly reduce the R&D risks and cost, shorten the R&D time [3–6].

Fuel Supply System is a crucial complex system involving mechanical, electrical and hydraulic technologies. According to CCAR 29 regulations, one of the main

L. Weihao (✉) · G. Yuqiang · C. Qipeng · Z. Hui
China Helicopter Research and Development Institute, Route Dongsan, Airport Economic Zone, Tianjin 300308, China
e-mail: Liuwh001@avic.com

G. Yuqiang
e-mail: Guoyq001@avic.com

© The Author(s), under exclusive license to Springer Nature Switzerland AG 2021
D. Krob et al. (eds.), *Complex Systems Design & Management*,
https://doi.org/10.1007/978-3-030-73539-5_24

311

functions of Fuel Supply System is to provide the Engine with sufficient fuel required under all operating and maneuvering conditions [7].

The present work is a practice to use a MBSE method for the development of helicopter Fuel Supply System. A system model is first architected with a practical modeling methodology and an efficient modeling tool based on the standard SysML. Then the model is used to make a trade-off analysis to choose an optimal solution architecture before subsequent detail design and manufacture begins.

2 Problem Description

According to the Engine Manual of Mi-26 helicopter, the pressure range at the Engine fuel inlet is so narrow that Fuel Supply System architecture design and booster pump selection will be complex (Fig. 1).

Generally, a simple and common civil helicopter Fuel Supply System model is presented in Fig. 2 considering the basic requirements from CCAR 29.951 & 953 & 955. Pressurized fuel is supplied to the engine by booster pump located in feeder tanks. There are also Non-Return Valve, Pessure transducer and SOV.

The challenge for Fuel Supply System design are from following requirements:

- According to the stakeholder's requirements, the height between the booster pump outlet and the Engine fuel inlet is 3.88 m, and the G-Load requirements is −0.5 to +3.5 g. In the fuel pipeline the pressure variation will reach up to 118.6 kPa due to G-Load only.

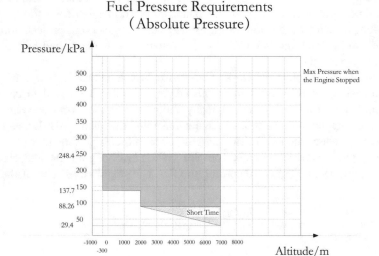

Fig. 1 Required fuel pressure (absolute pressure) by the Engine at different altitudes

Fig. 2 Fuel supply system general architecture

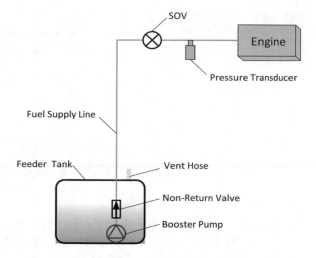

- Existing commercial off-the-shelf booster pumps possess a pressure variation of approximately 30 kPa due to engine consumption.
- Flow resistance due to various temperature, fuel type and velocity etc.

Thus, the total pressure variation will reach up to 148 kPa, which exceeds the Engine fuel pressure limitations, as shown in Fig. 3.

According to the Fuel Supply System architecture (Fig. 4), the fuel pressure at the Engine fuel inlet can be computed as followings:

1. H_1 is the plumb height between fuel surface and booster pump inlet.

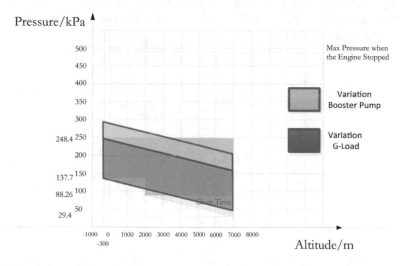

Fig. 3 Comparison of the required fuel pressure by the Engine with its variations caused by G-Load and booster pump characteristics

Fig. 4 General fuel supply
system schematic diagram

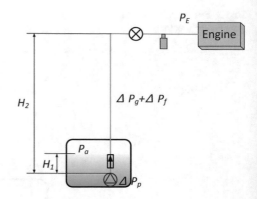

2. H_2 is the plumb height between Engine inlet and pump outlet.
3. P_a is ambient pressure associated with altitude h and relevant ambient temperature T. It will be referenced as $P_{a(h,T)}$.
4. ΔP_{g1} is the pressure of fuel height H_1. It is affected by G-Load factor δ, fuel density ρ and height H_1. It will be referenced as $\Delta P_{g1(\delta,\rho,H_1)}$.
5. ΔP_p is the booster pump pressure associated with pipeline fuel flow rate Q and fuel temperature T_f. It will be referenced as $\Delta P_{p(Q,T_f)}$.
6. ΔP_{g2} is the pressure variation mainly caused by the G-Load of helicopter. It is affected by G-Load factor δ, fuel density ρ, and height H_2, therefore will be referenced as $\Delta P_{g2(\delta,\rho,H_2)}$.
7. ΔP_f is the fuel pressure loss due to pipeline friction. It's a function of pipeline diameter ϕ, height H_2, flow rate Q, fuel density ρ and fuel temperature T_f, therefore will be referenced as $\Delta P_{f(\phi,H_2,Q,\rho,T_f)}$.
8. P_E is the Engine inlet fuel pressure. The limitation of P_E is effected by altitude h according to Engine manual, therefore the permitted engine inlet pressure will be referenced as $P_{E(h)}$.

Then, P_E can be expressed by the equation below:

$$P_E = P_{a(h,T)} + \Delta P_{g1(\delta,\rho,H_1)} + \Delta P_{p(Q,T_f)} - \Delta P_{g2(\delta,\rho,H_2)} - \Delta P_{f(\phi,H_2,Q,\rho,T_f)} \quad (1)$$

where:

$$\Delta P_{g1(\delta,\rho,H_1)} = \delta\rho g H_1 \quad (2)$$

$$\Delta P_{g2(\delta,\rho,H_2)} = \delta\rho g H_2 \quad (3)$$

Usually, T is approximately equal to fuel temperature T_f. Thus, from Eq. (1),

$$P_E - P_{a(h,T_f)} = \Delta P_{p(Q,T_f)} - \delta\rho g(H_2 - H_1) - \Delta P_{f(\phi,H_2,Q,\rho,T_f)} \quad (4)$$

Considering $P_{E(h)}$, following inequality should be satisfied:

$$\left| P_{E(h)} - P_{a(h,T_f)} \right| \geq \left| \Delta P_{p(Q,T_f)} - \delta \rho g (H_2 - H_1) - \Delta P_{f(\phi,H_2,Q,\rho,T_f)} \right| \quad (5)$$

Also δ has to be kept in accordance with the G-Load of the helicopter. Considering the extreme values of δ, following inequality should be satisfied:

$$\begin{cases} P_{E(h),max} - P_{a(h,T_f)} \geq \Delta P_{p(Q,T_f)} - \delta_{min} \rho g (H_2 - H_1) - \Delta P_{f(\phi,H_2,Q,\rho,T_f)} \\ P_{E(h),min} - P_{a(h,T_f)} \leq \Delta P_{p(Q,T_f)} - \delta_{max} \rho g (H_2 - H_1) - \Delta P_{f(\phi,H_2,Q,\rho,T_f)} \end{cases} \quad (6)$$

According to Eq. (6), $\Delta P_{p(Q,T_f)}$ and $\Delta P_{f(\phi,H_2,Q,\rho,T_f)}$ mainly depend on the inherent characteristics of booster pump, fuel supply hoses and fuel type. Thus, the relative pressure at the Engine fuel inlet $P_{E(h)} - P_{a(h,T_f)}$ will be impacted significantly by height $H_2 - H_1$.

According to Eq. (6), one possible solution is to reduce the height $H_2 - H_1$, which means the fuel supply line should be shorten by putting the feeder tank on the upper deck of the helicopter (Fig. 5). The feeder tank will be supplied by the Fuel Transfer System from the Auxiliary Tank. Float switch and actuated valve will detect the high fuel level and cut off the fuel supply from the auxiliary tank to the feeder tank in order to prevent overpressure. The feeder tank connects with the outside through vent hoses in order to keep ambient pressure. Fuel stored in the feeder tank is supplied to the Engine with the help of booster pump.

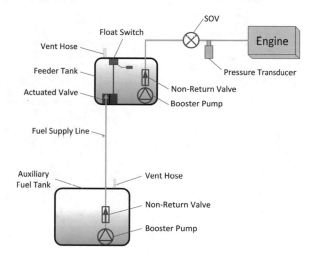

Fig. 5 Alternative fuel supply system architecture

3 System Architecture

Since the system model is the basis of MBSE activities, it is of significance to architect system with appropriate modeling language, tool and methodology. In previous work, the system architecture is based on the Object Management Group's (OMG) Systems Modeling Language (SysML) and produced by a commercial modeling tool Magic-Draw with a practical modeling methodology MagicGrid. As the present work is dedicated to trade-off analysis of system solution architectures, only the problem and the solution domains of MagicGrid methodology are adopted. The remaining sections in this chapter will present model artifacts produced in the modeling workflow.

3.1 Problem Domain

The purpose of the problem domain is to analyze stakeholder needs and refine them with SysML model elements to get a clear and coherent description of what problems the system of interest (SoI) must solve. From the black-box perspective, it focuses on how the SoI interacts with the environment without getting any knowledge about its internal structure and behavior.

Stakeholder needs of the helicopter Fuel Supply System are captured by a requirement table shown in Fig. 6. It includes airworthiness regulations, functional and performance requirements, which are derived from the Fuel Supply System's superior system, i.e. the fuel system. Non-functional stakeholder needs serve as high-level

#	△ Name	Text
1	⊟ Ⓡ SN-1 CCAR-29.955	The fuel system for each engine must provide the engine with at least 100% percent of the fuel required under all operation and maneuvering conditions to be approved for the rotorcraft.
2	Ⓡ SN-1.1 Fuel pressure	The fuel pressure corrected for accelerations (load factors), must be within the limits specified by the engine type certificate data sheet.
3	Ⓡ SN-1.2 Fuel level	The fuel level in the tank may not exceed that established as the unusable fuel supply for that tank under ccar 29.959, plus that necessary to conduct the test.
4	Ⓡ SN-1.3 Fuel head	The fuel head between the tank and the engine must be critical with respect to rotorcraft flight attitudes.
5	Ⓡ SN-1.4 Fuel flow transmitter	The fuel flow transmitter, if installed, and the critical fuel pump (for pump-fed systems) must be installed to produce (by actual or simulated failure) the critical restriction to fuel flow to be expected from component failure.
6	Ⓡ SN-1.5 Power resource	Critical values of electrical power or other source of fuel pump motive power must be applied.
7	Ⓡ SN-1.6 Fuel property	Critical values of fuel properties which adversely affect fuel flow are applied during demonstrations of fuel flow capability.
8	Ⓡ SN-1.7 Fuel Filter	The fuel filter required by CCAR.29.997 is blocked to the degree necessary to simulate the accumulation of fuel contamination required to activate the indicator required by CCAR.29.1305 (a)(17).
9	⊟ Ⓡ SN-2 Functional requirements	
10	⊟ Ⓡ SN-2.1 Supply Fuel to the Engine.	
11	Ⓡ SN-2.1.1 Fuel supply capacity	Supply to the engine with fuel according to the pressure, flow and temperature requirements in all flight envelopes.
12	Ⓡ SN-2.1.2 Fuel pressure monitoring	Monitor the fuel pressure supplied to the Engine.
13	Ⓡ SN-2.1.3 Pump status monitoring	Monitor the status of the pump.
14	⊟ Ⓡ SN-2.2 Shut off fuel supply	
15	Ⓡ SN-2.2.1 Fuel shut off capacity	Provide means to cut off the fuel supply to the engine.
16	Ⓡ SN-2.2.2 Shut-off valve monitoring	Monitor the status of the shut-off valve.

Fig. 6 Stakeholder needs

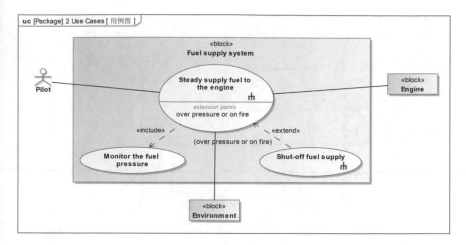

Fig. 7 Use cases

key performance indicators that should be checked within the solution domain model, such as the fuel pressure at the inlet of the engine. The functional stakeholder needs include "fuel supply to the Engine" and "cut off the fuel supply", which are two key functions that the Fuel Supply System must have.

Functional stakeholder needs are refined with use cases and use case scenarios. Three use cases in the system context are captured by a SysML use case diagram, as shown in Fig. 7. They are more precise than the stakeholder needs in telling what the pilots expected from it, what external systems it interacts with, and what they want to achieve by using it.

The primary use case scenario, "Steady supply fuel to the engine", is captured by a SysML activity diagram shown in Fig. 8. It can be found that an activity "Supply fuel to the engine" is allocated to the Fuel supply system by swimlanes. Moreover, the external interaction interfaces between the Fuel Supply System and the external systems can also be identified. Summarizing all use case scenarios, all required activities and external interaction interfaces can be obtained.

Then, the Fuel Supply System can be unfolded from the white-box perspective. The required activities for the Fuel Supply System are further refined with decomposed sub-functions, as shown in Fig. 9. Moreover, the logical subsystems and internal interaction interfaces can also be identified.

3.2 Solution Domain

Once the problem domain analysis is completed and stakeholder needs are transferred into the model, it is time to start thinking about the solutions. Figure 10 shows a system requirements specification described by a SysML requirement diagram. It is

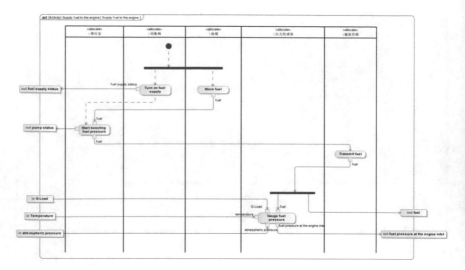

Fig. 8 Primary use case scenario

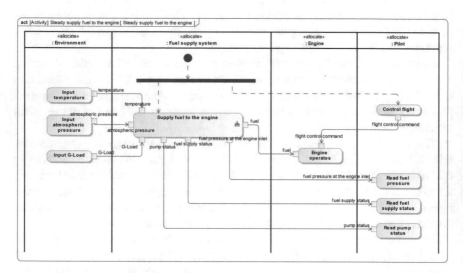

Fig. 9 Refinement of the primary activity

identified from the problem domain model to refine the activities, parts, interfaces
and ports captured in problem domain. Based on it, a high-level solution architecture
(HLSA) can be given by a SysML block definition diagram from the problem domain
model, as shown in Fig. 11.

According to the HLSA, two detailed solution architectures are considered in
present work. In the first one, the feeder tank is placed at the bottom of helicopter
(Fig. 2). It has been found that the fuel pressure at the Engine fuel inlet due to plumb

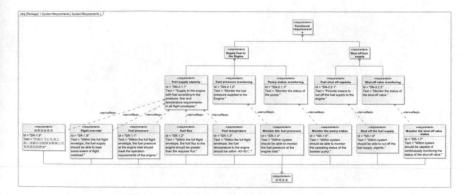

Fig. 10 System requirements specification

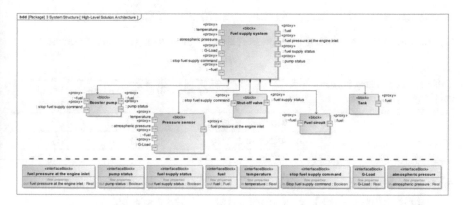

Fig. 11 High-level solution architecture

height of fuel supply line and G-Load will vary wildly. An alternative architecture puts the feeder tank at the height close to the Engine (Fig. 5). The calculation model for the former is described by a SysML parametric diagram shown in Fig. 12. With the altitude (A), fuel temperature (T), G-Load factor (G-Load), and fuel pressure at the outlet of the booster pump (Pp) as the input, the fuel pressure at Engine fuel inlet (Pe) can be immediately evaluated and verified against the required range of the Engine.

4 Trade-Off Analysis

Once detailed solution architectures are determined, a trade-off analysis can be performed to choose the optimal one. Moreover, the selection of existing commercial off-the-shelf booster pumps is also considered in this analysis. Table 1 gives six

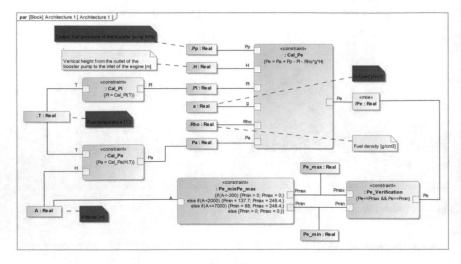

Fig. 12 Calculation model for the fuel pressure at the engine inlet in solution architecture 1

Table 1 Six practical solutions considered in present trade-off analysis

	Pump RLB-40	Pump RLB-48	Pump RLB-111
Architecture 1	(a)	(c)	(e)
Architecture 2	(b)	(d)	(f)

practical solutions from two solution architectures and three booster pumps.

A trade-off procedure is concluded in Fig. 13. For a specific flight mission, the altitude (A), fuel temperature (T), G-Load factor (G-Load), and fuel pressure at the outlet of the pump (Pp) vary within certain ranges. They are first discretized and combined into different operating points. Those operating points are inputted into the calculation model represented by the SysML parametric diagram, which integrates Matlab/Simulink models to calculate and verify the fuel pressure at the Engine fuel inlet (MoE). Only when all operating points pass the verification can a solution be asserted as feasible and safe for the flight mission. In present SysML modeling tool (No Magic MagicDraw), the failed operating points will be marked red in the right sidebar. It is obvious that only the solution (f), i.e. architecture 1 combined with pump RLB 111, passes the verification, while the other five solutions are failed.

In order to give a deeper analysis, the results are displayed graphically in Fig. 14. The outer frame represents the allowable pressure range at the Engine fuel inlet. The cubes distinguished with different color represent the actual pressure that supplied to the Engine. If the any part of the cubes exceeds the boundary of the outer frame, it means that the operating point relevant to the "part" cannot meet the pressure requirement of the Engine. Obviously, conclusions can be drawn as follows:

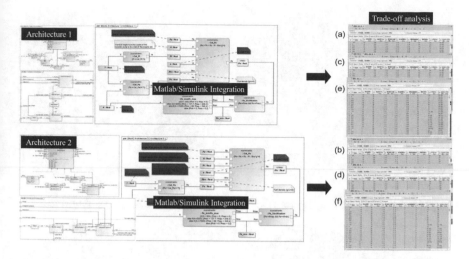

Fig. 13 Trade-off procedure

1. Among the 6 solutions, only the solution (f) "Architecture 2 + Pump RLB-111" passes the verification. It's found that architecture 2 can effectively reduce the pressure variation and still shows potentiality for better pump selections.
2. The fuel pressure variation at the Engine fuel inlet obtained by architecture 2 is significantly smaller than architecture 1 considering the same pump and same operating point. In architecture 1, the pressure is more sensitive to the G-Load factor because the plumb height between the outlet of the booster bump and the inlet of the Engine is much higher.

5 Conclusion

A complete process of developing a Fuel Supply System using MBSE method is presented in this paper, including requirements analysis, function analysis, system architecture definition and trade-off. It provides a capability that MBSE is a proper way to deal with the Fuel Supply System development.

Variable booster pump, which has the ability to adjust the output fuel pressure $(\Delta P_{p(Q,T_f)})$ accordingly, might be another possible solution for this problem. However, more detailed requirements shall be analyzed and more complex joint simulation between system and equipment model need further discussion.

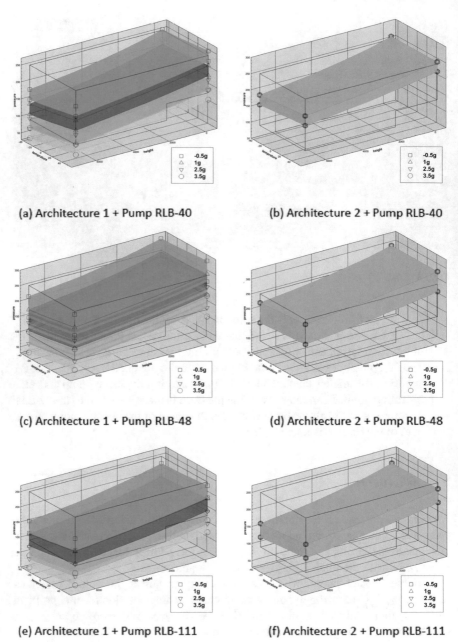

(a) Architecture 1 + Pump RLB-40   (b) Architecture 2 + Pump RLB-40

(c) Architecture 1 + Pump RLB-48   (d) Architecture 2 + Pump RLB-48

(e) Architecture 1 + Pump RLB-111   (f) Architecture 2 + Pump RLB-111

Fig. 14 Results of trade-off analysis for six practical solutions

References

1. INCOSE: INCOSE Systems Engineering Handbook, Version 4.0 (eds.) (2015)
2. ISO/IEC: ISO/IEC 15288: 2008 Systems and software engineering-System lifecycle process (2008)
3. Motamedian, B.: MBSE applicability analysis. Int. J. Sci. Eng. Res. (2013)
4. Arnould, V.: Using model-driven approach for engineering the system engineering system. In: 13th annual conference on system of systems engineering (SoSE) (2018), pp. 608–614
5. Zhang, S.J., Li, Z.Q., Hai, X.H.: Safety critical systems design for civil aircrafts by model based systems engineering (in Chinese) (2018), pp. 299–311
6. Qian, M.: Design method of civil aircraft functional architecture based on MBSE (2019)
7. CCAR- 29-R2 Airworthiness Standards Transport Category Rotorcraft.

Non-functional Attribute Modeling and Verification Method for Integrated Modular Avionics System

Peng Guo, Feiyang Liu, Na Wu, Yahui Li, and Ning Hu

1 Introduction

Integrated modular avionics (IMA) system has strict requirements for non-functional attributes such as real-time, safety, and reliability [1, 2]. At present, domestic and foreign industrial departments lack technical means to ensure the above-mentioned non-functional attributes requirements, resulting in high development costs, long development cycle and high failure rate [3, 4]. In 2002, the National Institute of Standardization and Technology (NIST) research pointed out that 70% of errors were introduced early in the life of the system, and 80% of them were not discovered at high repair costs until after comprehensive testing. In the final confirmation test stage, although 20.5% of errors could be found, but it took 300–1000 times the repair cost. At present, the development cost of integrated modular avionics systems has exceeded the upper limit of the industrial departments' affordability. The aviation industrial departments urgently need to change the traditional design methods, among which MBSE (model-based systems engineering) is an effective solution (Fig. 1) [5–7].

At present, the development of integrated modular avionics system based on MBSE has the following problems: (1) In terms of process, there is a lack of a set of development process suitable for integrated modular avionics system, covering the three stages of requirement analysis, architecture design and multi-dimensional analysis. (2) At the requirement level, the mainstream way in domestic and foreign

P. Guo (✉) · N. Wu · Y. Li · N. Hu
Aviation Key Laboratory of Science and Technology on Airborne and Missileborne Computer, School of Computer Science and Technology, Computing Technique Research Institute, AVIC, Nanjing University of Aeronautics and Astronautics, Room 15, JinYe2 Road, Xi'an, Shaanxi, China
e-mail: nwpu062475@163.com

F. Liu
Computing Technique Research Institute, AVIC, Room 15, JinYe2 Road, Xi'an, Shaanxi, China
e-mail: feiyang_liu2017@163.com

© The Author(s), under exclusive license to Springer Nature Switzerland AG 2021
D. Krob et al. (eds.), *Complex Systems Design & Management*,
https://doi.org/10.1007/978-3-030-73539-5_25

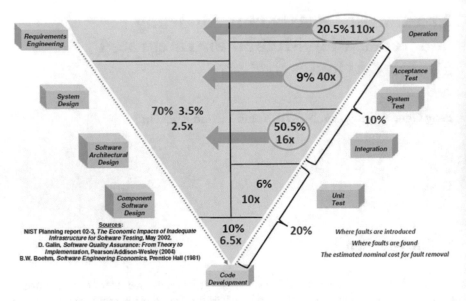

Fig. 1 Disadvantages of traditional development methods

is to use SysML (System Engineering Modeling Language) for requirement analysis to form a functional model of a comprehensive avionics system, and use AADL (Embedded System Architecture Analysis and Design Language) for system Architecture design and analysis form an architecture model. At present, most of them use manual secondary design to transform the functional model to the architecture model, which leads to inconsistencies between the functional model and the architecture model, and the engineering staff repeats their work. (3) At the system structure level, the existing AADL has insufficient descriptive capabilities in terms of nonfunctional attributes, and cannot describe information such as errors/faults to hazard propagation, hazard behavior, etc.; AADL lacks standardized analysis methods, such as scheduling analysis complying with ARINC 653 and safety analysis based on FMECA [8–11].

Aiming at the problems faced in the development of integrated modular avionics systems, this paper proposes non-functional attributes the modeling and verification method for integrated modular avionics systems, extracts a set of development processes suitable for integrated modular avionics systems, and put forward key technologies such as automatic model transformation, hazard sub-language extension based on AADL, AADL-based scheduling analysis for ARINC 653 system, and FMECA-based safety analysis [12–14].

2 Architecture-Centric Design Process

The model-based design process for IMA system follows the promotion of MBSE by Aviation Industry Corporation of China. The first step is to take the result of the requirement analysis as input, which is function architecture model, and use the SysML2AADL model transformation to automatically complete the 100% transmission of requirement information, transforming to the AADL initial architecture model; the second step is to refine the model based on the initial architecture model, and design the logical architecture model and physical architecture model of IMA system. The logical architecture model is mainly based on processes and threads. The main software task is composed. The physical architecture is mainly composed of the processor, memory, bus and peripherals. If applying ARINC 653 OS, the partition management, inter/intra-partition communication relationship should be described, and the logical architecture, physical architecture, ARINC 653 model Deployment requires additional interface models, including system data interfaces and electrical interfaces; the third step is to add non-functional attribute models based on the system architecture model, including security, reliability, safety and real-time models; last, based on the above complete architecture model, we take multi-dimensional verification of the system's non-functional attributes (Fig. 2).

3 Key Technology

This chapter focuses on the detailed introduction of the key technologies of the IMA system design process, including the automatic transformation of the SysML-based functional model to the AADL-based architecture model, hazard sub-language extension based on AADL, AADL-based scheduling analysis for ARINC 653 system, and FMECA-based safety analysis.

3.1 Model Transformation of Functional Model to Architecture Model

In order to realize the automatic transformation of IMA system requirements information, this paper proposes a transformation method from SysML-based functional models to AADL-based architecture models. At the functional model level, SysML BDD (block definition diagram) is mainly used to define the functional architecture of the system, IBD (internal block diagram) is used to define the relationship between functions, and activity diagrams are used to describe the functional logic of the system. At the system architecture level, AADL components are used to describe the composition of the system, connections are used to describe the relationships between components, and behavior state machines are used to describe the

Fig. 2 Architecture-based
airborne information system
design system

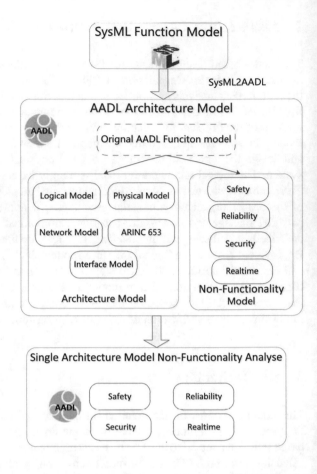

behavior of components. Figure 3 shows the model transformation framework, which extracts the SysML and AADL meta-models that meet the IMA system requirements. The ATL language is used to design SysML to AADL mapping rules, including parsing SysML XML files and unparsing AADL XMI files. Finally, an AADL model conforming to the OSATE tool is generated.

Table 1 shows the transformation rules from SysML BDD and IBD to AADL components. SysML Block is mapped into AADL abstract components. According to the architect's needs, AADL abstract can be refined into specific components such as processes, processors, device, and so on. The subordinate relationship of SysML BBD is mapped to AADL subcomponents, the port of SysML Block is mapped into AADL Port, and the relationship of Block in SysML IBD is mapped into AADL connection. Table 2 shows the rules for converting SysML activity diagrams into AADL behavior attachments, including elements such as initial nodes, activity nodes, activity endpoints, and control flow.

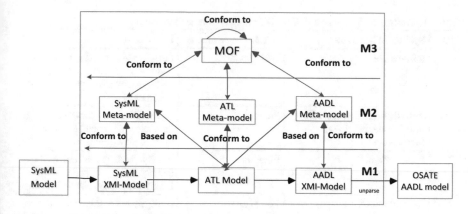

Fig. 3 Functional model to architecture model transformation framework

Table 1 Transformation of SysML2AADL system static structure

ID	BDD/IBD element	AADL element	Description
1	Block (system stereotype)	Abstract	Block (system) describes the system function module declaration, and maps to the abstract component declaration
2	BlockImpl (system stereotype)	Abstract implementation	Block (system) describes the realization of system function modules and maps to the realization of abstract components
3	Block (data Stereotype)	Data	Block (data) describes the declaration of custom data types and maps to data components
4	BlockImpl (data stereotype)	Data implementation	BlockImpl (data) describes the implementation of custom data types and maps to the implementation of data components
5	Part	SubComponent	Part represents the constitute of the Block, which is mapped as a subcomponent
6	Connector	Connection	Connector represents the abstract connection between Parts, which is mapped to Connection
7	FlowPort (dataEventPort stereotype)	DataEventPort	FlowPort (dataEventPort) describes the data event port of Part and maps to DataEventPort in the AADL component declaration feature

Table 2 Transformation of SysML2AADL system dynamic behaviour

ID	SysML ACT element	AADL BA element	Description
1	InitialNode	InitialState	InitialNode describes the initial node of the activity and is mapped to the InitalState in the behavior attachment
2	ActivityNode (not including InitialNode, FinalNode, MergeNode)	State	ActivityNode describe the node in the Activity, mapped to the state in the behavior annex
3	FinalNode	FinalState	FinalNode marks the end of the activity and is mapped to the FinalState in the behavior annex
4	AcceptEventAction	Port? (var)	AcceptEventAction is mapped into the receiving statement Port? (var)
5	CallBehaviorAction	SubProgram	CallBehaviorAction means that another behavior is triggered at startup, and CallBehaviorAction is mapped into a SubProgram
6	SendSignalAction	Port! (var)	SendSignalAction is mapped to the sending statement Port! (var)
7	ControlFlow	Transitions	ControlFlow is mapped to transitions in the behavior annex
8	Variable	Variable	The local variables used by the Activity are mapped to the local variables of the corresponding behavior annex
9	Guard	Condition	Guard is mapped to the trigger condition of behavior annex

3.2 Hazard Sub-language Extension Based on AADL

AADL was originally an architecture design language specifically for avionics and flight control. Although it has EMV2 that specifically describes safety and reliability, it has insufficient on hazard. Only error events and error states are defined in Error Model, and it is impossible to determine which events and states may cause harm; Error Model lacks a quantitative description of hazards; Hazard analysis is an important part of safety analysis, Error Model cannot perform hazard analysis on components and systems.

Based on the research of AADL technical standards and related safety analysis methods, the AADL Hazard Model Annex is proposed to solve the problem of insufficient modeling capabilities of the above error model. As shown in Fig. 4, we design AADL-Hazard meta-model. We use an AADL Hazard Model to describe the system hazards by adding dependencies, link it with the error model, and integrate the hazard behavior into the system architecture model. The error behavior state

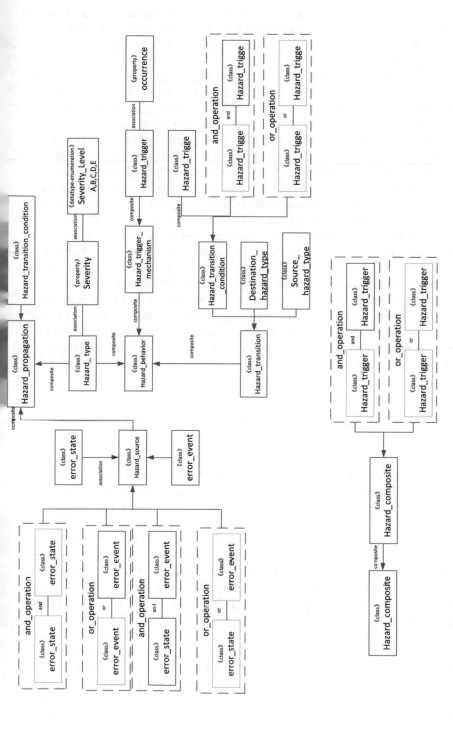

Fig. 4 AADL-Hazard meta-model

machine is essentially a state machine, and the hazard behavior state machine also describes the hazard behavior in the same way as the state machine. The error event, state, or their logical combination in the error model can be the hazard source in the hazard model.

The hazard meta-model is shown in Fig. 4, which establishes all hazards in the system, and specifies the severity of each hazard, which is divided into A, B, C, D, and E level. We establish the state of hazard behavior, including the type of hazard, the hazard trigger mechanism, and the hazard transfer. The hazard type refers to the hazard type. The hazard trigger mechanism includes multiple triggers, and each trigger has a probability of occurrence, including exponential, delayed, immediate and fixed probability. Hazard transfer consists of source hazard, destination hazard and hazard transfer conditions. Hazard transfer conditions are triggered by hazards, or two hazards triggered AND operations, or two triggered OR operations. Hazard propagation includes three parts, the source of the hazard, the conditions for the transfer of the hazard, and the hazard. The hazard sources of the system include two types, one is the event that triggers the system failure, the other is a certain failure state after the system fails. The hazard sources include failure state, failure event, failure state and operation, failure event and Operation, failure state and event AND operation, failure state and event OR operation. The hazard transfer conditions are as described. The hazard refers to the hazard type. The combined behavior of hazards includes multiple hazard combinations, each hazard combination is composed of a combined hazard expression and a combined hazard type, and each hazard expression is composed of inter-hazard and operation, inter-hazard or operation.

3.3 Safety Modeling and Analysis Based on FMECA

Failure Mode and Criticality Analysis (FMECA) is currently the main technical means for comprehensive avionics system safety analysis. The current analysis capability of AADL does not support FMECA. Based on AADL EMV2 and the proposed AADL-Hazard sub-language, this paper proposes FMECA-based safety modeling method.

Error models and hazard models involve multiple aspects of errors and hazards, but FMECA pays more attention to all possible failure modes of the product and their possible impact, and assigns the severity of each failure mode and its probability of occurrence. Therefore, FEMCA-based safety modeling needs to establish component error behavior model, hazard source model, hazard propagation model, combined error behavior model and other related attributes, as shown in Figs. 5 and 6.

The FMECA process is shown in Fig. 7. Based on the AADL architecture model, error model and hazard model, a system safety model is formed. Based on the safety analysis, qualitative FMEA analysis and quantitative CA analysis are carried out, and finally the FMECA model is formed. The structure of the entire FMECA analysis result is shown in the left part of Table 3, including failure mode and impact analysis and criticality analysis. FMEA uses partial qualitative analysis, CA uses partial

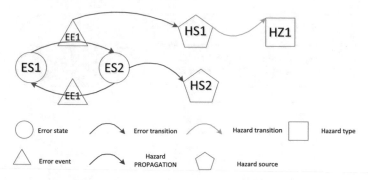

Fig. 5 AADL-Hazard model

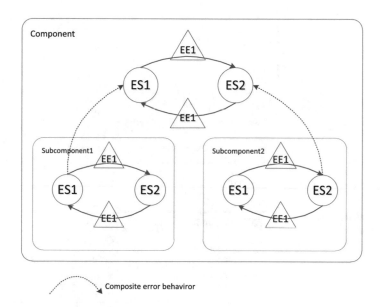

Fig. 6 AADL-Hazard composite model

quantitative analysis. The input of the entire FMECA comes from the AADL architecture model and the additional safety model, in which the units and functions are
extracted from the architecture model, the failure mode comes from the error model,
the failure effects come from error model composite behavior, severity type comes
from the hazard model, the probability of occurrence level comes from the probability calculation of the error model, and the detection difficulty is equivalent to the
probability calculation from the hazard model. The above calculation uses the DSPN
calculation tool TimeNet, and the risk priority number (RPN) comes from the above
three. The product of those is shown in Formula 1. Effect Severity Ranking (ESR)
represents the severity level of the failure mode, Occurrence Probability Ranking

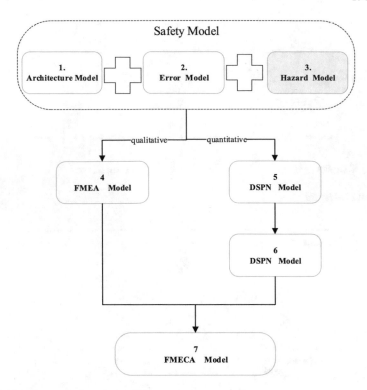

Fig. 7 FEMCA process based on AADL

Table 3 Relation between
FMECA and AADL

FMECA standard		AADL
Item		Component
Function		Property: description
Failure mode		Error state
Cause		Property: casuse
Failure effect	1st level	Hazard source
	2st level	Error model: composite error behavior
	3st level	Composite error behavior
Severity		Property:hazard::severity
OPR		DSPN Model for error model
ESR		Property:error::severity
DDR		DSPN Model for hazard model
RPN		Formula (1)

(OPR) represents the occurrence probability level of the failure mode, and Detection Difficulty Ranking (DDR) represents the detection difficulty level of the failure mode.

$$RPN = ESR * OPR * DDR \qquad (1)$$

3.4 Modeling and Scheduling Analysis for ARINC 653 System

At present, the mainstream IMA adopts the ARINC 653 standard to provide a partition isolation protection mechanism for application software. AADL has developed an annex to meet ARINC 653 and supports ARINC 653 in the architecture model. However, at present OSATE, which is the international mainstream AADL development environment, does not yet support ARINC 653 schedulable analysis. This paper presents the scheduling analysis for ARINC 653 system based on AADL.

Currently AADL ARINC 653 modeling supports partition management, partition modeling, multi-processing architecture modeling, process modeling, inter-partition communication modeling, intra-partition communication modeling, memory requirement modeling, health monitoring modeling, mode modeling, and Device driver modeling. The scheduling analysis adopts module, partition, process and system modeling. Table 4 shows the mapping relationship between AADL and ARINC 653, including module, partition and process parts. The input of scheduling analysis includes two parts, one part is task information, which supports periodic and event-type tasks. Periodic tasks need to be described the period, execution time range, deadline and priority of the task. Event-type tasks need to describe the task's activation time, execution time range, and deadline; the other part is ARINC 653 information, including major time frame, partition slot time, partition order, etc. The entire scheduling analysis process shows that the AADL model covering ARINC 653

	ARINC653	AADL	Property
Table 4 AADL and ARINC653 mapping relationship	Module	Processor	• ARINC653:: Module_Major_Frame • ARINC653:: Partition_Slots • ARINC653:: Slots_Allocation • Scheduling_Protocol
	Partitions	Virtual processor	• Scheduling_Protocol
	Process	Thread	• Deadline • Period • Priority • Compute_Execution_Time

Table 5 AADL component and Cheddar component transformation rules

AADL element		Cheddar element	
Component name	Property	Component name	Property
Processor	Scheduling Algorithm	Processor	Scheduler
Process	\	Address space	Stack size
Thread	Types of	Task	task_type
	bound processor name		cpu_name
	Cycle		Period
	Release time		start_time
	Execution time		Capacity
	Deadline		Deadline
	Priority		Priority

is converted to AAXL, and AAXL is converted to Cheddar XML as shown in Table 5, and then the simulation engine is performed to analyze the task Gantt chart and the conclusion of whether the task can be scheduled, as shown Fig. 8. The principle of the entire scheduling is divided into three steps, as shown Fig. 9. The first step is to determine which event occurs in each unit time, and store these events in a table.

Fig. 8 Scheduled analysis process based on AADL

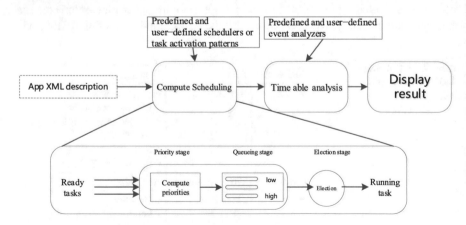

Fig. 9 Scheduling analysis principle

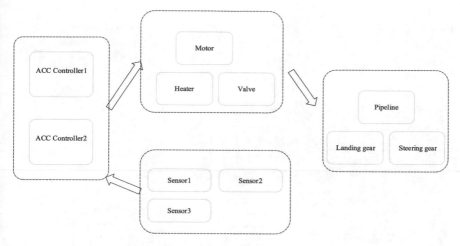

Fig. 10 Air compressor control system

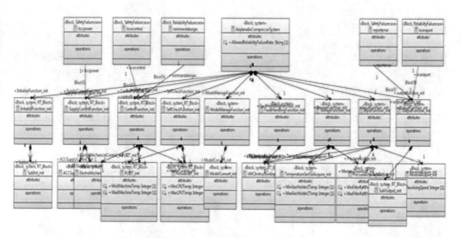

Fig. 11 BDD for air compressor control system

The second step is to perform analysis to view the changes in the attributes in the timetable; the third step is to display the analysis results.

4 Case Study

We choose an air compressor control system (ACCS) as a typical application to verify the correctness of the modeling and verification technology in this paper. The air compressor control system, as an airborne safety-critical system, drives the motor to drive the air compressor to work and output nitrogen energy, as shown in Fig. 10.

a

b

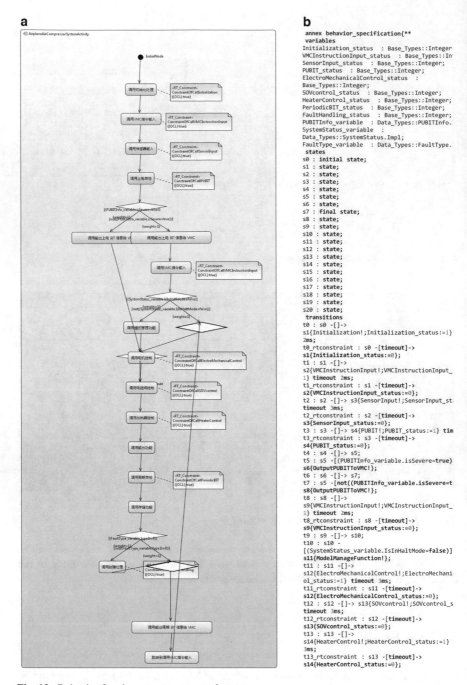

```
annex behavior_specification{**
  variables
Initialization_status  : Base_Types::Integer
VMCInstructionInput_status  : Base_Types::In
SensorInput_status  : Base_Types::Integer;
PUBIT_status  : Base_Types::Integer;
ElectroMechanicalControl_status  :
Base_Types::Integer;
SOVcontrol_status  : Base_Types::Integer;
HeaterControl_status  : Base_Types::Integer;
PeriodicBIT_status  : Base_Types::Integer;
FaultHandling_status  : Base_Types::Integer;
PUBITInfo_variable  : Data_Types::PUBITInfo.
SystemStatus_variable  :
Data_Types::SystemStatus.Impl;
FaultType_variable  : Data_Types::FaultType.
  states
s0 : initial state;
s1 : state;
s2 : state;
s3 : state;
s4 : state;
s5 : state;
s6 : state;
s7 : final state;
s8 : state;
s9 : state;
s10 : state;
s11 : state;
s12 : state;
s13 : state;
s14 : state;
s15 : state;
s16 : state;
s17 : state;
s18 : state;
s19 : state;
s20 : state;
  transitions
t0 : s0 -[]->
s1{Initialization!;Initialization_status:=1}
2ms;
t0_rtconstraint : s0 -[timeout]->
s1{Initialization_status:=0};
t1 : s1 -[]->
s2{VMCInstructionInput!;VMCInstructionInput_
1} timeout 2ms;
t1_rtconstraint : s1 -[timeout]->
s2{VMCInstructionInput_status:=0};
t2 : s2 -[]-> s3{SensorInput!;SensorInput_st
timeout 3ms;
t2_rtconstraint : s2 -[timeout]->
s3{SensorInput_status:=0};
t3 : s3 -[]-> s4{PUBIT!;PUBIT_status:=1} tim
t3_rtconstraint : s3 -[timeout]->
s4{PUBIT_status:=0};
t4 : s4 -[]-> s5;
t5 : s5 -[(PUBITInfo_variable.isSevere=true)
s6{OutputPUBITToVMC!};
t6 : s6 -[]-> s7;
t7 : s5 -[not((PUBITInfo_variable.isSevere=t
s8{OutputPUBITToVMC!};
t8 : s8 -[]->
s9{VMCInstructionInput!;VMCInstructionInput_
1} timeout 2ms;
t8_rtconstraint : s8 -[timeout]->
s9{VMCInstructionInput_status:=0};
t9 : s9 -[]-> s10;
t10 : s10 -
[(SystemStatus_variable.IsInHaltMode=false)]
s11{ModelManageFunction!};
t11 : s11 -[]->
s12{ElectroMechanicalControl!;ElectroMechani
ol_status:=1} timeout 3ms;
t11_rtconstraint : s11 -[timeout]->
s12{ElectroMechanicalControl_status:=0};
t12 : s12 -[]-> s13{SOVcontrol!;SOVcontrol_s
timeout 3ms;
t12_rtconstraint : s12 -[timeout]->
s13{SOVcontrol_status:=0};
t13 : s13 -[]->
s14{HeaterControl!;HeaterControl_status:=1}
3ms;
t13_rtconstraint : s13 -[timeout]->
s14{HeaterControl_status:=0};
```

Fig. 12 Behavior for air compressor control system

First, we select Papyrus SysML 1.4 to establish the functional model of the system, including BDD and Activity Diagram as shown in Figs. 11 and 12. Using SysML to AADL model transformation method, the generation AADL model is shown in Figs. 12, 13 and 14b. On this basis, the architecture design of ACCS is carried out, and then the non-functional attribute analysis is carried out, as shown in Fig. 14a.

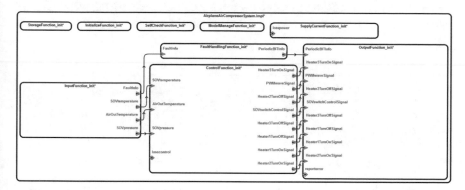

Fig. 13 Behavior for air compressor control system

a

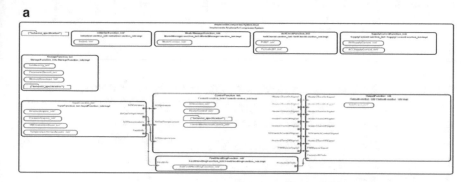

b

```
annex Hazard_Model{**
    use error behavior systemerrmodel;
    use hazard types accl_hazardmodel;
    --use hazard behavior accl_hebehavior;
    hazard sources
        hs1: error state error1;
        hs2: error state error2;
        hs3: error state error3;
    hazard trigger mechanism
        trigger1:trigger{Occurrence=>poisson 0.6};
        trigger2:trigger{Occurrence=>poisson 0.5};
        trigger3:trigger{Occurrence=>poisson 0.7};
    hazard propagations
        state hs1-[trigger1]->acclhazard1;
        state hs2-[trigger2]->acclhazard2;
        state hs3-[trigger3]->acclhazard3;|
    **};
end accl.impl;
```

Fig. 14 AADL architecture model

FMECA Result											
					Fault Effect				CA		
ID	Component	Funcition	Fault mode	1st Level Effect	2ed Level Effect	Final Effect	Severity	Severity Leve	Probabilit	Test Leve	RPN
1	accl system	output nitrogen	accl.error1	acclhazard1	acclhazard1	acclhazard1	C	3	9	3	81
2	accl system	transfer status da	accl.error2	acclhazard2	acclhazard2	acclhazard2	A	5	9	3	135
3	accl system	transfer error dat	accl.error3	acclhazard3	acclhazard3	acclhazard3	B	4	9	3	108
4	input process	receive command	input.nondatacmd	inputhazard1	acclhazard1	acclhazard1	B	5	8	3	120

Fig. 15 Safety model based on AADL

Figure 16 shows the FEMCA analysis results. ACC adopts ARRINC 653 standard, which has 7 partitions and. The scheduling of partition tasks on ACC is shown in Fig. 17.

FMECA分析结果

					故障影响				危害性分析			
编号	构件名	功能描述	故障模式	故障原因	局部影响	高一层次影响	最终影响	严酷度类别	软件严酷度等级	软件发生概率等级	软件故障模式被检测难度等级	风险优先数
1	acc1 system	正常输出氮气	acc1.error1	systemFail1事件导致	acc1.error1 危险 acc1hazard1	acc1.error1 危险 acc1hazard1	acc1.error1 危险 acc1hazard1	C	3	9	3	81
2	acc1 system	正常上传空压机状态信息	acc1.error2	systemFail2事件导致	acc1.error2 危险 acc1hazard2	acc1.error2 危险 acc1hazard2	acc1.error2 危险 acc1hazard2	A	5	9	3	135
3	acc1 system	正常上报错误信息	acc1.error3	systemFail3事件导致	acc1.error3 危险 acc1hazard3	acc1.error3 危险 acc1hazard3	acc1.error3 危险 acc1hazard3	B	4	9	3	108
4	input process	正常接收vmc指令	input.nondatacmd	inputFailed事件导致	input.nondatacmd 危险 inputhazard1	acc1.error1 危险 acc1hazard1	acc1.error1 危险 acc1hazard1	B	5	8	3	120

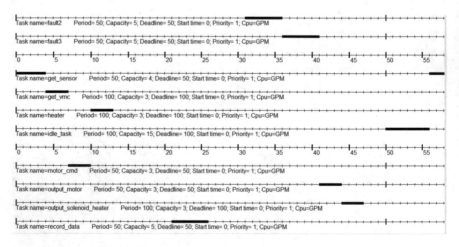

Fig. 16 AADL-FMECA model

Task name=fault2 Period= 50; Capacity= 5; Deadline= 50; Start time= 0; Priority= 1; Cpu=GPM
Task name=fault3 Period= 50; Capacity= 5; Deadline= 50; Start time= 0; Priority= 1; Cpu=GPM
0 5 10 15 20 25 30 35 40 45 50 55
Task name=get_sensor Period= 50; Capacity= 4; Deadline= 50; Start time= 0; Priority= 1; Cpu=GPM
Task name=get_vmc Period= 100; Capacity= 3; Deadline= 100; Start time= 0; Priority= 1; Cpu=GPM
Task name=heater Period= 100; Capacity= 3; Deadline= 100; Start time= 0; Priority= 1; Cpu=GPM
Task name=idle_task Period= 100; Capacity= 15; Deadline= 100; Start time= 0; Priority= 1; Cpu=GPM
0 5 10 15 20 25 30 35 40 45 50 55
Task name=motor_cmd Period= 50; Capacity= 3; Deadline= 50; Start time= 0; Priority= 1; Cpu=GPM
Task name=output_motor Period= 50; Capacity= 3; Deadline= 50; Start time= 0; Priority= 1; Cpu=GPM
Task name=output_solenoid_heater Period= 100; Capacity= 3; Deadline= 100; Start time= 0; Priority= 1; Cpu=GPM
Task name=record_data Period= 50; Capacity= 5; Deadline= 50; Start time= 0; Priority= 1; Cpu=GPM

Fig. 17 Gantt chart of Cheddar

5　Conclusion

Based on AADL, this paper proposes a non-functional attribute modeling and verification method for IMA system. We refined a set of design procedures suitable for IMA system, realized the automatic transformation of integrated modular avionics system functional model to architecture model, proposed a safety-oriented AADL-Hazard sub-language, and designed it based on AADL Safety analysis method and scheduling analysis of ARINC 653 system based on AADL.

For future work, there are several directions to be explored. (1) At the system level, form a scientific deployment method from task to platform; (2) At the OS level, form a special AADL model that satisfies the our OS product of our institute; (3) At the network level, a dedicated AADL avionics model that meets the backbone network of our actual products FC, AFDX, TTE, etc. (4) At the scheduling analysis level, the single-processor scheduling analysis is upgraded to cover the distributed network System scheduling analysis.

References

1. Baldonado, M., Chang, C.-C.K., Gravano, L., Paepcke, A.: The Stanford digital library metadata architecture. Int. J. Digit. Libr. **1**, 108–121 (1997)
2. Aeronautical Radio Incorporated: Avionics Application Software Standard Interface (2013)
3. Delange, J., Feiler, P., Gluch, D.P.: AADL Fault Modeling and Analysis Within an ARP4761 Safety Assessment. Software Engineering Institute Carnegie Mellon University (2014)
4. Chilenski, J.J., Kerstetter, M.S.: SAVI AFE 61S1 Report. Aerospace Vehicle Systems Institute (2015)
5. SAE International Group: SAE Architecture Analysis and Design Language (AADL) Annex Volume 2: Annex F: ARINC653 Annex (2011)
6. SAE International Group: SAE Architecture Analysis and Design Language (AADL) Annex Volume 3: Annex E: Error Model Annex (2013)
7. Singhoff, F., Legrand, J., Nana, L., Marcé, L.: Cheddar: A Flexible Real Time Scheduling Framework. University of Brest
8. OMG: Systems Modeling Language (2016). https://www.omg.org/spec/SysML/20161101
9. Feiler, P., Rugina, A.: Dependability Modeling with the Architecture Analysis & Design Language (AADL). Software Engineering Institute (2007, July)
10. Robati, T., El Kouhen, A., Gherbi, A., Hamadou, S., Mullins, J.: An Extension for AADL to Model Mixed-Criticality Avionic Systems Deployed on IMA Architectures with TTEthernet. ACVI at MoDELS (2014)
11. Lafaye, M., Gatti, M., Faura, D., Pautet, L.: Model driven early exploration of IMA execution platform. In: Digital Avionics Systems Conference (DASC), IEEE/AIAA 30th (2010)
12. SAE International: ARP-5580—Recommended Failure Modes and Effects Analysis (FMEA) Practices for Non-automobile Applications. SAE International (2001)
13. SAE International: ARP-4761—Guidelines and Methods for Conducting the Safety Assessment Process on Civil Airborne Systems and Equipment. SAE International (1996)
14. Yang, Z., Hu, K., Ma, D., Bodeveix, J.-P., Pi, L., Talpin, J.-P.: From AADL to timed abstract state machines: a verified model transformation. J. Syst. Softw. (2014)

Ontology for Systems Engineering Technical Processes

Chuangye Chang, Gang Xiao, and Yaqi Zhang

1 Introduction

Nowadays, the complexity of engineered system has increased dramatically and will continue to increase. To deal with the complexity, International Council on Systems Engineering (INCOSE) promotes Model Based Systems Engineering (MBSE) as the state-of-the-practice to support system specification, definition, analysis, verification and validation activities through the formalized application of system modeling and simulation, beginning in the conceptual design phase and continuing throughout development and later life cycle phases [1].

System model is the main artifact of MBSE for specifying, analyzing, designing, and verifying systems, and is fully integrated with other engineering models [2]. System model is composed of interconnected model elements that represent key system aspects, including requirements, behavior, structure and parametric. To satisfy system modeling requirements, Object Management Group (OMG) defined Systems Modeling Language (SysML), which is a profile of Unified Modeling Language (UML) [3]. SysML provides a good start to model a system, but there are still two gaps. First, the classification of concepts in SysML is ambiguous, which impedes stakeholders to differentiate concepts in specific domains. Second, there is lack of rigorous semantic definition, which is impossible to check the completeness, consistency, correctness of the system model. Therefore, although Systems Engineering processes have been defined, there are still insistent demands to specify which kinds of concepts and relationships need to be modeled with which kinds of metamodels from SysML.

C. Chang (✉)
Institute of Unmanned System, Beihang University, Beijing, China
e-mail: changchy@buaa.edu.cn

G. Xiao · Y. Zhang
National Key Laboratory for Complex Systems Simulation, Beijing, China

© The Author(s), under exclusive license to Springer Nature Switzerland AG 2021
D. Krob et al. (eds.), *Complex Systems Design & Management*,
https://doi.org/10.1007/978-3-030-73539-5_26

Works exist to define ontologies for Systems Engineering processes [4–6] which focus on the standardized information model to facilitate interoperability between projects and parties. However, those ontologies are based on the former edition of Systems Engineering processes with coarse granularity and there is no relationship between ontologies and SysML metamodels. Furthermore, a formal approach to combine domain ontologies and standard metamodels has been defined based on Category Theory to ease integration of a large audience in the modeling process through adapted viewpoints [7–9]. This paper utilizes this approach to define ontologies with fine granularity for Systems Engineering technical processes and combines these ontologies with SysML metamodels.

According to the practices from INCOSE, MBSE is heavily applied in the first four technical processes recursively and iteratively in the early lifecycle phases. Business or Mission Analysis Process and Stakeholder Need and Requirement Definition Process are used to frame problems, whilst System Requirement Definition Process and Architecture Definition Process are used to define solutions. Business Requirement Specification (BRS), Stakeholder Requirement Specification (StRS), System Requirement Specification (SyRS) as well as Component Requirement Specification (CRS) [10] could be generated from system models constructed in each process respectively. Besides, System Analysis Process is used to analyze the system models based on simulation technology. Figure 1 is the overall relationships between these processes and corresponding artifacts. It is worth to mention that these four processes are not necessary to be sequential. There are intensive iterative and incremental cycles during system development.

This paper extracts main concepts from the four processes in terms of requirements, structures, behaviors, parametric, etc. The semantic relationships between these concepts are defined rigorously and explicitly, which also indicate the modeling steps for each process. Different colors are used for the four processes, which are

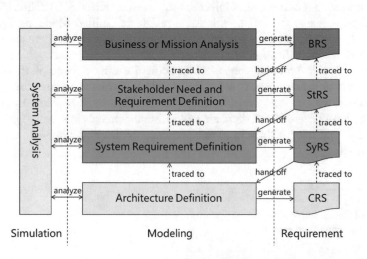

Fig. 1 Overall relationship between the main technical processes

consistent with the ontology definition in Sect. 2. For every concept and relationship, dedicated SysML constructs and diagrams are assigned to normalize artifacts modeling due to the stereotype mechanism of SysML. In each process, specific system models are constructed with different viewpoints and fidelity. In system models, constraints between system properties are also specified as a causal equations. These equations are handled through external solvers and facilitate the analysis of alternatives. Consequently, requirements are specified and verified through system models. Requirement specifications are generated from corresponding system models and handed off to the upcoming engineering process. In addition, traceability between system models as well as requirements are maintained to perform change and impact analysis.

This paper is organized as follows. Section 2 illustrates the ontologies for the four technical processes as well as mappings from ontologies to SysML metamodels. Section 3 discusses methods about functions identification and definition, transformation from logical architecture to physical architecture. Section 4 concludes this paper and outlines the future work.

2 Ontology Definition

According to the ontology definition from NASA JPL, there are three kinds of ontologies: foundation ontologies, discipline ontologies, and application ontologies. Foundation ontologies define fundamental terms used in all projects, disciplines, and applications, such as mission, project, stakeholders, requirements, functions, components, interfaces, analysis, etc. [11]. This section will focus on the foundation ontologies definition dedicated for the first four technical processes from INCOSE Systems Engineering Handbook Edition 4.0.

2.1 Ontology for Business or Mission Analysis Process

In this process, business or mission is defined, major stakeholders are identified, problem or opportunity is addressed, possible solutions are characterized, preliminary Measure of Effectiveness (MOE) and validation criteria are proposed. Finally, BRS is generated and analyzed as the base for the upcoming engineering efforts. Figure 2 depicts the ontologies for this process.

Mission is the starting point of the engineering processes. In military domain, a mission is an operational task with combat objectives, usually assigned by higher headquarters. Mission could be represented as *Block* in SysML *Block Definition Diagram (BDD)*.

The success of mission is measured by several MOEs. MOEs are a collection of operational measures that indicate how well the solution achieves the intended

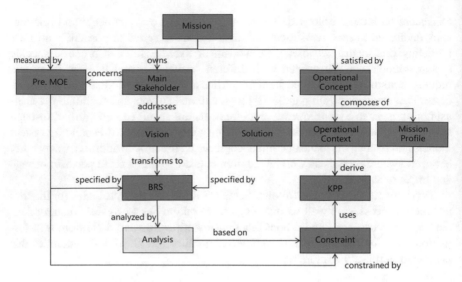

Fig. 2 Ontology for business or mission analysis process

purpose. MOE could be represented as *Value Property* of mission *Block* in SysML *BDD*.

Major stakeholders are identified at the business level, who may be people from marketing, operational, engineering department, representatives from users, regulatory agencies, and governments, etc. Stakeholder could be represented as *Block* with *«actor»* stereotype in SysML *BDD*.

The major stakeholders will address the target markets, sound Return of Investment (ROI), competence constraints, feasible solutions, acceptable risks, as well as the obliged regulations, which are documented as visions. Vision describes the future state of the enterprise, without regard to how it is to be achieved. Vision could be represented as *Block* in SysML *BDD*. Traceability between major stakeholders and visions should be maintained.

The mission is accomplished by operational concept at the business level, which includes operational context, solution, and mission profile. Operational context includes the external entities that will interact with the solution. Solution is taken as a black-box in this process. Mission profile describes the dynamic aspect of the operational concept, while operational context and solution describe the static aspects. Through the interaction between operational context and solution, main features of the solution are characterized. Operational context and solution could be represented as *Block* in SysML *BDD*. Mission profile could be represented as *Interaction* in SysML *Sequence Diagram (SD)*.

Based on the operational concept, Key Performance Parameter (KPP) is derived. KPPs play as the factors that will affect the accomplishment of MOEs. Several equations are formalized to express constraints between MOEs and KPPs. KPPs could be represented as *Value Property* of operational context and solution *Block* in

SysML *BDD*. Equations could be represented as *Constraint Block* in SysML *BDD*, and *Constraint Parameters* in *Constraint Block* should be bound with MOEs and KPPs in SysML *Parametric Diagram (PAR)*.

After the relevant efforts have completed, BRS is generated to specify all the requirements from the business aspect. Meanwhile, BRS is analyzed based on the constraints between MOEs and KPPs. Business requirement could be represented as *Requirement* in SysML *Requirement Diagram (REQ)*. It is worth to mention that MOEs may not appear in all business requirements, but most of them.

2.2 Ontology for Stakeholder Need and Requirement Definition Process

In this process, System of Interest (SoI) is defined as a black-box from the life-cycle points of view. MOEs and validation criteria for the mission are determined, all the stakeholders are identified and corresponding needs are elicited, lifecycle concepts are elaborated based on use case and scenario approach, service functions and external functional flows are identified and synthesized, operational modes are defined and integrated with service functions. Finally, StRS is generated and traced back to BRS. Figure 3 depicts the ontologies for this process.

MOEs and validation criteria are complemented and determined. The number of MOEs should be limited, since each MOE has high monitoring cost. KPPs are refined and allocated to operational context as Context Parameters (CP) and SoI as preliminary Measure of Performance (MOP). CP could be the properties of external entities. Then constraints are updated between MOE, MOP and CP. MOE, MOP, CP, as well as the constraints will inherit artifacts from the previous process.

The boundary of SoI becomes explicit and clear due to solution and operational context. Considering the lifecycle of SoI, five lifecycle concepts are recommended to identify all the stakeholders and elicit the relevant needs, which are acquisition concept, deployment concept, operational concept, support concept, and retirement concept [1]. SoI will inherit solution *Block* from the previous process.

Regarding each lifecycle concept, use cases are defined to specify functionalities that SoI provides to its users. Use cases should be independent, which helps different team work on different use case agilely and simultaneously. There are many roles that interact with SoI through use case. These roles are modeled as actors, which could be played by any stakeholders or entities from operational context. Use cases could be represented as *Use Case* in SysML *Use Case Diagram (UC)*, while actor could be represented as *Actor* in SysML *UC*.

Each use case could be refined into 3–50 scenarios [9], which describe the possible sequences of interactions between SoI and actors [12]. There are two kinds of scenarios, which are "sunny day" and "rainy day". Each scenario represents a fragment of interaction between SoI and actors which are related with the corresponding use case. Scenario approach is applied to explore the potential service functions,

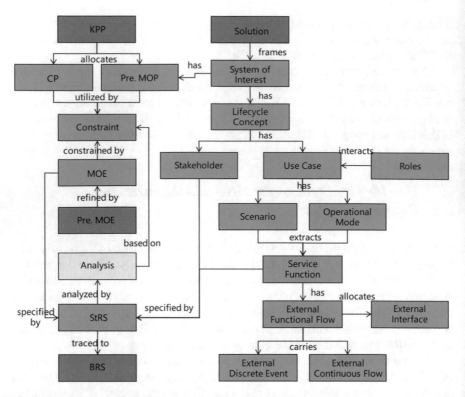

Fig. 3 Ontology for stakeholder need and requirement definition process

which are synthesized and organized according to use cases. Scenario could be represented as *Interaction* in SysML *SD*, while service functions could be represented as *Activity* in SysML *Activity Diagram (ACT)*.

Operational modes are defined during use case analysis and service functions are allocated to operational modes to specify the function validity over time. Operational modes are triggered by discrete events and transit from one mode to another. Operational mode could be represented as *State Machine* in SysML *State Machine Diagram (STM)*.

External functional flow acts as the input and output of service functions. External functional flow carries external discrete event or continuous flow, which could be material, energy, information, or construction etc. Based on external functional flow, external interface is identified. External functional flow could be represented as *Proxy Port* of SoI *Block* in SysML *BDD*. External interface could be represented as *Interface Block* in SysML *BDD*, which is utilized to type *Proxy Port*. External discrete event and continuous flow could be represented as *Reception* and *Flow Property* of *Interface Block* in SysML *BDD*.

After the relevant efforts have completed, StRS is generated to specify all the stakeholder requirements from the operational aspect. Meanwhile, StRS is analyzed based on the constraints between MOE, MOP and CP. Stakeholder requirement could be represented as *Requirement* in SysML *REQ*.

2.3 Ontology for System Requirement Definition Process

In this process, service functions are transformed into technical functions based on the domain knowledge without considering the implementation technology. Technical functions should be refined recursively down to the leaf technical functions which could be allocated to a component. Usually, these technical functions could be abstracted to several layers, such as system, subsystem, and component. When all the technical functions have been defined, a functional architecture needs to be constructed, which is an arrangement of functions and their subfunctions and interfaces (external and internal) that defines the execution sequencing, conditions for control or data flow, and the performance requirements to satisfy the requirements baseline [13]. Meanwhile, system states should be defined based on the operational modes, and integrated with technical functions. Finally, SyRS is generated and traced back to StRS. There are 13 kinds of requirements in SyRS, however it is mainly composed of technical functions with MOP, as well as constraints from SoI [14]. Figure 4 depicts the ontologies for this process.

In the previous process, service functions have been identified and organized according to the relevant use cases. In this process, these service functions should be orchestrated, which become the use case specification respectively. Since use cases

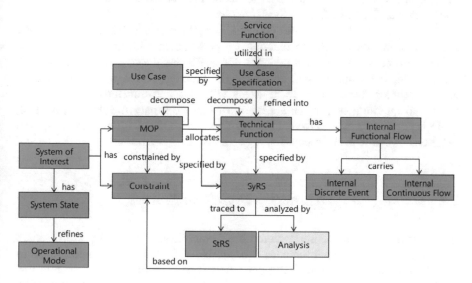

Fig. 4 Ontology for system requirement definition process

are independently, use case specifications could be engineered simultaneously. Use case specification could be represented as *Activity* in SysML *ACT*.

Based on use case specification, service functions with coarse granularity are decomposed into technical functions, and technical functions could also be refined recursively down to a level where either function specification is clear enough for allocation or realization without further decomposition, or there are identified products that realize it [9]. It is possible that there are several alternatives for one function decomposition and a trade-off study is needed to determine a preferred one. Technical functions could be represented as *Activity* in SysML *ACT*.

Along with decomposition of technical functions, MOPs are also decomposed and flowed down to lower level functions. Equations are formalized to express constraints between MOPs at different layers. MOPs could be represented as *Value Property* of SoI *Block* in SysML *BDD*. Equations could be represented as *Constraint Block* in SysML *BDD*, and *Constraint Parameters* in *Constraint Block* should be bound with MOPs in SysML *PAR*.

During the decomposition, interactions between technical functions are defined as internal functional flow. Internal functional flow carries internal discrete event or continuous flow, which represent input and output of technical functions and is independent of platform. Internal functional flow could be represented as *Control Flow* or *Object Flow* in SysML *ACT*. Internal discrete event could be represented as *Send Signal Action* or *Accept Event Action* in SysML *ACT*. Internal continuous flow could be represented as *Object Node* in SysML *ACT*.

Operational mode is refined and complemented by system state. Technical function integrates with system state and transition to define the function validity over time. This is a very important aspect for technical function. System state could be represented as *State Machine* in SysML *STM*.

After the relevant efforts have completed, SyRS is generated to specify all the system requirements from the functional aspect. The establishment of functional architecture provides a solid foundation for SyRS definition. Based on functional architecture, the input, output, control logic, hierarchy level, state, performance etc. of technical function could be easily extracted and transformed into system functional requirements, which are the most important part in SyRS. Meanwhile, SyRS is analyzed based on the constraints between MOPs at different level. System requirement could be represented as *Requirement* in SysML *REQ*.

2.4   Ontology for Architecture Definition Process

In this process, functional architecture from the previous process is transformed into system architecture, which includes two phases, logical architecture composed of logical components and physical architecture composed of physical components. First, functions are grouped and allocated to logical components layer by layer with a top-down pattern. Therefore, it is highly recommended that functional architecture definition and logical architecture definition are conducted concurrently. Second, when logical architecture reaches the lowest level, which means that logical component has reached physical domain and is ready to start discipline engineering, or a Commercial Off the Shelf (COTS) exists to satisfy this logical component, it is time to start the definition of physical architecture with a bottom-up pattern. During physical architecture definition, implementation technologies, such as mechanical, electrical, software, thermal, etc., are selected and relevant features as well as preliminary Technical Performance Measure (TPM) are defined. Meanwhile, configuration item is identified from physical architecture. Finally, CRS is generated from physical architecture and distributed to discipline engineers or corresponding suppliers. CRS is traced back to SyRS to make sure all the requirements are satisfied, both functional and non-functional. Figure 5 depicts the ontologies for this process.

System architecture is the fundamental concepts or properties of a system in its environment embodied in its elements, relationships, and in the principles of its design and evolution [15]. System components and their interconnections are the basic parts that compose of system architecture. System component undertakes one or multiple technical functions, and is implemented by proper technology with TPMs. Therefore, system component has two phases, which are logical and physical, reflecting the maturity of engineering efforts. System components could be represented as *Block* in SysML *BDD*. TPM could be represented as *Value Property* of component *Block* in SysML *BDD*.

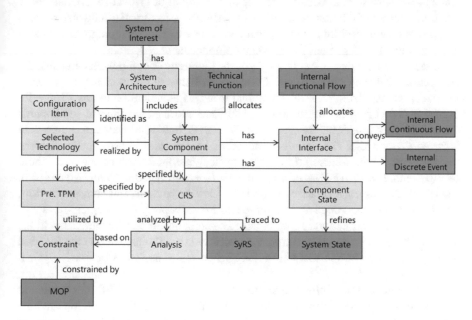

Fig. 5 Ontology for architecture definition process

Logical component is the abstraction of physical components that perform the system functionality without imposing implementation constraints [16]. The selection of logical component is based on qualitative or quantitative criteria such as modularity or reuse of existing components [10]. Logical architecture describes solution in terms of logical components, relationships, states, performance etc. [17]. Comparing with the direct allocation from functional architecture to physical architecture, there are three benefits to define logical architecture. First, logical architecture could serve as an intermediate level of abstraction between functional architecture and physical architecture to mitigate the impact of requirements changes and technology evolutions. Second, it can serve as a reference architecture for a family of products that support different physical implementations in terms of production line engineering. Third, it could help separate of concerns and divide complex system with good modularity to collaborate efforts from different expert teams. *Allocation* in SysML could be used to represent the traceability from technical function to logical component.

Physical component is the substantiation of logical component that is implemented through technology. Usually, logical component could be mapped to several physical components, and vice versa. Physical architecture defines a specific design implementation corresponding to a particular logical architecture. Physical architecture should answer all requirements, not only functional, but also non-functional. Considering non-functional requirements, here are a few examples, redundancy of some block with several occurrences, spatial isolation of critical functional flows, additional sensors for better monitoring, etc. *Allocation* in SysML could also be used to represent the traceability from logical component to physical component, as well as from software component to hardware component.

System architecture not only represents the hierarchy of components, but also the connections between components. Connections between components are internal interfaces derived from internal functional flows. If technical functions F1 and F2 exchange object flow or control flow and they are allocated to different components C1 and C2, then functional flow between F1 and F2 shall be allocated to an interface between C1 and C2. The definition of internal interface will also go through the transformation from logical interface to physical interface. Internal interface could be represented as *Proxy Port* of component *Block*, and typed by *Interface Block* in SysML *BDD*. *Interface Block* conveys internal discrete event and continuous flow, which could be represented as *Reception* and *Flow Property* of *Interface Block*.

System state is refined and complemented by component state. Technical functions at lower levels integrate with component states and transitions to define the function validity over time. Component state could be represented as *State Machine* in SysML *STM*.

After the definition of physical architecture have completed, CRS is generated to specify component requirements from the logical and physical aspects. Meanwhile, CRS is analyzed based on the constraints between MOP and TPM. Component requirement could be represented as *Requirement* in SysML *REQ*. Finally, CRS becomes the contract to conduct detailed design and implementation of real physical components.

3 Discussion

During ontologies definition and MBSE practice for the first four technical processes, there are two interesting topics worth to discuss.

The first topic is about function identification and definition. In this paper, two kinds of functions are defined separately, which are the user-oriented service function and engineering-oriented technical function. Figure 6 shows the method how these functions are identified and defined. Regarding the service function, there are two approaches, which are usage-driven approach and feature-driven approach [17]. Usage-driven approach ensures functional requirements are traced directly to the user's operational requirements, which in turn ensures the design is influenced foremost by the end-user's needs. While in the feature-driven approach, desired features, functions, or capabilities for a system are listed by domain experts and engineers in consultation with the end-users. In this paper, usage-driven approach was selected to identify service functions due to the scenario description. During the practice, SysML *SD* is recognized as a good candidate to describe different scenarios, which is easy to communicate with stakeholders. It is worth to mention that the same service function could appear in different scenarios, and a consolidation is needed to identify the minimum set of service functions. Finally, SysML *ACT* is modeled to elaborate the use case specification composed of service functions.

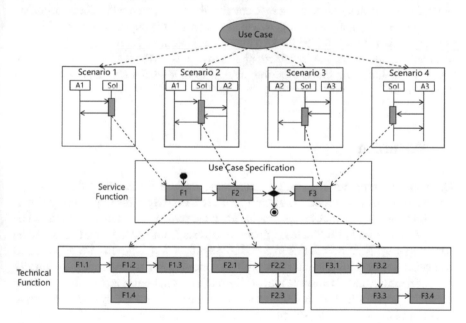

Fig. 6 Function identification and definition

The second topic is about transformation from logical architecture to physical architecture. In this paper, three architectures are defined, which are functional architecture, logical architecture, and physical architecture. Regarding the three architecture, there are numerous definitions from different points of view. Since aviation product is safety-critical, ARP-4754A is selected as the reference standard, where functional architecture is separated from the system architecture definition [18]. First, functional architecture is transformed to logical architecture through grouping and allocating functions until physical domain is reached or COTS exists. Then transformation from logical architecture to physical architecture starts. Nowadays, most of the products have the nature of Cyber Physical System (CPS), which are composed of two main parts, controller with strong communication and computation capability, and controlled object with multi-physical sensing and actuating capability. According to the classification of models from INCOSE, there are two kinds of models, which are descriptive model and analysis model. Many Domain Specific Modeling Languages (DSML) exist to deal with the modeling efforts. In this paper, SysML is selected as the language for descriptive model. Regarding the analysis model, Architecture Analysis and Design Language (AADL) from Society of Automotive Engineering (SAE) is selected as the language for embedded real time system modeling and simulation, while Modelica from Modelica Association is selected as the language to perform modeling and simulation of multi-physics system. To facilitate the co-simulation among different engineering disciplines, Functional Mockup Interface (FMI) from Modelica Association is selected as the interface standard for information exchange. Finally, analysis results will feed and enrich the physical architecture, and physical architecture is traced back to logical architecture. There are many research work and standards for the transformation between SysML, AADL, and Modelica. Figure 7 illustrates the transformation method from logical architecture to physical architecture.

4   Conclusion

This paper presents the motivation, scope, and method for ontology definition for the first four technical processes aligning with INCOSE Systems Engineering Handbook Edition 4.0. The ontologies provide accurate semantic to elaborate on the technical processes and coordinate engineering efforts from different parties. In addition, mappings from ontologies to SysML constructs and diagrams are proposed to normalize the modeling methods as well as formal artifacts. Particularly, a further step has been advanced to enable the transformation from logical architecture to physical architecture, which involves information exchange between system descriptive models and discipline analysis models. According to the practice of a UAV system, system models based on the ontologies and SysML have adequate capabilities to represent requirements specification from business, operational, functional to logical and physical layers, which enable itself the single source of truth.

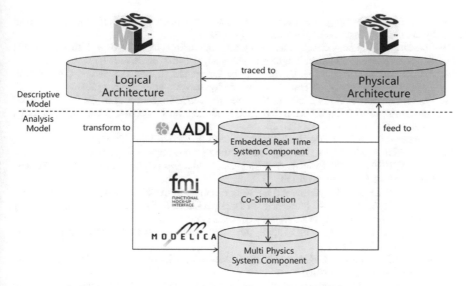

Fig. 7 Transformation from logical architecture to physical architecture

In this paper, ontologies are still expressed in text. A formal representation of ontologies will be established in the future. Web Ontology Language (OWL) from World Wide Web Consortium (W3C) is a good candidate to represent rich and complex concepts about things, groups of things, and relations between things [19]. SysML is a graphical modeling language for representing systems engineering concepts, while OWL is a formal language for expressing ontologies using a logical formalism. Reasoning could be conducted for SysML models based on OWL models. MBSE methodology leverages both OWL and SysML [20, 21].

References

1. International Council on Systems Engineering (INCOSE): Systems Engineering Handbook V4.0 (2015)
2. International Council on Systems Engineering (INCOSE): Systems Engineering Vision: 2025 (2014)
3. Object Management Group (OMG): Systems Modeling Language (OMG SysML™) V1.4/ Available at: https://www.omg.org/spec/SysML/
4. van Ruijven, L.C.: Ontology and model based systems engineering. In: Conference on Systems Engineering Research (CSER), St. Louis, MO, USA (2012)
5. van Ruijven, L.C.: Ontology for systems engineering. In: Conference on Systems Engineering Research (CSER), Atlanta, GA, USA (2013)
6. van Ruijven, L.C.: Ontology for systems engineering as a base for MBSE. In: INCOSE International Symposium (2015)
7. Ernadote, D.: An ontology mindset for system engineering. In: IEEE 1st International Symposium on Systems Engineering, pp. 454–460 (2015)

8. Ernadote, D.: Ontology reconciliation for system engineering. In: IEEE 2nd International Symposium on Systems Engineering, pp. 1–8 (2016)
9. Ernadote, D.: Ontology-based pattern for system engineering. In: ACM/IEEE 20th International Conference on Model Driven Engineering Languages and Systems (2017)
10. Andrianarison, E., Piques, J.-D.: SysML for Embedded Automotive Systems: A Practical Approach, Embedded Real Time Software and System. Toulouse, France (2010)
11. Bayer, T., Dvorak, D., Friedenthal, S., Jenkins, S., Lin, C., Mandutianu, S.: Model-Centric Engineering, Part 2: Introduction to System Modeling. Jet Propulsion Laboratory, California Institute of Technology, USA (2012)
12. Cockburn, A.: Writing Effective Use Cases. Addison-Wesley, USA (2000)
13. IEEE 1220: Application and Management of the Systems Engineering Process (2005)
14. ISO/IEC/IEEE 29148: Systems and Software Engineering-System Life Cycle Processes-Requirements Engineering (2011)
15. ISO/IEC/IEEE 42010: Systems and Software Engineering-Architecture Description (2011)
16. Friedenthal, S., Moore, A., Steiner, R.: A Practical Guide to SysML: The Systems Modeling Language, 3rd edn. Morgan Kaufmann, USA (2015)
17. Pearce, P., Hause, M.: ISO-15288, OOSEM and Model-Based Submarine Design. SETE/APCOSE (2012)
18. Society of Automotive Engineers (SAE): Guidelines for Development of Civil Aircraft and Systems, ARP-4754A (2010)
19. World Wide Web Consortium (W3C): Web Ontology Language (OWL). Available at: https://www.w3.org/2004/OWL/
20. Douglass, B.P.: Agile Systems Engineering. Morgan Kaufmann, USA (2016)
21. Tang, J., Zhu, S., Faudou, R., Gauthier, J.-M.: An MBSE framework to support agile functional definition of an avionics system. In: Proceedings of the 9th International Conference on Complex Systems Design & Management (2018)

Proposal of Assurance Case Description Method in Design for Environment (DfE) Process

Kenichi Shibuya, Nobuyuki Kobayashi, and Seiko Shirasaka

1 Introduction

In project management, Gantt charts and flow charts are used to manage projects, however it is pointed out that these methods do not contain accurate information for decision making [1]. Individual designers involved in product development may be aware of the design process for which they are responsible, however may not be fully aware of the design process outside their responsibilities. As a result, an improper design process causes a major design rework, and as a result of accumulating component designs, it is difficult to meet product requirements simply by matching existing technologies [2]. The reason is that 80% of all life-cycle costs of the product is decided at the initial design stage [3]. Therefore, in order to meet product requirements, it is necessary for the project manager to have each individual designer involved in product development understand the design process outside the scope of their responsibility at the initial design stage. As an approach to Sustainable Development Goals (SDGs), it is necessary to implement the design for environment (DfE) based on the guidelines [4, 5], therefore, we investigated DfE process (DfEP) in this study.

K. Shibuya (✉)
Graduate School of Science and Technology, Keio University, 3-14-1, Hiyoshi, Kohoku, Yokohama 223-8522, Kanagawa, Japan
e-mail: kenichi.shibuya@keio.jp

N. Kobayashi
The System Design and Management Research Institute of Graduate School of System Design and Management, Keio University, 4-1-1 Hiyoshi, Kohoku, Yokohama 223-8526, Kanagawa, Japan
e-mail: n-kobayashi5@a6.keio.jp

S. Shirasaka
Graduate School of System Design and Management, Keio University, 4-1-1 Hiyoshi, Kohoku, Yokohama 223-8526, Kanagawa, Japan
e-mail: shirasaka@sdm.keio.ac.jp

The purpose of this study is to have each individual designer involved in product development to understand the design process outside the scope of their responsibility in the DfEP from the requirement definition to the prototyping at the phase of feasibility study stage based on V model approach [6] of system engineering. The proposal of this study is the assurance case (AC) description method (DM) which is in order to visualize the DfEP. As for the evaluation method in this study, a questionnaire survey is conducted before and after seeing an AC whether the designer be able to understand the processes outside the scope of his/her responsibility.

Next, we describe the novelty of this study. From the viewpoint of project management, Levardy and Browning [7] proposed a modeling framework for process models that support product development. Browning [8] pointed out that the design development process is complicated by the independent design activities and the design information flow with many starting points, and proposed to analyze the complicated process using DSM (Design Structure Matrix). Browning et al. [9] showed that it is possible to support restructuring from the viewpoint of "As Is" and "To Be" by using the model of the product design process and the development organization for which DSM is applied. Nakazawa and Masuda [10] pointed out that the main part of the product development process is an indigenous person, and an RDC model was proposed as a methodology for explicitly describing the implicit design process. However, these studies do not refer to the dependability and operation guarantee of the methodology. Focus on the methodology, AC [11] DM was proposed for solving communication challenges in business [12]. Kobayashi et al. [13, 14] also showed that using ACs increases the feasibility of accomplishing management vision and management strategy. However, these previous studies do not mention the communication and consensus in DfEP. Scoping to DfEP and consensus building, Ameknassi et al. [15] showed that multi stage methodology for defining, modeling, and solving DfE problems. This integrated approach includes LCA, checklist and QFD in order to support designers in implementing DfE activities effectively. Ramanujan et al. [16] indicated the visual analytics system which is called VESPER, for generating contextual DfE principles in sustainable manufacturing. The tasks involved can be divided into following steps (i) data gathering and preprocessing, (ii) interactive visual exploration, and (iii) DfE principle(s) generation. In order to carry out design cooperatively, Takechi et al. [17] proposed a method for linking process information to resolve inconsistencies and a method for changing process information to resolve conflicts, based on the modeling of the product and the design process. Ignacio et al. [18] proposed the consensus model for heterogeneous group decision making problems guided also by the heterogeneity criterion. However, these studies do not refer these studies do not satisfy both of consensus and accountability. Therefore, the novelty of this study is to propose the AC DM which is applied for the DfEP, in order to visualize the DfEP and also assure the dependability and accountability. Furthermore, the feature of AC is defined as "a documented body of evidence that provides a convincing and valid argument that a system is adequately dependable for a given application in a given environment". AC is required to submit to certification bodies for developing and operating safety critical systems, e. g., automotive, railway,

Table 1 Six nodes in ACs

Node	Figure	Explanation
Goal		Goal describes what to assure, with a combination of a subject and predicate
Strategy		Strategy describes how to break down the goal into sub-goals leading to the lower layer
Context		Context describes the state, or environment and conditions of the system, and shows ways to lead to the goal and strategy
Evidence		Evidence eventually assures that we can reach the goal, and shows ways to lead to the goal
Monitoring		Monitoring is intended to represent evidence available at runtime, corresponding to the target values of in-operation ranges
Undeveloped		Undeveloped shows the status that there is no evidence or monitoring, or discussion supporting the goal

defense, nuclear plants and oil production platforms [19]. This research attempts to apply AC to the environment-friendly design of automobile development.

This paper is organized as follows. We explain the previous studies of this proposal in Sect. 2. We show the proposed AC DM for DfE in Sect. 3. Then, we show evaluation method, results of evaluation, and discussion in Sect. 4. Finally, we conclude this paper and state the future research topics in Sect. 5.

2   Previous Studies

The AC [20] extends the scope of discussion to the overall quality with an acceptable level of quality among stakeholders, including the "safety" that was targeted in the Safety Case [21]. In this study, D-Case (Dependability-Case) [22], which is an extension of the DM called GSN (Goal Structuring Notation) proposed by Tim Kelly [23], is used as a DM. D-Case is used instead of GSN in this study because it assumes the operation stage and adopts a node called "Monitor", which is prepared only for D-Case. This study uses six nodes in Table 1.

3   Proposed Assurance Case Description Method for DfE

In this chapter, we propose the method for describing the AC which is applies the argument decomposition pattern [24, 25] for DfE. On the assumption that it will be confirmed by all members involved in DfE, D-Case of the DfEP is described based on the V model [6] in system engineering.

In the requirement definition, based on IEEE1220 [26], the process was shortened in consideration of time constraints [25]. Including the product design including UI/UX, the customer's requirements and related laws and regulations are clarified, product planning is performed, and it is defined as a requirement specification by combining with the requirement quality development table by QFD [27].

In concept design, it is divided into the selection of environmentally friendly parts and the creation of the concept design document incorporating the parts. When selecting environmentally friendly parts, HAZOP analysis was used to extract hazards from the user's perspective (external analysis), and FTA was used to extract hazard factors from the developer's perspective (internal analysis). FMEA is applied to the function of parts, weighting is applied by the degree of risks, and then, environmentally friendly parts and application destinations are selected from the viewpoint of 3R (Reduce, Reuse, Recycle). Finally, the concept design specifications is comprehensively created.

In the detailed design, the structural design of the automobile interior parts is performed based on the concept design specifications which is created in the concept design. It is required that physical property values are obtained from parts manufacturers and also refer Ashby maps [28]. Next, in layout design, functional parts and structural members are laid out in view of interference of parts and design tolerances according to production equipment and quality standards, and finally 3D CAD is presented. At this stage, the feasibility is confirmed to be consistent with the required specifications and concept design documents.

Finally, we describe the DM of AC by D-Case [29] for system dependability. In addition, the present study provides a limitation to the DM [30] for solving the known problem regarding the DM of the AC. First, we set the business system that you want to achieve the top goal, and then, we divide it into "Attractive quality system" and "Must be quality system" in the strategy node. In this study, the requirement definition and the concept design are defined as "Attractive quality system", and the detailed design is defined as "Must be quality system" [31, 32]. In "Attractive quality system", the work for the purpose of value co-creation is the center, and it is described by the evidence node. On the other hand, in "Must be quality system", the monitoring node is applied to the goal node. The reason for this is to confirm "Attractive quality system" as the implementation of the customer satisfaction value. Therefore, the evidence node is not used for the goal node in "Must be quality system". As shown in Fig. 1, 3D CAD data refers to the requirement specifications and concept specifications in "Attractive quality system". The monitoring node in "Must be quality system" is connected to the evidence node in "Attractive quality system" (Fig. 2, dotted arrow). The reason is to judge whether to achieve the top goal based on the monitoring results. However, this DM has the limitation described later. First, design tolerances are not constant because they vary depending on production equipment and quality standards required. Second, the subject being monitored is not always constant. Third, the monitored results are not always necessary for evidence in "Attractive quality system". Proposed AC in this study has these limitations. Additionally, proposed

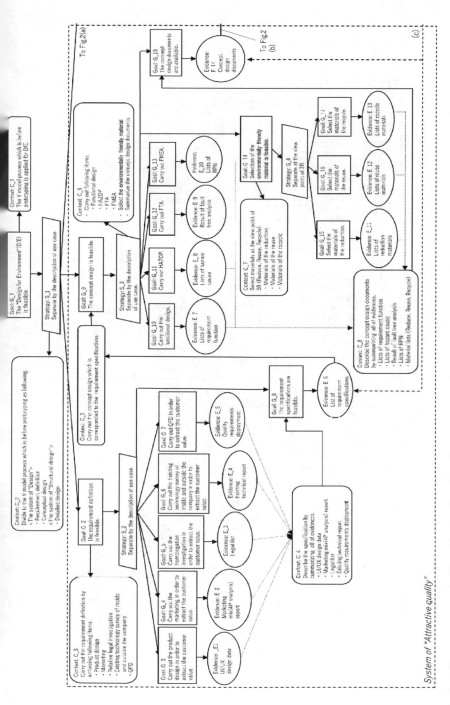

Fig. 1 DfE assurance case in "Attractive quality system"

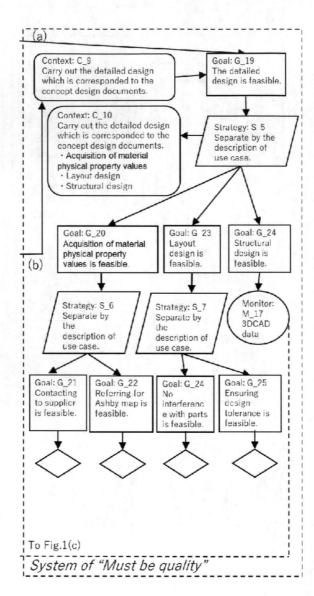

Fig. 2 DfE assurance case in "Must be quality system"

Table 2 The proposed method of AC description

Step 1	Set goal node
Step 2	Set context node as sub-goals by dividing the goal into sub-goals. In addition, set priorities, if any, in the context node when the priority of sub-goals is important
Step 3	For strategy node, divide the goal into sub-goals (in prioritized order, if any)
Step 4	Set the sub-goals (in prioritized order if any) underneath the Strategy node. In addition, set evidence nodes if the sub-goals need to be prioritized
Step 5	Assume Step 4 to be Step 1, and repeat this process until sub-goal nodes are completely deconstructed
Step 6	Set evidence node or monitoring node until you cannot divide
Step 7	Link multiple evidence nodes to context nodes, and context nodes are linked to goal nodes, in order to express the procedure to be executed on the premise of the output of multiple procedures
Step 8	Summarize evidence node with context node of the next procedure, in order to synthesize the result of multiple procedures
Step 9	Confirm the relationship between the monitoring node and the evidence node
Step 10	Connect the dotted arrow from the monitoring node to the evidence node according to the result of Step 9

AC in this study is guaranteed by the agreement of the project member participants. Next, the procedure applied to the DM of the DfEP is described. Proposed method of AC description is showed in Table 2.

As a result, it is possible to measure the design achievement of a product by using the 3D CAD data which monitored in "Must be quality system" for the required specifications including the design created in "Attractive quality system".

4 Evaluation

4.1 Evaluation Method

We conducted a survey for DfE project members in order to verify the effect of applying the proposed AC at the initial design and development stage. At that time, the target persons were designers, structural design engineer, product planner, and quality control engineer. This study confirms three questionnaires in Table 3. Evaluation methods are independent sample t-test and open cording for each questionnaire, before and after applying proposed AC. The reason is to confirm whether each individual member in charge is aware of the design process outside the scope of his/her responsibility. Responses were given on a five-point ordinal scale, ranging from -2—"disagree," to $+2$—"agree," with 0 representing "neither agree nor disagree." Scores from $+1$ to $+2$ were assumed to be valid for business improvement. With regard to the qualitative data analysis, free descriptive answers were used as the data,

Table 3 The result of group statistics and independent sample t-test

No.	Questionnaires	Applying AC	Mean value	Standard deviation	Significance (both sides)
Q1	Do you understand the DfEP?	Before	1.9333	0.96115	0.000
		After	4.2000	0.56061	
Q2	Do you understand the relevant divisions in the DfEP?	Before	2.1333	1.06010	0.000
		After	3.9333	0.59362	
Q3	Do you understand the relationship between the DfEP and related divisions?	Before	2.0000	0.84515	0.000
		After	3.9333	0.59362	

Table 4 Open coding procedure

Step 1	The free description of the questionnaire survey and the verbatim comments are extracted, and the viewpoint of categorizing the Affinity Diagram is determined by using the KJ method [33]
Step 2	Based on the viewpoint set in Step 1, sort the comments of the respondents according to the Affinity Diagram
Step 3	As a result of open coding, summarize the main points of the group and describe the title of each group
Step 4	Count the number of descriptions related to open coding results

and analyzed by the following procedure in Table 4, using qualitative coding methods [34]. We set that the viewpoint in step 1 is to relate to the design process outside the scope of their responsibility, in order to confirm the purpose of this study which described in Chap. 1.

4.2 Result of Evaluation

Total of 15 respondents who were involved in the project were evaluated by conducting a questionnaire survey before and after applying the AC. Table 5 shows the respondents' years of work experience in the project. The reason for describing the number of years of experience is that it is possible to judge whether or not the proposal in this study is effective and that the degree of understanding of the proposed method depends on the work experiences. In this study, the target was the members who actually participated in the DfE project and became the person in charge. Table 5 shows the results of group statistics and independent sample t-test.

Table 5 The work experience years

The work experience years	Number of respondents
1–5	5
6–10	3
11–20	4
21–	3
Total	15

Table 6 The results of open cording

No.	Open cording results	Counts
1	The overall of the DfEP can be seen visually	27
2	Personal work is based on a well-defined workflow so that the person in charge can understand and move	17
3	Could discuss the business processes required for the DfEP	16
4	Visualized the relationship with other organizations in the DfEP	14
5	Proposed AC could be used as a common language	8
6	DfEP is divided for each design phase, so that it is able to aware of the schedule of tasks in the DfEP	6

As shown in Table 3, before and after applying the AC, the mean value increased. The significance probability was 0.000, which was confirmed to be significant. In other words, by applying the AC, it is possible to understand the DfEP and to understand the related divisions and the relationship between the DfEP and related divisions. Table 6 shows the results of open coding. By applying proposed AC from the open coding results, the purpose of this study described in Chap. 1, "Each individual designer involved in product development grasps the design process outside the scope of their responsibility" is achieved. In order to ensure the reliability of the open coding results when confirming the analysis results in Table 6 with the respondents, all agreed with the analysis results [35]. In addition, the reliability of the analysis results was confirmed by confirming it with one expert researcher who had verification experience using the qualitative survey method in the study of applying the AC [36]. Furthermore, we confirmed whether respondents could perform a repeat and point after applying the proposed AC for respondents [14]. As a result, all 15 respondents were able to repeat and point respectively. From this result, it became clear that it was possible to recognize the DfEP defined in proposed AC.

4.3 Discussion

From the results of independent sample t-test in Table 3, the proposed AC DM is "each individual designer can understand the DfEP", "each individual designer can

understand the related department", "Individual designers can understand the relationship between the DfEP and related departments." The results will be considered
below. Kobayashi et al. [12] indicated that the person in charge recognizes the task
importance proposed by Hackman et al. [37] by applying the AC. In fact, in this
study, from the open coding result of No. 1 in Table 6, the respondents answered,
"The overall of the DfEP can be seen visually." In order to realize DfE, it is possible
to utilize the AC as a tool for grasping the whole process and share the position of
one's own work, which is considered to contribute to the recognition of task importance. In addition, the open coding result of No. 2 in Table 6 indicates that the tacit
knowledge which is the personalized experience in the DfE is conversed to formal
knowledge owing to proposed AC. Specifically, The open coding result of No. 3
in Table 6 shows that the persons in charge are able to reconcile the recognition of
tasks in the DfE. In other words, we consider that it contributes to the recognition
of task importance. Moreover, the open coding result of No. 4 in Table 6 indicates
that in DfE, each individual designer is able to recognize design processes that are
out of his or her scope. The open coding result of No. 5 in Table 6 indicates that
in communication, they answered that they can be used as a common language for
carrying out the DfE, and as mentioned by Kobayashi et al. [14], they could share
their consciousness by referring to the AC. Finally, the open coding result of No.
6 in Table 6 indicates that the person in charge is aware of each design phase, and
thus can have the consciousness of the time axis in the development process. That is
because each individual designer understands the phase of overall development and
the position of each person's work. It contributes the consciousness for the schedule
driver in order to success the project management [38] and also the task importance. The result which project members understand the position of their work in the
design phase indicates that they are aware of the development time schedule which
the project manager controls.

From the above, it is important for each individual person to understand the DfEP
for the purpose of grasping the design process outside the scope of responsibility
and to share the purpose and the awareness of the time schedule of the design phase
among the stakeholders. Visualization of the DfEP using proposed AC is effective for
the realization of DfE. The reason is that proposed AC is used for the consciousness
of the members for the task importance and the time schedule.

5 Conclusion

This study proposed an AC DM on DfEP, for being able to visualize the DfEP
procedure that assuring a task. Based on the independent t-test result, the proposed
DM was judged effective compared with current practice. Furthermore, from the open
coding results, the results suggested why the proposed DM is effective. This study
can be seen from the difference of each average value shown in Table 3, there is a
possibility that the effectiveness may be further improved. Improving the expression
method therefore is possible in the future. Future research topic includes increasing

the number of applications. This study is related with AC notation for quality of DfEP. This study is related with AC notation for quality of DfEP. In the future research, it is necessary to develop a method to assure the actual work by linking it with the management strategy and management vision, which are the superordinate concepts of environmentally friendly design. The reason why future research needs to link it with the management strategy and management vision, which are the superordinate concepts of environmentally friendly design is that project members perform the DfEP in order to accomplish the superordinate purpose related the management. Additionally, future research needs a way to assure that person in charge concretely performs development processes with high quality. The reason why needs a way to assure that person in charge concretely performs development processes with high quality is to improve the feasibility of accomplishing the superordinate concepts (e.g. management vision, management strategy). The previous study states that providing managers with a method to translate management vision and strategy into action is a key parameter that will lead to a successful project [39]. In addition, the previous study shows that improving the feasibility of accomplishing management vision and management strategy is desirable for organizations [13, 40–43].

References

1. Browning, T.R.: On the alignment of the purposes and views of process models in project management. J. Oper. Manag. **28**(4), 316–332 (2010)
2. Koga, T., et al.: Design process guide method for minimizing design loops and design conflicts. J. Adv. Mech. Des. Syst. Manuf. **3**(3), 191–202 (2009)
3. Fabrycky, W.J., Blanchard, B.S.: Life-Cycle Cost and Economic Analysis. Prentice Hall International Series in Industrial and Systems Engineering (1991)
4. United Nations: The Sustainable Development Goals Report 2019 (2019)
5. METI, Japan: The Guide for SDG Business Management (2019, May)
6. Harold, M., Forsberg, K.: Visualizing System Engineering and Project Management as an Integrated Process. Wiley, Hoboken (2014) provided with permission
7. Levardy, V., Browning, T.R.: An Adaptive Process Model to Support Product Development Project Management, forthcoming (2009)
8. Browning, T.R.: Process integration using the design structure matrix. Syst. Eng. **5**(3), 180–193 (2002)
9. Browning, T.R., Fricke, E., Negele, H.: Key Concepts in modeling product development processes. Syst. Eng. **9**(2), 104–128 (2006)
10. Nakazawa, T., Masuda, H.: Requirement-definition-confirmation modeling approach for identifying uncertainties in product design processes. In: Design Engineering Technical Conference, Philadelphia, PA (CD-ROM) (2006)
11. ISO15026-2-2011: Systems and Software engineering Part2: Assurance Case
12. Kobayashi, N., Kawase, M., Shirasaka, S.: A proposal for an assurance case description method—aiming to tackle challenges in work-related communication. J. Jpn. Assoc. Manag. Syst. **33**, 91–107 (2016)
13. Kobayashi, N., Nakamoto, A., Kawase, M., Sussan, F., Shirasaka, S.: What model(s) of assurance cases will increase the feasibility of accomplishing both vision and strategy? Rev. Integr. Bus. Econ. Res. **7**(2), 1–17 (2018)

14. Kobayashi, N., Nakamoto, A., Kawase, M., Sussan, F., Ioki, M., Shirasaka, S.: Four-layered assurance case description method using D-case. Int. J. Jpn. Assoc. Manag. Syst. **10**(1), 87–93 (2018)
15. Ameknassi, L., Ait-Kadi, D., Keivanpour, S.: Incorporating design for environment into product development process: an integrated approach. IFAC **49**(12), 1460–1465 (2016)
16. Ramanujan, D., Bernstein, W.Z., Totorikaguena, M.A., Ilvig, C.F., Ørskov, K.B.: Generating contextual design for environment principles in sustainable manufacturing using visual analytics. J. Manuf. Sci. Eng. **141**(2), 021016 (2019)
17. Takechi, S., Sawda, K., Aoyama, K., Nomoto, T.: A study on design process information for collaborative engineering. J. Soc. Naval Architects Jpn. **190**, 377–386 (2000)
18. Pérez, I.J., Cabrerizo, F.J., Alonso, S., Herrera-Viedma, E.: A new consensus model for group decision making problems with non-homogeneous experts. IEEE Trans. Syst. Man Cybern. Syst. **44**(4) (2014, April)
19. Matsuno, Y., Yamamoto, S.: An implementation of GSN community standard. In: 2013 1st International Workshop on Assurance Cases for Software-Intensive Systems, San Francisco, CA, USA (CDROM) (2013)
20. Menon, C., Hawkins, R., McDermid, J.: Defence standard 00-56 issue 4: towards evidence-based safety standards. In: Proceedings of the Seventeenth Safety-Critical Systems Symposium, pp. 223–243 (2009)
21. Kelly, T., Weaver, R.: The goal structuring notation—a safety argument notation. In: Proceedings of the Dependable Systems and Networks 2004 Workshop on Assurance Cases (2004)
22. Matsuno, Y., Takamura, H., Ishikawa, Y.: A dependability case editor with pattern library. In: IEEE 12th International Symposium on High Assurance Systems Engineering, pp. 170–171 (2010)
23. GSN Community: GSN Community Standard Version 1. Origin Consulting (York) (2011)
24. Bloomfield, R., Bishop, P.: Safety and Assurance Cases: Past, Present and Possible Future—An Adelard Perspective (2010)
25. Yamamoto, S., Kaneko, T., Tanaka, H.: A proposal on security case based on common criteria. In: Proceedings of the 2013 International Conference on Information and Communication Technology, pp. 331–336 (2013)
26. IEEE 2005: IEEE Std. 1220-2005 Systems Engineering: Application and Management of the Systems Engineering Process (2005)
27. Akao, Y.: New product development and quality assurance—quality deployment system. Stand. Qual. Control **25**(4), 7–14 (1972)
28. Srikar, V.T., Mark. S.: Materials selection in micromechanical design: an application of the Ashby approach. J. Microelectromech. Syst. **12**(1) (2003)
29. Matsuno, Y., Yamamoto, S.: A new method for writing assurance cases. Int. J. Secure Software Eng. **4**(1), 31–49 (2013)
30. Yamamoto, S.: An evaluation of argument patterns based on data flow. In: Proceedings of ICT-EurAsia 2014, Bali, Indonesia, pp. 432–437 (2014)
31. Berger, C., Blauth, R., Boger, D.: Kano's methods for understanding customer defined quality. Center Qual. Manag. J. (Fall) 3–35 (1993)
32. Ohtomi, K.: Kansei modeling for delight design based on 1DCAE concept. In: Proceedings of the 11th International Modelica Conference, Versailles, France, issue 118, article no. 087, pp. 811–815 (2015)
33. Scupin, R.: Method: a technique for analyzing data derived from Japanese ethnology. Hum. Organ. **56**(2) (1997)
34. Strauss, A., Corbin, J.: Basics of Qualitative Research: Techniques and Procedures for Developing Grounded Theory, 3rd edn. Sage Publications, Londo (2008)
35. Kawase, M.: Crafting selves in multiple worlds: a phenomenological study of four foreign-born women's lived-experiences of being "foreign(ers)". VDM Verlag Dr. Müller (2008)
36. Golafshani, N.: Understanding reliability and validity in qualitative research. Qual. Rep. **8**(4), 597–607 (2003)

37. Hackman, J., et al.: Work Redesign. Addison-Wesley, Reading, MA (1980)

38. Yaghootkar, K., Gil, N.: The effects of schedule-driven project management in multi-project environments. Int. J. Project Manag. **30**(1), 127–140 (2012)

39. Epstein, M.J., Manzoni, J.-F.: The balanced scorecard and tableau de bord: translating strategy into action. Manag. Acc. **79**(2), 28–36 (1997)

40. Masumoto, M., Tokuno, T., Yanamoto, S.: A method for assuring service grade with assurance case—an experiment on a portal service. In: IEEE International Symposium on Software Reliability Engineering Workshops (ISSREW) (2013)

41. Kobayashi, N., Yamamoto, S.: The effectiveness of D-Case application knowledge on a safety process. Procedia Comput. Sci. **60**, 908–917 (2015). (19th International Conference on KBIIES)

42. Kobayshi, N., Hikishima, E., Nakamoto, A., Inayama, T., Shirasaka, S.: A proposal of assurance case description method using process of ISO 15288. Int. J. Jpn. Assoc. Manag. Syst. **11**(1), 139–146 (2019)

43. Kobayashi, N., Kawase, M., Shirasaka, S.: A Proposal of assurance case description method for sharing a company's vision. J. Jpn. Assoc. Manag. Syst. **34**(1), 85–94 (2017)

Quality Infrastructure System in China: An Agent-Based Model

Shiying Ni, Liwei Zheng, and Lefei Li

1 Introduction

Quality Infrastructure (QI) is a complex system, which comprises the organizations, policies, legal and regulatory framework, and practices for supporting and enhancing the quality, safety, and environmental soundness of goods, services, and processes. The system relies on five key elements, namely metrology, standardization, accreditation, conformity assessment, and market surveillance [1]. It is believed to improve product quality, ensure public safety, optimize the business environment, and foster international trade, especially in developing countries [2].

China has made great efforts in QI construction and obtained great progress. According to a report by PTB in 2011 [3], China ranked 19/53 in a quantitative evaluation of QI development level, which is measured by composite indicators of QI elements. Moreover, China was the best among countries that received WTO official development assistance. However, China is still suffering from quality issues constantly. According to a report by the State Administration for Market Regulation of China [4], 10.1% of all batches of products are found not qualified in selective examinations, and over 15% of all types of products have a qualification ratio lower than 80% in 2018. This situation indicates that the efficacy of the QI system, especially in its goal to improve product quality in China, is far away from desirable. Therefore, figuring out the internal working mechanism, that is, how key elements

S. Ni (✉) · L. Li
Department of Industrial Engineering, Tsinghua University, Beijing 10084, China
e-mail: nsy17@mails.tsinghua.edu.cn

L. Li
e-mail: lilefei@tsinghua.edu.cn

L. Zheng
Key Laboratory of Quality Infrastructure Efficacy Research, AVIC China Aero-Polytechnology Establishment, Beijing 100028, China
e-mail: xing_cun@sina.com

© The Author(s), under exclusive license to Springer Nature Switzerland AG 2021
D. Krob et al. (eds.), *Complex Systems Design & Management*,
https://doi.org/10.1007/978-3-030-73539-5_28

in the QI system act and interact to influence the market, will help the government to make better QI resource allocation and thus to achieve better efficacy of the QI system.

There are already research interests paid to QI system, including theoretical or empirical QI impacts study [2, 5–9] and QI development level evaluation by composite indicators [3, 10, 11]. However, they are not able to depict the complex interaction among QI elements sufficiently and answer how the system works intrinsically and how each element influences the efficacy of the whole system well. To address these questions, a model representation for the complex system should be done. There are several kinds of models, including the system dynamic model, discrete-event model, agent-based model, as well as the structural equation model. In the QI system, there are multiple heterogeneous actors, such as enterprises with different cost structures and operation strategies and consumers with various preferences. Besides, the relationship between actors is complex which cannot be described by certain simple functions. Additionally, in the real market environment, game-theoretic thoughts are involved in actor behaviors. In the consideration of the reasons above, we choose Agent-Based Modeling (ABM) for the QI system, for its advantages in modeling heterogeneous actors with complex behaviors. SysML modeling tools are used in system analysis and design step. This paper aims at building an agent-based model for the QI system, depicting its key elements and their interactions, providing an effective model framework for further research on QI efficacy.

The following sections will be organized as follows. In Sect. 2, we will introduce some related works of the QI system and ABM. Analysis of the existing QI system in China will be discussed in Sect. 3. In Sect. 4, the conceptual model, simulation model together with experimental results are illustrated. We give a brief conclusion about our work in Sect. 5.

2 Related Works

2.1 Quality Infrastructure

As a complex system, the concept of QI receives plenty of revision and refinement in practice. In the latest definition of QI adopted in June 2017 by INetQI and the World Bank, QI is the system comprising the organizations (public and private) together with the policies, relevant legal and regulatory framework, and practices needed to support and enhance the quality, safety and environmental soundness of goods, services, and processes. It relies on metrology, standardization, accreditation, conformity assessment (testing, inspection, and certification), and market surveillance [1]. The QI system has a 5-level structure including governance, quality infrastructure institutions, quality infrastructure services, enterprises, and consumers [12]. There are several research interests regarding the QI system and its key elements, including impacts analysis and evaluation methods.

The impacts of the QI system and its key elements are discussed both qualitatively and quantitatively. Gonçalves and Peuckert [2] summarizes activity, main functions, main beneficiaries, and main impacts of standardization, metrology, conformity assessment, and accreditation from literature. Ntlhane [8] uses a questionnaire to compare QI construction in Sweden and South Africa. Results turn out that the QI system reduces inferior products and has a positive impact on the economy. In the standardization field, empirical researches report positive impacts of international harmonized standards on China's exports [5–7], and EU's agricultural exports [9]. Given the positive impacts of the QI system, some researchers focus on the evaluation of the QI level to support construction decisions of governments. PTB [3] designs an index system with widely attainable indices, including total ISO9001 issues, total accredited bodies, total technical committees participation, etc. Measurements are carried out in 53 WTO members and China ranked 19 of all countries, 1 of countries who received WTO official development assistance. Choi et al. [10] propose a capability assessment framework based on the TQM model. Three key pillars, including standardization, conformity assessment, and metrology are assessed in 7 assessment categories, namely laws, systems, and institutions, strategies and implementation plans, stakeholders, infrastructure, human resources, process, and outcome. Pilot implementation conducted in four countries demonstrates that the framework helps to recognize the overall development level. Feng et al. [11] build an index system of 43 basic indicators in 5 first-level indicators including metrology, standardization, certification and accreditation capability, inspection and testing, and quality management capability.

There are sufficient evidences to show that the QI system plays a significant role in society and the economy. Figuring out the working mechanism of QI elements inside the QI system will further provide decision supporting for governments to better allocate QI resources and improve QI efficacy. Unfortunately, there are few literatures modeling all QI elements in one system, but capturing partial components of the QI system. Moljevic [13] provides an idea of modeling the impact of quality by a system dynamic model. Quality is connected to several aspects, including the rate of consumption, demand, price, etc. The complex interactions are well represented by nodes and relations, and finally, have an overall impact on regional developments. Ohori and Takahashi [14] build an agent-based model concerning standardization issues on market. Both consumers and firms are regarded as agents, which captures the heterogeneity of agents and micro-interactions in the real market. This model supports the decision making of the method or timing for the standardization by enabling analysis of possible market changes following different policies and market mechanisms behind the changes.

2.2 Agent-Based Modeling

Considering a complex system with multiple heterogeneous agents with complex interactions and behaviors we are facing, we use ABM to model the QI system. In

ABM, a system is modeled as a collection of autonomous decision-making entities called agents [15]. Agents are diverse, heterogeneous, and dynamic in their attributes and behavioral rules, and therefore are able to model complex systems with interdependencies and imperfect assumptions [16]. ABM is widely used in a variety of domains, including social science [17], marketing [18], supply chain [19], transportation system [20], etc.

In our problem, the configuration of elements in the QI system is similar to policy or regulation implementation, and the efficacy of the QI system can be depicted by performance indices in the market. Hence, it can be treated like a policy or regulation impact evaluation problem on market. In this topic, Zhao et al. [21] analyze the effectiveness of various policies on the proper growth rate of distributed photovoltaic (PV). The adoption behaviors of PV systems are influenced by various factors including household income, word-of-mouth effect, advertisement effect, etc., which depicts part of consumer behaviors in the real market. Silvia and Krause [22] assess the performance of different policy interventions to encourage the public to adopt plug-in electric vehicles. Policies are interpreted as specific model settings, including vehicle price, public charging network, and the number of government-owned electric vehicles. Leal and Napoletano [23] study the effects of a set of regulatory policies towards high-frequency trading. Regulations are treated as different parameter settings such as minimum resting times, or different model structures such as the added circuit breakers.

3 Quality Infrastructure in China

3.1 Overview Framework

Figure 1 shows an overall framework for the QI system with a five-level structure. The dotted line with an arrow represents a weak (usually undirect) relation between elements while the solid line describes a tighter relation. Grey block stands for the key element of QI defined in [1]. Firstly, the governance level is the foundation of the QI system, which comprises quality policy and regulatory framework, to ensure the legitimacy of organizations and processes in this framework. The following is the QI institutions level. Institutions for three core elements, namely standardization, accreditation, and metrology, are established. They are usually large public institutions, providing other components of the QI system with the necessary resources and guides to operate optimally. For example, enterprises can consult standards for product, process, or system by national standardization body to set up their business with better effectiveness. Below is the QI services level. This level includes all services in conformity assessment and calibration services in metrology. Services provided at this level are more directly related to the public for quality promotion. Enterprises at the enterprise level will be able to provide services and products with necessary certificates or reports provided by the QI services level. The last level with

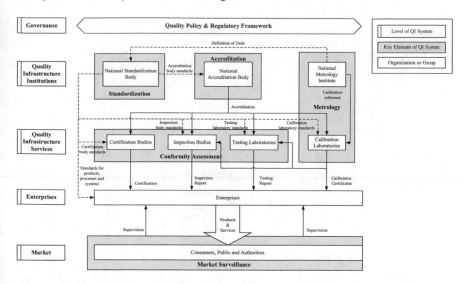

Fig. 1 Framework of QI system

the widest range is the market level. The products and services provided by enterprises are competing in the market. Consumers will select desirable ones, thus have an impact on enterprises in turn. Besides, the public and authorities will supervise the market by reporting quality issues and conducting quality selective examinations. These elements collaborate to ensure enterprises to provide better products and services to the market.

3.2 Process of Interest

There are multiple missions of the QI system, including improving product quality, ensuring public safety, improving the business environment, and fostering international trade. For different missions, the QI system works differently. We focus on the mission of improving product quality in the domestic market. Hence, we extract relevant entities, interactions, and activities, mainly concerning enterprises' activities in product production, circulation, and supervision. We must highlight that the entities, interactions, or activities that are not discussed in our work are never meant to be negligible in the QI system. The simplification is done to enable us to focus on the key processes we are interested in.

Figure 2 shows how the QI system works to enhance the product quality in the market. The national metrology body first determines the metrology level for the whole system. Metrology is the basic element in the QI system that has multiple impacts on other elements. We now concentrate on its impacts on accreditation because metrology will influence the instrument precision, thus having further

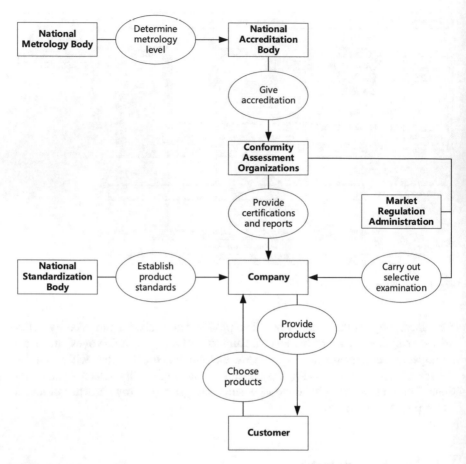

Fig. 2 Flowchart for QI system

impacts on the accreditation process. The national accreditation body will then give accreditation to conformity assessment organizations. Note that we do not distinguish different services in conformity assessment because their roles and activities are similar.

The company is the center of the whole flowchart, which has several connections with other parts. For companies to produce products and sell them to customers, they must follow certain product standards and get production permits. The role of the national standardization body here is to establish product standards. In fact, the body will establish a series of standards such as managerial standards and safety standards, for companies to follow, the simplification here mainly concerns our focus on product quality. Companies are able to provide products to the market when they obtain production permits. The application for production permits requires several documents, among which certificates and reports of conformity assessment are essential. Hence, we make conformity assessment organizations give production permits to companies for convenience, but the Quality and Technology Supervision Bureau should play this role in reality. After that, companies provide products to customers and customers select products from companies. When companies are competing in the market, the market regulation administration will conduct selective examinations to see if the products are qualified. Conformity assessment organizations will provide technical support in selective examinations. By all these means, the QI system enhances quality promotion in the market.

There are several potential risks that prevent the QI system from reaching better efficacy. Enterprises' real quality level may be lower than their announced quality level because of quality fluctuation in daily production or saving cost by purpose. Zhang et al. [24] point out there are divergences in the capability of conformity assessment organizations, resulting in mistakes or even fraud in conformity assessment. This phenomenon will result in unqualified products in the market. Besides, standardization will also play a significant role. Too strict standards make small enterprises hard to survive, resulting in a shortage of market supply. Too loose standards lead to a relatively low overall quality level of products in the market. Moreover, the cycle and ratio of the selective examinations will also influence the overall quality performance.

4 Agent-Based Modeling for QI System

4.1 Conceptual Model

We carry out the conceptual design for the QI system before modeling and simulation. There are several useful tools in system analysis and design, including Unified Modeling Language (UML) for object-oriented system design in software engineering, Gaia for agent-oriented analysis and design, and Systems Modeling Language (SysML) for Model-Based Systems Engineering (MBSE). In the consideration of model generalization ability and expansibility, we choose SysML in our research. Sha et al. [25] propose a 5-step approach of ABM development using several SysML diagrams, such as Requirement Diagram, Use Case Diagram, Block Definition Diagram, etc. Concerning the specific problem we are addressing, we utilize 3 types of SysML diagrams, including Block Definition Diagram for agents, attributes and relations description, Sequence Diagram for events and behaviors description, and State Machine Diagram for agent state description.

Although there are only five key elements in the definition of QI, we extend the concept of the QI system with enterprises and consumers because they are significant stakeholders in the system. A Block Definition Diagram (BBD) for the QI system by SysML is shown in Fig. 3. It depicts the component blocks of the QI system and various relationships among them. Main parts, values, and operations are defined in the blocks, too. There are 8 types of blocks, including the block of the whole QI system and its seven components. Seven composition lines with solid diamonds

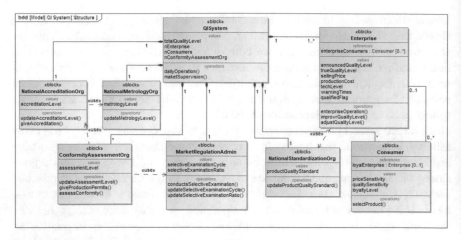

Fig. 3 Block definition diagram for QI system

demonstrate the composition relationship between the QISystem block and other blocks. Besides, there are also reference associations between Enterprise block and Consumer block. One enterprise may have several consumers and each consumer may show his or her loyalty to an enterprise. Dashed lines with arrows show dependency relationships between blocks. For example, enterprises should rely on national standardization organizations to determine a minimum quality level. Once the quality standard changes, enterprises should adjust their quality level accordingly.

Sequence Diagram (SD) (Fig. 4) depicts the main dynamic interactions inside the

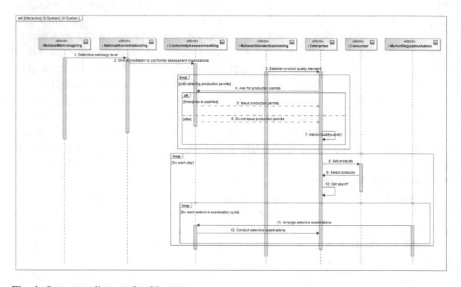

Fig. 4 Sequence diagram for QI system

QI system, focusing on events and sequences. NationalMetrologyOrg block first determines the metrology level. Next, the NationalAccreditationOrg block gives accreditation to ConformityAssessmentOrg. And then, the NationalStandardization block establishes a product quality standard. Enterprise block will go through several loops. They should first ask for production permits and obtain production permits, or once they failed, they improve product quality level and try again. In the following daily loop, Enterprise block sells products to Consumer block, Consumer block select products from enterprises, and Enterprise block gets its payoff for the following operation. The selective examination will be carried out every selective examination cycle, involving the interactions of MarketRegulationAdmin block, ConformityAssessmentOrg block, and Enterprise block.

State Machine Diagram (SMD) in Fig. 5 describes the life cycle of the enterprise by different states and the state transition rules. Enterprises are first in setting up state. Only when they obtain production permits, they are allowed to compete in the market. There are potential risks in daily operation. The first one is financial risks. Enterprises

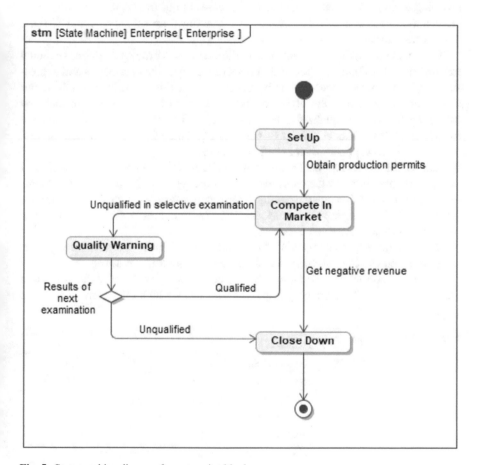

Fig. 5 State machine diagram for enterprise block

will go bankrupt when their revenues become negative. They close down and end their life cycle. Another risk is that when they get unqualified in the selective examination, they will receive a quality warning and lose part of their loyal consumers. Authorities will check the quality level again in the next selective examination. If enterprises are qualified, they will return to the normal state of competing in the market, or they are forced to close down by market regulation administration.

4.2 Experiments and Results

Although there are already several attempts [25, 26] in automatic simulation model generation from the conceptual model in SysML, transforming a complex real-world system model into a simulation model automatically is still a challenge [26]. Luckily, with a well-defined conceptual model, we are able to build the corresponding agent-based model manually with ease. In our model, classes and agents are defined according to BBD, events, and functions are set by SD, and SMD is directly transformed into the state chart in Anylogic.

It is worth noting that our main contribution here is to provide an effective model framework of the QI system for further research. Hence, the parameters and expressions are determined randomly or by experience in this work. Nevertheless, our experiments demonstrate that our model is consistent with the real situation, and they can also provide some qualitative insights for us. With more realistic data obtained for this model, the model can carry out quantitative analysis or optimization and provide decision support for real-world problems.

Some of the model parameters are shown in Table 1. To minimize the influence of randomness and increase the credibility of results, we run 10 repeated experiments for each setting and average the results. Consumers select products from enterprises according to the Discrete Choice Model, with individual-specific utility functions based on product quality and cost.

To verify the negative impacts of unqualified conformity assessment organizations proposed in [24], we carry out simulations to investigate it. Figure 6a shows how average quality level changes with different assessment levels. It illustrates that the average quality level roughly improves over time, which QI system is effective

Table 1 Model settings

Parameter	Value
Number of enterprises	100
Number of consumers	20,000
Number of conformity assessment organizations	20
Runs of each experimental setting	10
Product quality standard	3
Initial average of the quality level of enterprises	3.5

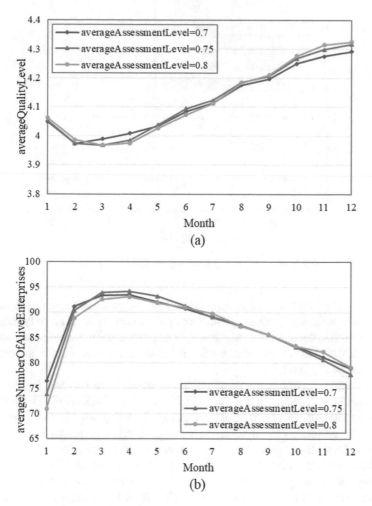

Fig. 6 **a** Average quality level with different assessment levels **b** average number of alive enterprises with different assessment levels

in improving product quality level in the market. A quality performance is better in a system with a higher average assessment level. It is probably because when conformity assessment organizations have a higher assessment level, unqualified enterprises are harder to enter the market, or they will be uncovered and get punishment more easily. This result is consistent with common sense. It is noted that there is a slight decline in the average quality level at the beginning of the simulation. It may be caused by the entrance of enterprises who did not obtain production permits in month 1 and improved their quality level later to get permits. Although they are qualified, their quality levels are not so good as the ones who are qualified in the very beginning. Graph of the average number of alive enterprises in Fig. 6b supports this

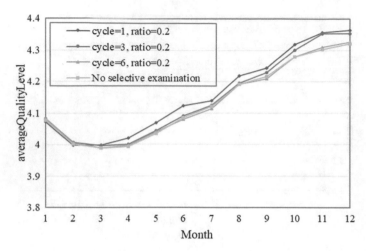

Fig. 7 Average quality level with different selective examination cycles

hypothesis. The number of alive enterprises first increases because qualified enterprises enter the market gradually, resulting in a decrease in the average quality level. The number then decreases because enterprises close down when revenue becomes 0 or get unqualified in two successive selective examinations, leading to improvement of average quality level.

The selective examination is an important measure to regulate the market by market regulation administration. We can figure out the impacts of selective examination by simulations. As it is shown in Fig. 7, the quality performance of the system with selective examinations is better than that without selective examinations. It indicates that selective examination helps in the quality promotion. Besides, the difference of performance with different selective examination cycles is small in the beginning and gets larger as time goes by. Generally, more frequent selective examinations result in a better average quality level in the market when selective examination ratios are the same. It is reasonable because unqualified enterprises are more likely to be chosen for selective examination. However, it takes time and money to carry out selective examinations. Hence, it can be an optimization problem to determine the cycle and ratio of selective examinations.

5 Conclusion

This paper proposes an agent-based model framework for exploring the complex internal mechanism of the QI system. This model is based on the abstraction of the existing QI system, focusing on its goal to improve product quality in the market. SysML modeling tools, including Block Definition Diagram, Sequence Diagram, and State Machine Diagram, are used in building conceptual models. The successful transformation from a conceptual model in SysML to an executable agent-based model in Anylogic shows the effectiveness and promise of SysML in complex system modeling. Two sets of simulation experiments are carried out with our model. Results show that our model works in analyzing how QI elements interact to influence product quality. Although the settings of parameters and expressions need to be studied and designed more sufficiently, we still highlight that the proposed framework has the potential to provide a model framework for further research in QI system modeling. With fine data collected, more researches can be done. For example, researchers can optimize the allocation of QI construction resources concerning cost and benefits. New policies or new processes can be studies by modifying or extending this framework.

Acknowledgements This study was co-supported by the Key Laboratory of Quality Infrastructure Efficacy Research (grant number KF20180401) and the National Key R&D Program of China (grant number 2017YFF0209400).

References

1. UNIDO: Quality Infrastructure: UNIDO's Unique Approach (2018)
2. Gonçalves, J., Peuckert, J.: Measuring the Impacts of Quality Infrastructure: Impact Theory, Empirics and Study Design. Physikalisch-Technische Bundesanstalt (2011)
3. Harmes-Liedtke, U., Di Matteo, J.J.O.: Measurement of Quality Infrastructure. Physikalisch Technische Bundesanstalt (PTB), Braunschweig (2011)
4. Announcement of the State Administration for Market Regulation on the State Supervision and Spot Check of Product Quality in 2018 (2019). Available from: https://gkml.samr.gov.cn/nsjg/zljdj/201904/t20190426_293185.html
5. Mangelsdorf, A.: The role of technical standards for trade between China and the European Union. Technol. Anal. Strateg. Manag. **23**(7), 725–743 (2011)
6. Li-juan, Y.: Trade effects of Chinese standards: empirical research based on standard category in 33 ICS sectors. In: 2013 International Conference on Management Science and Engineering 20th Annual Conference Proceedings. IEEE (2013)
7. Mangelsdorf, A., Portugal-Perez, A., Wilson, J.S.: Food standards and exports: evidence for China. World Trade Rev. **11**(3), 507–526 (2012)
8. Ntlhane, M.D.: Comparison of Quality Infrastructure of the Republic of South Africa and Sweden. University of Johannesburg (2015)
9. Shepherd, B., Wilson, N.L.: Product standards and developing country agricultural exports: the case of the European Union. Food Policy **42**, 1–10 (2013)
10. Choi, D.G., et al.: Standards as catalyst for national innovation and performance—a capability assessment framework for latecomer countries. Total Qual. Manag. Bus. Excell. **25**(9–10), 969–985 (2014)
11. Feng, L., Liao, J.X., Huang, J.X.: The evaluation method and empirical study of national quality infrastructure capability index. In: Huang, D., et al. (eds.) Conference Proceedings of the 6th International Symposium on Project Management, pp. 1322–1329. Aussino Acad Publ House, Marrickville (2018)
12. UNIDO: Quality Policy: Technical Guide (2018)
13. Moljevic, S.: Influence of quality infrastructure on regional development. Int. J. Qual. Res. **10**(2) (2016)

14. Ohori, K., Takahashi, S.: Market design for standardization problems with agent-based social simulation. J. Evol. Econ. **22**(1), 49–77 (2012)
15. Bonabeau, E.: Agent-based modeling: methods and techniques for simulating human systems. Proc. Natl. Acad. Sci. **99**(suppl 3), 7280–7287 (2002)
16. Macal, C.M., North, M.J.: Agent-based modeling and simulation. In: Proceedings of the 2009 Winter Simulation Conference (WSC). IEEE (2009)
17. Desmarchelier, B., Fang, E.S.: National culture and innovation diffusion. Exploratory insights from agent-based modeling. Technol. Forecast. Soc. Chang. **105**, 121–128 (2016)
18. Rand, W., Rust, R.T.: Agent-based modeling in marketing: guidelines for rigor. Int. J. Res. Mark. **28**(3), 181–193 (2011)
19. Long, Q.Q., Zhang, W.Y.: An integrated framework for agent based inventory-production-transportation modeling and distributed simulation of supply chains. Inf. Sci. **277**, 567–581 (2014)
20. Zhang, G., et al.: Agent-based simulation and optimization of urban transit system. IEEE Trans. Intell. Transp. Syst. **15**(2), 589–596 (2013)
21. Zhao, J., et al.: Hybrid agent-based simulation for policy evaluation of solar power generation systems. Simul. Model. Pract. Theory **19**(10), 2189–2205 (2011)
22. Silvia, C., Krause, R.M.: Assessing the impact of policy interventions on the adoption of plug-in electric vehicles: an agent-based model. Energy Policy **96**, 105–118 (2016)
23. Leal, S.J., Napoletano, M.: Market stability vs. market resilience: regulatory policies experiments in an agent-based model with low-and high-frequency trading. J. Econ. Behav. Organ. **157**, 15–41 (2019)
24. Zhang, L., Huang, H.J., Cai, L.Y.: On the reform and development of quality infrastructure under the new situation in China. Qual. Explor. **16**(02), 60–67 (2019)
25. Sha, Z., Le, Q., Panchal, J.H.: Using SysML for conceptual representation of agent-based models. In: ASME 2011 International Design Engineering Technical Conferences and Computers and Information in Engineering Conference. American Society of Mechanical Engineers Digital Collection (2011)
26. Maheshwari, A., Kenley, C.R., DeLaurentis, D.A.: Creating executable agent-based models using SysML. In: INCOSE International Symposium. Wiley Online Library (2015)

Realizing Digital Systems Engineering—Aerospace and Defence Use Case

Eran Gery

In defense, complex requirements are evolving to meet the demands of advanced battle fields where there will be greater numbers of autonomous devices operating under a command and control mesh. Historically these devices were dominated by electro-mechanical systems under human operator control and their design fell under a mechanical/aeronautical or electrical engineering discipline. Today's devices are now dominated by software and present increasing design complexity with the advent of technologies like IoT, AI, and microprocessors [1, 2].

In addition, new challenges are rapidly emerging across both defense and aerospace industries. On the defense side, increasing commoditization of advanced technologies such as UAVs, rocket technologies and electronics creates a new set of threats that must be designed for. On the civil side forces like globalization, industrialization, and porous technology creates an increasingly competitive marketplace that has to be balanced with growing pressures around regulation and functional safety.

Therefore, the traditional system engineering approaches that have been key to sustaining both defense and aerospace industries fall short in effectively addressing the emerging challenges: addressing growing complexity, technological choices, and all that with growing time constraints. These dynamics have triggered systems engineering initiatives such as the DoD digital engineering strategy and the INCOSE 2025 vision [3, 4].

The emerging digital ssystem engineering approach focuses on digital practices. Designing and analyzing systems starting with a functional perspective which is not locked to a discipline. Digitally transforming the engineering environment so teams can leverage capabilities like digital thread and digital twin to improve system design and time to market.

In this paper we introduce the digital systems engineering model, how it leverages advances in engineering lifecycle management to address the complexity and the time

E. Gery (✉)
New York, USA

© The Author(s), under exclusive license to Springer Nature Switzerland AG 2021
D. Krob et al. (eds.), *Complex Systems Design & Management*,
https://doi.org/10.1007/978-3-030-73539-5_29

to market challenges. We will discuss concepts like digital threads, model-based engineering and how AI and machine learning can assist in efficient deliveries. We will also illustrate how key activities are conducted using an A&D example.

1 Challenges in A&D Systems

The nature of A&D systems has always been about designing systems that will be leadership for years if not decades in the future, which implies delivery of more functionality while still meeting the same quality and schedule constraints. Increasingly faster technology introductions coupled with broader global competition is driving the A&D industry to rethink development models to address growing cost pressures and time to market demands.

Two massive trends are putting new pressure on innovators, engineers and planners—one is the way that information is increasingly available [i.e. Cloud, Mobile, Internet of Things (IoT)] and usable (i.e. Apps, API's, AI) in ways that A&D systems haven't had to generally deal with before and secondly that the way that technology evolves is increasingly unpredictable. These trends mean that A&D companies and systems need to be responsive and open to change in wholly new ways. Responsive across capabilities, architectures, organizations, systems, processes, methods, tools—which is counter to the prevailing industry culture. Scoping any system boundary or responsibility now requires openness because innovation arises everywhere, but so do threats. This creates business opportunities which can only be addressed by embracing new efficiency given older models and processes can't scale. Speed, agility and flexibility are the main drivers to take the industry to a new level—with smarter ways of achieving this level with security and safety—these factors change everything. The companies that excel with these characteristics will be the ones that set the pace for the next decade.

Last but not least, this industry typically develops product families, for different customers and purposes. Developing an effective reuse culture across related programs is a major opportunity for significantly improving key performance indicators (KPI's) such as cost and schedule.

2 Role and Value of Systems Engineering

Systems engineering (SE) has a strong track record as an interdisciplinary approach and means to enable the realization of successful systems. Successful systems must satisfy the needs of their customers, users and other stakeholders, and be delivered within expected time and budget ranges.

Due to the highly dynamic nature of the business and operating environment, tomorrow's Systems Engineering needs to better tie into the business or mission opportunity management.

The following study done by NASA and published in SeBok shows the correlation between systems engineering activity and reduced project overruns:

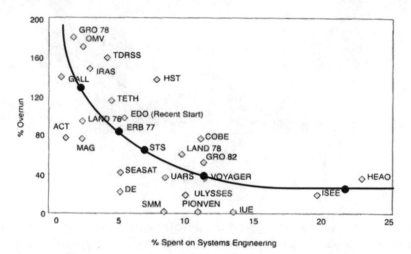

The above figure clearly shows that applying more systems engineering practice reduces project overruns.

Key systems engineering activities are in two main areas: technical processes and technical management processes, as specified by ISO 15288 and the INCOSE Handbook:

Technical processes

- Mission analysis
- Managing requirements
- System analysis—functional specification
- Architecture definition and allocation
- Verification and validation
- Operation and maintenance.

Technical Management processes

- Configuration and data management
- Variant management and PLE
- Change management—Analyzing the Impact of Change
- Planning and tracking
- Quality assurance.

The technical processes are typically overlaid on a "v" lifecycle model, where the analysis and design activities are on the left arm of the "v", and those are then validated by processes on the right arm of the "v".

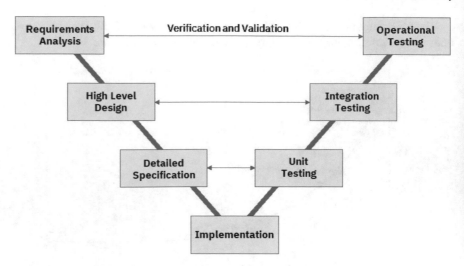

Systems engineering is strongly correlated to the system lifecycle process models. The common lifecycle process models are iterative, but they widely vary from incremental and iterative (IID) practices to lean/agile practices.

Today's dominant systems engineering (SE) practice is commonly referred to as "document centric". It relies on work products which are typically textual documents focused around system requirements, interface control documents, and various studies and analysis. The focus is mostly on the technical processes, more specifically requirements management. There is also some level of adoption of model-based techniques to specify the functionality and architecture of the system. There is also some base practice to specify the validation procedures, usually as a special set of requirements. While there is some use of digital tools (for example to manage requirements), the various artifacts are not digitally related and the typical SE baseline is eventually a set of documents, some of which are produced from specialty tools, like requirements management tools (Fig. 1).

As for technical management processes, those get much less focus by systems engineers, probably due to the heritage of waterfall processes that did not require planning in the context of daily life, but part of large-scale planning done by project management professionals. Therefore, while planning and configuration management are very common among software developers, they are rarely practiced by systems engineers. The fact that planning and the various aspects of data management are not practiced hinders the aim and need to have highly effective and efficient overall engineering practices.

Such environments that are based on engineering baselines which are sets of disparate documents, lack digital connectivity, or traceability, to track and manage the dependencies across the various artifacts. It also lacks means for visibility to effectively comprehend the implication of changes, and the overall integrity of the system. It also does not have the means to effectively manage complex engineering projects.

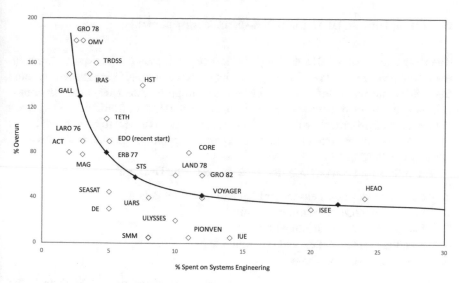

Fig. 1 Document centric systems engineering

This has several implications:

- Difficulty to adapt to technological innovation or business model disruption fast enough due to lack of visibility and effective change management. In general, such environments do not provide the means for agility and plan changes.
- Expensive rework due to late discovery of issues: document centric environments do not provide means to validate the systems engineering artifacts until realization phases where rework is much more expensive.
- Inefficient supplier collaboration: collaborating with suppliers is also based on document exchange, which significantly slows down the process. Creating and receiving artifacts from suppliers is a process that requires extra manual steps until the cycle is stable and validated.
- Non-optimal designs due to early design lock: such systems with ineffective change mechanisms gravitate towards minimizing change. This results in processes that prioritize less change and early design locks, as change is expensive. The implication is compromised performance of the resulting system.
- High costs of regulatory compliance activities: the core of regulatory requirements for safety and maturity relies on repeatability, evidence, and traceability. In document centric environments demonstrating traceability, collecting testing evidence, and demonstrating well specified, repeatable process require dedicated costly effort as it is not inherently supported as part of the engineering process.

3 The Premise of Digital Systems Engineering

The DoD called out in 2018 the Digital Engineering Strategy. The expectation from digital engineering is to "conduct engineering in more integrated virtual environments to increase customer and vendor engagement, improve threat response timelines, [..], reduce cost of documentation and impact sustainment affordability [...] Such engineering environments will allow DoD and industry partners to evolve designs at conceptual phase, reducing the need for expensive mockups, premature design lock, and physical testing."

The pillars for such environments, as stated by the DoD are:

1. Formalize development integration and use of models
2. Provide authoritative source of truth
3. Incorporate technical Innovation
4. Establish infrastructure and environments
5. Transform the workforce.

Digital systems engineering focuses on the scope of systems engineering in the context of a digital engineering framework, and it is part of the digital transformation wave being adopted by the industry. As opposed to the document centric approach, digital systems engineering relies on a digital representation of all the artifacts within the scope of systems engineering, in the context of a lifecycle information model. The information model realizes a complete digital continuity across systems engineering artifacts. It consists of artifacts such as requirements, functional and architectural elements (MBSE), V&V elements such as test plans and representation of implementation artifacts provided by implementation disciplines. The information model is complemented with lean/agile planning support and analysis capabilities to support effective change management, and overall visibility into various aspects of the system under designs. This infrastructure supports:

- Consistency across engineering and resiliency to changes
- Agile processes with the necessary tracking and visibility that enable higher velocity
- Effective fulfillment of industry compliant processes, such as ARP4754 and DO178
- Collaboration across the wider team and the supply chain.

4 Key Imperatives for Effective Digital Systems Engineering

As discussed above, digital systems engineering aims to enable higher precision and agility to cope with the technological challenges of tomorrow's systems. As part of that we see five key imperative to enable effective digital processes, that will be discussed in the next paragraphs:

- Digital continuity is the foundation of modern engineering practices. It enables a digital information model across the engineering lifecycle activities. Such information models are also referred to as digital threads.
- Leveraging information through technology advancements such as data analytics and machine learning to support decision making, consistency and compliance.
- Foster model-based systems engineering as a core lifecycle activity.
- Integrated planning that facilitate lean and agile processes.
- Enable global configuration management and reuse.

All the above principles are implemented by the IBM's Engineering Lifecycle Management (ELM) platform. See [5] for details about IBM ELM platform.

5 Digital Continuity and Digital Threads

Digital information models are a foundation for digital engineering processes. They specify the necessary artifacts and relationships across them. Artifacts may reside in multiple repositories as there are various specialty tools used by systems engineers and also by the implementation disciplines. Requirements are related to various artifacts such as tests, and also to analysis and architecture artifacts such as system functions as part of a functional model (e.g. activity diagram) and solution architecture elements represented by architectural blocks in block definition diagrams. Downstream into the implementation disciplines architectural artifacts as well as requirements relate to software or hardware implementation artifacts, represented by specialty tools like BOMs in a PLM system. Figure 2 illustrates an information model, in this case the one mandated by DO-178C. It shows the key artifacts and the necessary relationships.

Such information models enable some key capabilities:

- Completeness and coverage analysis: every requirement is implemented by the system
- Consistency: if a requirement is modified the downstream artifacts such as implementation and validation artifacts have been updated accordingly
- Change management: guiding impact of change analysis based on information model relationships.

A key challenge is how to implement digital continuity in heterogeneous multi-disciplinary engineering environments. In such environments there are various specialty tools used by systems engineers that also needs continuity to specialty tools used by various implementation disciplines such as software, mechanical, and electrical. There have been many attempts to address this integration challenge. One class of solution is based on synchronizing (copying) the data from the various tools to a central repository, where traceability is managed. Data replication approaches have always been challenging and consistently fail. An approach which has proven as a breakthrough and pragmatic one is the linked data approach, originated from

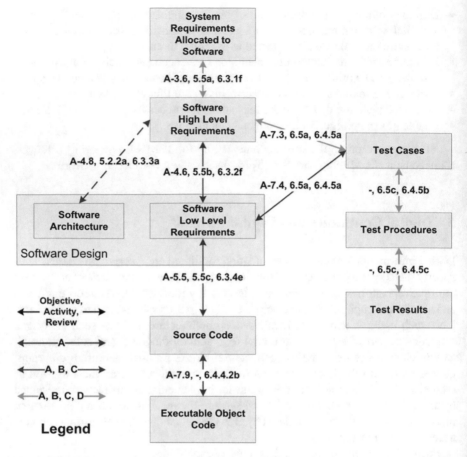

Fig. 2 DO178C lifecycle information model (*Source* Wikipedia)

the w3c internet stack. Linked data is the basis for the OASIS/OSLC [6] standard for lifecycle integration.

Figure 3 shows an impact analysis study based on concrete set of artifacts of a UAV system, based on implementation of OSLC with the IBM ELM system. The different ovals represent different types of artifacts across the lifecycle, such as change requests, requirements, analysis and design, and test artifacts.

6 Leveraging Analytics, Machine Learning and AI

A key factor to a successful delivery of a system is related to decision making, both in engineering management and core engineering activities. Visibility and aggregation of data, assessment of progress and deviations are key enablers to support such

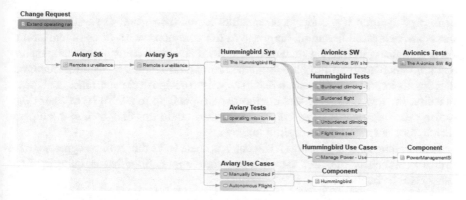

Fig. 3 Graphical impact analysis based on digital threads

engineering decisions. Providing relevant analytics to systems engineers as part of technical management processes enable better decisions which are based on actual data. Machine learning can further support such assessments based on historical and accumulated data throughout the project. AI assisted assessments of project health, impact of changes, risks, can leverage data and identify patterns that otherwise are not visible to the engineering managers.

Figure 4 shows an analytical view that depicts the level of requirements and test coverage across the UAV system, based on the digital information instances across the system.

A considerable amount of systems engineering information is found in "unstructured" format, more specifically, natural language. Stakeholder descriptions, such as

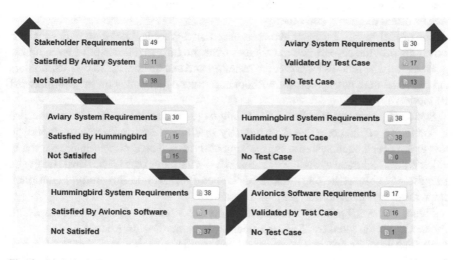

Fig. 4 Analytical view

concept of operation ("Conops"), stakeholder requirements, and also system requirements are expressed in natural language as the common means of communication.
Also, technical information such as descriptions and characteristics of components
and implementation artifacts are expressed in natural language. Such unstructured
data entails important information that translates to design decisions, reuse, and functionality. Surveys also show that engineers can spend up to a third (!) of their time
looking for information or repeating things they could not find, as non-intelligent
search does not result in the desired findings.

Natural language processing (NLP) can automate tasks the require engineers time
to deal with such unstructured data. Here are some applications that are being applied
using NLP technology:

- checking quality of requirements based on guidelines, such as INCOSE guidelines
- transforming unstructured information like Conops to structured model representations
- intelligent searching of matching data, like tests verifying requirements
- auto completion assistants, that enforce correct usage of domain terminology.

7 Model Based Analysis and Design

Model-based systems engineering (MBSE) is the formalized application of modeling
to support system requirements, design, analysis, verification and validation activities
beginning in the conceptual design phase and continuing throughout development
and later life cycle phases (INCOSE SE Vision 2020) (INCOSE-TP-2004-004-02,
Sep 2007). Such activities are still widely performed using unstructured and informal
means, such as natural language enriched with various types of graphical notations
such as block diagrams and others.

The model-based approach, which is based on a graphical notation with specified
semantics provide several key advantages over the informal descriptive techniques:

Precision: the models have precise meaning and can be consumed by stakeholders
and implementers with precision without complete dependency on the originator of
the model.

Standard practices: standard MBSE languages are based on immense expertise
of practicing engineers as well as industry guidelines and reference examples on
how to approach such systems engineering activities. The common language today
which is also endorsed by INCOSE is SysML—a variant of the OMG/UML language
for systems engineering purposes. SysML specify dedicated Structural, Functional,
Behavioral, and parametric views of a system.

Figure 5 shows 3 primary SysML diagrams functional, structural, and behavioral.
The model was created by IBM Engineering Systems Design Rhapsody.

Verifiability: Models can be analyzed and verified which means that many errors
that originated in otherwise ambiguous and erroneous specifications are discovered
at early stages. This has significant implication on overall cost of engineering as many
later discovered issues originate from specification and requirement errors. Model

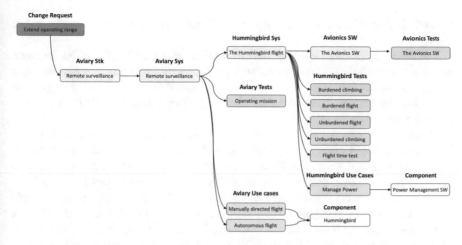

Fig. 5 Functional, structural, and behavioral views of the UAV

Base Testing (MBT) leverages modeling constructs to describe test scenarios. One of the use cases for MBT is for early verification of functional, or architectural models before they are handed off for implementation.

Figure 6 shows early verification of the specification model, what is also referred to as the "small v", versus the "big v" that represents an entire implementation and

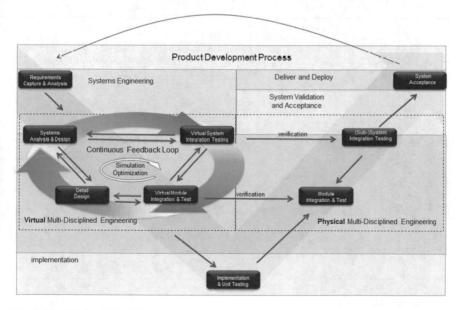

Fig. 6 "Small v" and big "V"

validation cycle. The small "v" can leverage validation techniques using model-based testing.

Optimization: models can be used to perform early trade-off studies between solution architecture. It is much cheaper to consider different solution architectures within virtual models than with physical implementations. Physical design solutions too often require "early design lockout" creating suboptimal KPIs in terms of reliability and operational costs.

However, it turns out that adoption of MBSE is not straightforward. It is key that a model-based approach is deployed with well-defined practices relevant to an organization and a domain. Reference implementations and templates can also greatly help engineers that are used to traditional unstructured practices. Another key aspect is that modeling artifacts and activities shall be part of the entire lifecycle information model and the digital threads. Isolated models provide very limited value and in many such cases the return on investment is questionable. Figure 7 shows how an analysis artifact is related to requirements which originate from a requirements management system (DOORS next generation).

Hence it is key that models are well integrated into the overall practice to effectively produce relevant deliverables which are consumable by downstream implementation teams.

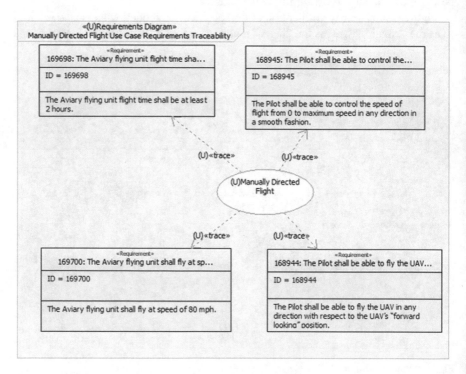

Fig. 7 Traceability from analysis (use cases) to requirements

8   Lean and Agile Work Management

Lean and agile approaches offer effective management of complex projects with continuous steering of the project based on actual progress. Scaled agile (SAFe) practices originate from agile software practices and lean engineering principles as introduced by Toyota production systems and is used by large, multi-team and multi-disciplinary projects. SAFe is a formal framework that maintains a community that evolve practices based on lean/agile principles: economic driven decisions, systems thinking, embracing variability and preserving options, iterative and incremental progress with objective measures of working systems, good control and visibility for work in progress (WIP), embracing innovation and decentralized decisions [7].

Studies made by scaled-agile (scaledagile.com) show 30–75% faster time to market, 25–75% defect reduction which demonstrates improvement in delivery and quality.

To properly enable SAFe for complex solutions there needs to be a digital supporting system for planning and tracking, that enables visibility and flow of information from top system level owners to subsystem and component implementation teams. The planning and tracking system shall be integrated with the digital information model with traceability from plan items to relevant engineering artifacts such as requirements, use cases, and test cases.

Analysis is core to successful conduct of scaled agile projects. Model based systems engineering fosters analysis of use cases and features that provide the basis for iterative planning. IBM's Harmony agile model-based systems engineering (aMBSE) is a practice that leverages the benefits of MBSE to drive iterative large-scale projects, based on a use case centric approach.

Figure 8 shows how use cases specified during model-based analysis are translated to solution epics in an agile plan.

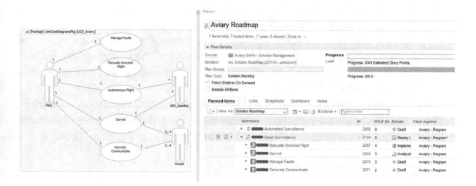

Fig. 8   From use cases to solution epic

9   Global Configuration, Reuse, and Variant Handling

An important part of SE "technical management" is managing the configuration of engineering data so it is consistent, as required by industry standards. The digital representation of a system is a set of inter-related artifacts across disciplines and tools. One of the fundamental challenges is how to support consistent baselines in such an environment. Configurations are also essential to facilitate reuse across programs. INCOSE 2025 Vision recognizes "product architecture reuse" as one of six primary concerns of systems engineers in the defense sector for the upcoming five years. It is very common that defense systems are delivered to multiple customers and purposes, essentially resulting in a family of products. The key challenge is how to effectively manage engineering multiple programs for multiple customers and/or purposes. This also implies support of concurrent configurations (or branches).

Global configuration management is an approach to manage configurations across a set of engineering repositories. It is part of the Oasis OSLC specification, and it is implemented by IBM Engineering Lifecycle Management solution and other engineering tools providers. Global configuration is a hierarchical structure that corresponds to the logical product breakdown structure of a system, starting from its root node and then hierarchically decomposed into a set of subsystems and next level subsystems, down to components. Each such node contains a set of engineering artifacts related to it, such as requirements, design models, V&V plans, etc. Each such node is going through a set of configurations that evolve as the lifecycle progresses. Figure 9 depicts a hierarchical breakdown structure of a UAV system, which contains a set of hierarchical global configurations of subsystems, in addition to artifact configurations of requirements, architectures, and tests.

As we described above, configurations of a system may evolve not only sequentially but also in parallel. Parallel configurations are often referred to as branches—a common paradigm that has existed in the software domain for many years. One of the important use cases is to support the reuse of a system across multiple programs. With configuration branches we can manage engineering assets across multiple programs, where programs can share the same version of a system node or maintain a variant of a node with a set of changes specific to a program. One of the important capabilities is that changes can be delivered across branches. So, a change that was applied in one program can be later delivered to another program.

Figure 10 shows a branch structure representing different variants of an aircraft platform. This figure also depicts how variants are updated from the primary platform, and how updates to variant2 are retrofitted back to the primary platform design.

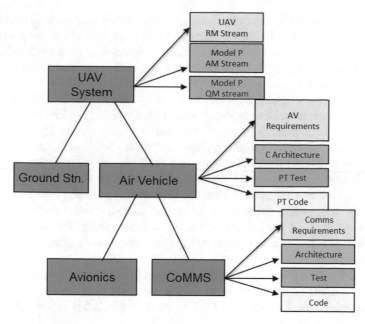

Fig. 9 PBS represented as a global configuration

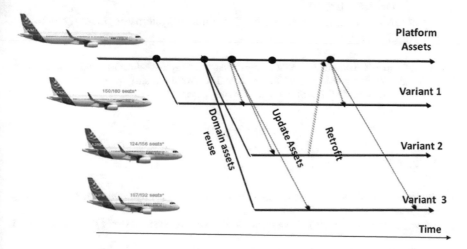

Fig. 10 Parallel configurations

10 Summary

The A&D industry faces increasing challenges from major technological disruptions in areas of connectivity, artificial intelligence, new materials and energy, and evolution of computing technology. That results in further increasing complexity and the need to rapidly adopt innovation. Document and discipline centric approaches do not scale and are subject to individual expertise and fragility of manual processes.

Technology advancements enable creation of effective digital engineering platforms that realize digital lifecycle information models. augmenting the traditional document centric approach.

Digital systems engineering enables visibility into the various aspects of the system development and fosters consistency of engineering artifacts and effectively supporting decision making by visualizing and analyzing system dependencies and impact of change. Digital systems engineering coupled with the Scaled Agile Framework (SAFe) supports lean/agile lifecycle models that harvest relevant concepts to enable effective practices that can accommodate change and better support the needs of the customer, and effectively steer the engineering activities based on concrete realities as they emerge throughout the project. Model based analysis and design practices enable early verification of requirements and design decisions, as well as more effective communication with the implementation teams using precise handoff specifications.

IBM ELM offers a leading integrated solution to improve your systems and software development lifecycle today as well as establish your foundation for digital systems engineering. IBM ELM is deployed at major A&D enterprise in North America, Europe, and Asia.

Visit the IBM web site to learn more about our ELM solution or to schedule a consult with an IBM Expert.

https://www.ibm.com/internet-of-things/solutions/systems-engineering?lnk=hpmpr_iot&lnk2=learn.

Acknowledgements I would like to thank my colleagues Gray Bachelor, Graham Bleakly, Daniel Moul, and also Kim Cobb for their useful inputs and suggestions related to this paper, and special thanks to Brian Sanders for his editorial update of this paper.

References

1. SEBok: https://sebokwiki.org/wiki/Guide_to_the_Systems_Engineering_Body_of_Knowledge_(SEBoK)
2. Incose Handbook: https://www.sebokwiki.org/wiki/INCOSE_Systems_Engineering_Handbook
3. DoD digital engineering initiative. https://fas.org/man/eprint/digeng-2018.pdf
4. INCOSE systems engineering vision 2025. https://www.incose.org/products-and-publications/se-vision-2025
5. ELM on Jazz.net. https://jazz.net/products/elm/
6. OSLC on Oasis website. https://www.oasis-oslc.org
7. SCALED AGILE: https://www.scaledagileframework.com/

Research on Behavior Modeling and Simulation of Complex UAV Based on SysML State Machine Diagram

Yuanjie Lu, Zhimin Liu, Zhixiao Sun, Miao Wang, Wenqing Yi, and Yang Bai

1 Introduction

The operating environment of the large UAV is complex, the interaction with external information is complex, and the internal function logic is complex. At the same time, the development of large UAV involves the integrated application and optimization of the aerodynamic, strength, structure, control, avionics, electromechanical, software and so on. Therefore, the system development of large UAV brings a new challenge to the design capability, methods and means. It is urgent to solve the problems of imperfect system design, poor traceability, inadequate verification, long system iterative design cycle and high outfield failure rate.

Model-based Systems Engineering (MBSE) apply modeling methods to support requirements, design, analysis, verification and validation activities that begin at the conceptual design stage and run through the development process and subsequent life cycle stages [1]. MBSE supports model-driven, fast iterative verification to make requirements traceable, express unambiguously, advance iterate verification to design stages and reduce physical changes to ensure quality, shorten cycle and reduce costs. SysML is commonly used to create models of system architecture, behavior, requirements, and constraints [2]. SysML provides activity diagram, sequence diagram, and state machine diagram that express the continuous and concurrent behavior of the

Y. Lu (✉) · M. Wang · Y. Bai
INCOSE CSEP, Shenyang Aircraft Design and Research Institute (SADRI), AVIC, No. 40 Tawan Road, Shenyang, Liaoning, China
e-mail: plough5221@sina.com

Y. Bai
e-mail: baijordon@163.com

Z. Liu · Z. Sun · W. Yi
Shenyang Aircraft Design and Research Institute (SADRI), AVIC, No. 40 Tawan Road, Shenyang, Liaoning, China
e-mail: lzm_w650910@sohu.com

© The Author(s), under exclusive license to Springer Nature Switzerland AG 2021
D. Krob et al. (eds.), *Complex Systems Design & Management*,
https://doi.org/10.1007/978-3-030-73539-5_30

system, in which state machine diagram is suitable for behavior modeling for discrete, event-driven systems [3, 4]. Based on the SysML state machine diagram, this paper studies the modeling of the behavior of the complex UAV system, and verifies the functional requirements and logical correctness of the UAV.

2   SysML State Machine Diagram

The SysML state machine diagram is used to describe the dynamic behavior of the system during its life cycle, as shown by the state sequence experienced by the system, the behavior in a particular state, the conditions that cause state transfer, and the response to an exception. The state machine diagram focuses on how the structure in the system changes state based on events that occur over time. The SysML state machine diagram generally consists of five main elements, defined as follows [5, 6].

State: An element persists in a situation of discernible, uncorrelated, and orthogonal over a considerable period of time. State is the abstract representation of the requirements at the highest level of the system, according to the functions currently configured and provided, describing the conditions necessary or permissible or prohibited for the system. State consists of simple state, composite state, and final state.

Event: The occurrence of a related event that might trigger a transfer.

Guard: Is a Boolean expression that determines which transfer condition to take. If the guard is true at the time of the event, a transfer will be taken; if the guard is false, the event will not be executed.

Behavior: SysML offers two kinds of behavioral features: operation and reception. Operation can be refined into entry, exit, and so on. Entry behavior is a list of operations performed whenever a state is entered. Exit behavior is a list of operations taken whenever a state is left. Reception means to take the action associated with the event without causing a state change.

Transition: The response to a related event causes the instance to move from one state to another. State transition execution sequence: Receive trigger events, judge conversion conditions, perform source state exit operations, perform state transition operations, and perform target state entry operations.

3   UAV Behavior Modeling

3.1   Modelling Methods

System modeling runs through requirements analysis, use case design, function analysis, architecture design, and other design and development stages. Generally there are two typical development methods. The first method allows for system modeling based on mature methodologies such as Harmony SE [7], CESAMES Systems Architecture Method [8], and so on. For example, using Harmony SE requires the modeling and analysis of system use case diagram, activity diagram, sequence diagram, block definition diagram, internal block diagram, state machine diagram of black box and white box behavior. The advantage of the first development approach is that the system can be designed more comprehensively by following rigorous modeling processes and complete modeling elements. In the second way, if the developer has rich experience in similar product development, the system modeling and simulation can be carried out directly based on the SysML state machine diagram. The advantage of the second development method is that the design and development efficiency of the system can be greatly improved. Based on the second method, this paper study the behavior modeling and simulation of UAV.

3.2 UAV State Definition

Before defining the state of the UAV, a series of preparations are required, including the system requirements capture, requirements analysis, function analysis, and mapping between requirements and functions. The focus of requirements analysis is to form a system specification. The focus of functional analysis is to form a function tree (functional breakdown structure).

To analyze the state of UAV, the most important work is to select the appropriate angle for decoupling and classification. It can be divided, for example, by in the air or on the ground, by power on or by power off, or by static or dynamic. For a typical complex UAV system, five states, such as Mission preparation, Flight, Mission execution, Post-flight inspection and Parking, can generally be defined, as shown in Fig. 1.

Mission preparation refers to the state in which the UAV is prepared to meet the flight conditions after receiving the mission. It mainly includes mission planning, oil and gas filling, taxiing, refueling, power on and self-checking, starting engine and so on.

Flight refers to the state of the UAV following the planned route to complete take-off, climb, cruise, landing and other behaviors. It mainly includes controlling surface, retracting landing gear, controlling engine, braking, and so on.

Mission execution refers to the state of the UAV entering the mission area and performing a specific mission. It mainly includes situational awareness, danger avoidance, mission replanning, and so on.

Post-flight inspection refers to the UAV after the completion of the mission to carry out the relevant inspection work to confirm the system in good condition. It mainly includes consumable residue check, data download, and so on.

Parking refers to a state in which the UAV is stored in a certain way within a certain time. It mainly includes inspection, storage, external protection, recovery and so on.

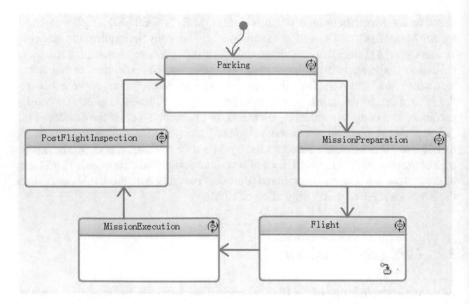

Fig. 1 State definition

3.3 Black Box Modeling

Behavior comes from function and can define two types of behaviors for the system: operation and reception. Operation can represent the behavior performed within the system. Reception can represent the behavior of the system interacting with the outside. Reception in black box modeling can represent the behavior of interaction between the system and external actor. In white box modeling, it can represent the behavior between the system and the external actors, but also the interaction between different subsystems. So the reception can be understood as an interface, divided into external interface and internal interface. Black box modeling can follow the process:

- According to the function breakdown structure, each function is defined as a behavior.
- The nested relation of composite state is defined, the sub-state is divided reasonably, and the logical relation between sub-states is defined. The level of the sub-state is refined so that it can exactly correspond to a series of operations.
- The transitions between states is defined, including triggers, events, guards, operation lists, etc., and the guard definition is leaded to attribute. Reception is associated with a trigger and event in a state diagram.
- According to the system operation of concept, the state machine diagram is debugged and executed according to a certain process, and the sequence diagram is a scene used by the system. In the simulation verification to be mentioned later, each test case is also a scenario used by the system.

3.4 White Box Modeling

In the system architecture design stage, according to the function clustering, the system is divided into a number of sub-systems, typically the UAV can be divided into the airframe, vehicle management system, integrated management system, the electromechanical system, the propulsion system and other sub-systems. Based on the system architecture, the white-box state machine model can create multiple blocks, each of which corresponds to a subsystem. However, the white box state machine diagram is not a simple "split" of the black box state machine, because the trigger between the states is transformed into an interface between subsystems.

4   Model Simulation and Verification

4.1 Purpose of the Verification

According to the requirements verification and traceability matrix (RVTM) of UAV system specification, the requirement items in UAV system specification need to be verified by simulation. Based on test case and external excitation, the model of black box and white box function of UAV is simulated to verify the compliance of the functional requirements of UAV system, including:

- The model covers all the functional requirements of the system.
- The external interface meets the requirements for interaction between the system and external actors.
- The internal interface satisfies the interaction between the internal subsystems.
- The UAV overall system function logic runs correctly.

4.2 Verification Criteria

The verification criteria contain the requirements specification verification and the overall system operation criteria, as shown in Fig. 2.

- According to the UAV system specification and the aircraft black box and white box model, the behavior, trigger, event and attribute corresponding to this requirement are confirmed.
- When all the system requirements are verified, according to the system operation of concept, the state of the system in the verification state diagram can be properly stimulated and transformed.

Fig. 2 Schematic diagram
of verification criteria

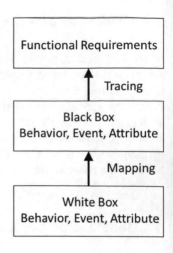

4.3 Test Case Design

Test cases are sets of test items used to meet the requirements of simulation and
verification and to ensure the correctness and completeness of verification, as shown
in Fig. 3. These items, which have clear input and output, are combined in a certain
way. The test case of the state machine model based on SysML is to test whether
the model can run exactly as intended by the designer. The processes to design test
cases include:

- System requirements analysis: Analyze the system requirements, extract the key
  points of functions, and generate functional test list to verify.
- Usage scenario analysis: According to the design process based on MBSE, use
  case describes the application scene. Through use case analysis, the typical application scenarios are extracted. These scenarios can be tested separately, or partly
  tested with other scenarios.
- Operation process analysis: The ideal operation flow of the system is analyzed
  according to the system operation of concept.
- Test case design: According to the operation of concept, the test items generated
  by the function points are effectively combined to form ordered set of cases under

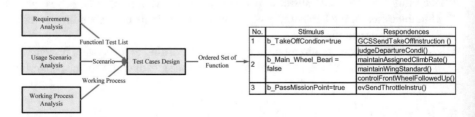

Fig. 3 Test case design process

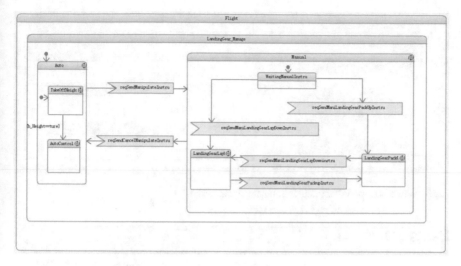

Fig. 4 Black box model

different usage scenarios. Each scenario can be executed sequentially and the functions can be verified in turn.

5 Examples

Taking the management behavior of UAV landing gear as an example, the feasibility of the system behavior modeling and simulation method based on SysML state machine diagram is verified. The modeling tool is Rhapsody software. The state machine black box modeling is shown in Fig. 4.

In the state machine white box modeling, behaviors are assigned to the integrated management system and the vehicle management system. The main behavior of the integrated management system is to receive and send ground instructions, and the main behavior of the vehicle management system is to execute according to instructions. See Figs. 5 and 6 for schematic diagrams.

Finally, by designing test cases and using state machine diagrams to run simulations, the compliance of the UAV system functional requirements and the correctness of internal logic operation were verified.

6 Conclusion

This paper has analyzed and described the system behavior modeling and simulation methods for complex UAV systems based on the SysML state machine diagram.

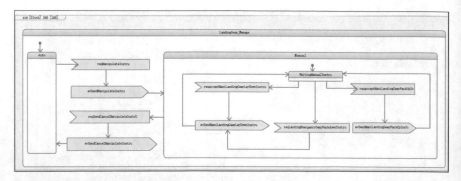

Fig. 5 White box model of integrated management system

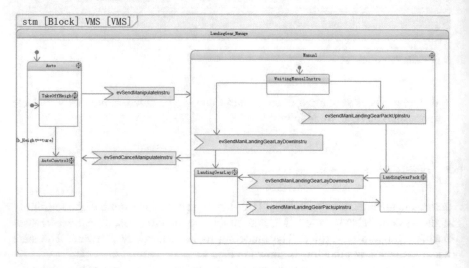

Fig. 6 White box model of vehicle management system

Taking UAV landing gear management behavior as an example, a SysML state machine diagram behavior model was established, and simulation verification was performed. Research and practice show that the UAV behavior model based on SysML state machine diagram in this paper can better reflect the real behavior of UAV system in operation and has strong engineering application value.

References

1. Estefan, J.A.: Survey of Model-Based Systems Engineering (MBSE) Methodologies, pp. 1–47. INCOSE MBSE Focus Group (2008)
2. Holt, J., Perry, S.: SysML for Systems Engineering, pp. 27–46. Institution of Engineering and Technology, UK (2008)
3. Mann, C.J.H.: A practical guide to SysML: the systems modeling language. Kybernetes **38**(1/2), 989–994 (2009)
4. Huang, E., Ramamurthy, R., McGinnis, L.F.: System and simulation modeling using SYSML. In: Simulation Conference, 2007 Winter IEEE, pp. 796–803 (2008)
5. Delligatti, L.: SysML Distilled: A Brief Guide to the Systems Modeling Language, pp. 89–176. Addison-Wesley, Boston (2013)
6. Douglass, B.P.: Agile Systems Engineering, pp. 5–36. Morgan Kaufmann, Burlington (2015)
7. Hoffmann, H.: SysML-based systems engineering using a model-driven development approach. In: INCOSE International Symposium, vol. 16, no. 1, pp. 804–814 (2006)
8. C.E.S.A.M.E.S.: CESAMES Systems Architecting Method: A Pocket Guide, pp. 3–125. CESAM Community (2017)

Research on the Method of Association Mapping Between Service Architecture Design and Service Component Development

Bing Xue, ZhiJuan Zhan, and YunHui Wang

1 Introduction

The development characteristics of network, information and system lead the technological change of all walks of life. In the field of avionics, avionics system architecture has experienced the development process of vertical, joint and integrated architecture, and gradually evolved to multi-platform avionics, "avionics cloud" and other horizontal avionics SoS architecture. Architecture is an abstraction of complex system. It provides a mechanism to understand and manage complex system by defining the composition structure and interaction relationship of system at the top level, hiding the local details of system components [1]. At present, the architecture standard models adopted in the military field mainly include the Zachman General Information System Architecture Framework [2], the US C4ISR Architecture Framework [3], the US DoD Architecture Framework (DoDAF) [4], and the British MoD Architecture Framework (MoDAF) [5]. Taking the most widely used DoDAF as an example, the framework has experienced version change from 1.0, 1.5 to 2.0, and its architecture development idea has also changed from product centered to data centered, and the service-oriented idea has been introduced to increase the service view and other products. The service architecture design based on the above methods has been widely studied in China. The University of National Defense Science and Technology introduces the service view in detail in the literature [6, 7], and studies the elements and development sequence of service architecture design, but lacks dynamic simulation verification and case application. In another paper [8], the author applied service architecture design to multi-platform avionics architecture, and completed

B. Xue (✉) · Z. Zhan · Y. Wang
Key Laboratory of Science and Technology on Avionics System Integration, Shanghai 200030, China
e-mail: xueicb16@126.com

Z. Zhan
e-mail: zzjaldm@aliyun.com

the simulation verification of executable model, realized the application of service architecture design method in avionics field.

Undertake the requirements of avionics architecture obtained from top-down analysis of service architecture design, carry out the development and implementation of Service-oriented Architecture (SOA), so as to complete the "design-development" penetration research of avionics architecture. SOA as a new architecture, uses the combination of components (services) with loose coupling and unified interface definition to form system functions. These components are published in the form of services, thus realizing the separation of technology and business. Among them, the development of service components has become the focus of SOA architecture implementation, including service template design, service encapsulation, service composition, service management and other key technologies. In literature [9], the development and implementation technology of SOA and service components are introduced in detail. The theoretical basis of SOA has been mature research and widely used, but there is no specific implementation case in the field of avionics. At the same time, there is a fault phenomenon from the top-level service architecture requirements to the development of service components. The development and composition of the bottom-level service components cannot meet the top-level architecture requirements in real time.

Therefore, aiming at the practical implementation of avionics domain architecture, this paper studies the mapping method between service architecture design and service component development, so as to realize the "top level avionics architecture design demand transfer service component development" through connectivity design.

2   Service Architecture Products for Multi-platform Avionics [8]

Based on the service view design idea in DoDAF 2.0, combined with the characteristics of multi-platform avionics architecture, the service architecture design of avionics system is carried out and the service architecture requirements are refined. Because this paper focuses on the relationship mapping between service architecture products and service component technology, we will not carry out the service architecture development and modeling in detail here. In reference [8], we elaborate the method flow and case analysis of service architecture design.

The service architecture include following product models: service interface representation model (SvcV-1), service resource flow representation model (SvcV-2), service and system matrix (SvcV-3a), service function representation model (SvcV-4), service and activity tracking matrix (SvcV-5), service event tracking representation model (SvcV-10c) and service state transition representation model (SvcV-10b). The specific products are as follows.

1.   Service Interface Representation Model (SvcV-1)

Svcv-1 describes the services, service items and their interconnection relations, which is used to define service concept, define service options, obtain service resource flow requirements, prepare capability integration plan, manage service integration transactions and prepare task plan (define capabilities and actors). Svcv-1 takes on the OV-2 resource flow description view, determines the services and service interfaces contained in all resources, and finally outputs the service interfaces and services by analyzing the resources involved in the task.

2.   Service Flow Representation Model (SvcV-2)

Svcv-2 describes the resource flows that need to be exchanged between services, which is used to standardize resource flows. Svcv-2 takes on OV-2 resource flow view and Svcv-1 service interface view. Through analyzing the interaction relationship of resource flow, it determines the resource interaction relationship between service interfaces residing on resources, and finally outputs the resource interaction relationship between service interfaces.

3.   Service and System Matrix (SvcV-3a)

SvcV-3a describes the relationship between various systems and services in a specific "architecture description", which is used to summarize the interoperability characteristics of system and service resource interaction, interface management, and comparison solution options. SvcV-3a takes on SvcV-1, analyzes the resource composition of specific architecture, and deploys services. It is the bridge between service layer and resource layer, and finally outputs the mapping matrix between services and system.

4.   Service Function Representation Model (SvcV-4)

SvcV-4 describes the functions implemented by services and service data flows, which are used to describe task job flows, identify service function requirements, decompose service functions, and associate relevant personnel and service functions. SvcV-4 takes on SvcV-1, analyzes the service functions that each service can realize according to the professional knowledge, and finally outputs all the service functions.

5.   Service and Activity Tracking Matrix (SvcV-5)

SvcV-5 describes the mapping between each service (activity) and the task activity (activity) they support, which is used to track service functional requirements and user requirements, track solution options and requirements, identify overlaps or gaps. SvcV-5 takes on OV-5 and SvcV-4, automatically generates the column vector of the matrix according to the activities in OV-5, automatically generates the row vector of the matrix according to the service function in SvcV-4, fills the mapping relationship between the activities and the service function into the matrix elements, and finally outputs the mapping relationship between the activities and the service function.

6.   Service Event Tracking Representation Model (SvcV-10c)

SvcV-10c describes the precise order of specific events of services, which is used to analyze resource events affecting tasks (business), analyze (dynamic) behaviors, and identify non-functional system requirements. SvcV-10c takes on OV-6c, SvcV-2 and SvcV-4. According to the event sequence in OV-6c and based on the resource interaction relationship in SvcV-2, the service function in SvcV-4 is refined into the sequence diagram of service access, and finally the service function call sequence based on service access is output.

7.   Service State Transition Representation Model (SvcV-10b)

Svcv-10b describes the response of services to events, which is used to define states, events and state transitions (behavior modeling) and identify constraints. SvcV-10b takes on SvcV-10c. Based on the sequence of service function calls, it draws the state machine conversion of each service access, which drives the dynamic simulation of the model, and finally outputs the state machine conversion model of each service access. According to all combat activities in svcv-5, a service template library is formed, and the full coverage of combat activities can meet the requirements of combat tasks. At the same time, it serves as the basis for subsequent system resource encapsulation to form sub services, and lays the instantiation foundation for sub service encapsulation.

3   Key Technologies of Service Component Development

Combined with avionics, the development of avionics service component includes two parts: Avionics resource encapsulation and Avionics service manager. In the avionics resource encapsulation, firstly, the resource service template is created according to the requirements of avionics service architecture, then the instantiation and encapsulation of the template is completed by extracting information from specific resource entities to form a sub service set, and then the sub service combination is completed based on the system functions according to the corresponding combination rules and methods to form a service set, finally, the formed services are described in the form of unified structure and call specification. The management and call of services are completed in the service manager. The services output from the resource encapsulation are registered in the service list first. The service management and service monitoring module will maintain the service list in real time. When the external task needs to call related services, the priority policy module of service call is combined with the service list, service management and service monitoring module to push services and provide task layer call. The whole process of service component development includes service template design, service encapsulation, service composition, service management and other key technologies.

- Service template design

The design of service template must meet the following four requirements:

1. Abstract—avionics resource service template does not involve specific resource objects, but is an abstract description of similar resources;
2. Normalization—the service template of avionics resources needs to be relatively stable in a certain range or a certain period of time, which has certain regulatory constraints on the development of avionics resources;
3. Generality—avionics resource service template is a description of similar avionics resources, which can be repeatedly applied to the instantiation of similar avionics resources;
4. Evolutionism—with the deepening of people's understanding of the essence of avionics resources, the service template of avionics resources will be summarized, defined and developed in practice.

- Service encapsulation

The specific resource entity with its own information parameters has the minimum granularity. As the object of the resource encapsulation, it brings its own performance indicators and model parameters into the corresponding type of service template for instantiation and encapsulation to form a sub service set. All input resources are encapsulated in the above process to form a sub service set, which lays the foundation for sub service composition to form services.

- Service composition

The sub service encapsulated by each function module resource of avionics system can complete the function with certain performance index. The system function with certain performance index can be completed by the service formed by different combinations of multiple sub services. The combination of sub services needs to follow the combination principle based on the requirements of performance indicators in accordance with the combination method based on the constraints of avionics system workflow. The sub services with the same or similar functions must be optimized by users on the basis of comprehensive measurement of Quality of Service (QoS), which can not be treated randomly and indiscriminately, which leads to the emergence of the optimization problem of service composition based on QoS.

- Service management

Service manager includes three parts: service list module, service management and service monitoring module, and service call priority policy module. The service formed by the resource encapsulation is output after standard description and registered in the service list module. The service list displays the usage status, calling interface, function type and performance index of each service. The service management and service monitoring module provides operations such as service start, service pause, service recovery and service stop, which is the management of avionics service when performing tasks. Service start is to start the specified avionics service, publish the interface of avionics service to the corresponding service request end, and complete the start of avionics service. Service pause is to set the specified avionics service as the pause service stage, which does not accept the call request of the task. Service recovery is to recover the specified avionics service from the suspended service state, so that the service can restart receiving the call request of the task. Service stop is to destroy the specified avionics service and unregister the service from the service platform. The service management and service monitoring module needs to update and manage the service status in the service list in real time. Service call priority policy module is the bridge between service and task mapping. When external task demands call a certain type of service, priority policy is assigned based on the performance index constraints of the service in the service list and the status of the service in the service management and service monitoring module to determine the specific service ID to be called, and then proceed to the next step call. The service manager breaks through the two-way process from top-down task requirements to service invocation, and from bottom-up service encapsulation to service registration, decouples the one-to-one mapping of avionics system resources and tasks, so as to realize the intelligent and autonomous operation of avionics system.

4 Research on the Method of Association and Mapping Between Service Architecture and Service Components

Through the service architecture design and service component development association mapping method, service template design based on service and combat activity tracking matrix, sub service encapsulation based on service function and system mapping matrix, service composition based on service rule model and service management and verification based on executable model are carried out to form "architecture design component development" integrated through service component development process. The overall idea is shown in Fig. 1.

1. **Design of service template based on service and operational activity tracking matrix**

First, based on the service and combat activity tracking matrix (SvcV-5) in the avionics system service architecture, a coarse-grained service template is designed, which includes the functional requirements and performance requirements options for completing combat activities, and does not involve specific performance indicators. The implementation steps are as follows:

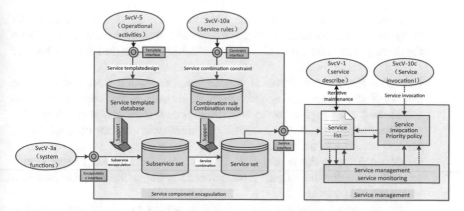

Fig. 1 Association and mapping between service architecture and service components

1.1　Design common service template definition. Service template elements include identity, type, function attribute, performance attribute, usage attribute, configuration information, etc. The Common Service Template (CST) is shown in Fig. 2.

CST = *<ServiceID, Type, FunctionPro, ParameterPro, UsingPro, DeploylmentInfo>*

ServiceID is the unique identity of service template. Each kind of service template has only one *ServiceID*;

Type is the type of this service template;

FunctionPro is the function attribute of service template, that is, the detailed function that can be completed by specific service under such service template;

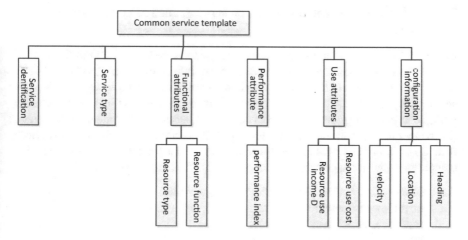

Fig. 2 Common service template properties

Parameter Pro is the performance attribute of the service template, which describes the performance index that the specific service can achieve to achieve the corresponding function, as one of the constraints when the service requester applies for service call;

UsingPro is the use attribute of service template, which is expressed as the *Use Cost* and *Income* of specific service. As one of the constraints when service requester applies for service invocation, *UsingPro* and *ParameterPro* are combined to determine whether a system resource can be used as the support resource of a service.

DeploymentInfo is the deployment configuration information of the service template, including the velocity, location and heading of the platform deployed by the specific service.

Using the service template can simplify the virtual encapsulation of system resources. The next step is to customize services that are more suitable for the abstract expression of system resources based on the service template according to the characteristics of different types of system resources, extract the description and implementation of resources and regulate their operation, so that system resources can reflect the attributes of system resources after encapsulation as services.

1.2   Define a set of operational activities. Take on the service architecture model, capture all the operational activities involved in refining avionics system functions from the svcv-5 view, merge them into categories, and generate a set of operational activities.

1.3   Design domain service template. According to the definition of common service template, each activity in the collection generated in step (1.2) is assigned the attribute of service template element, and the domain service template library is generated.

2.   Sub service encapsulation based on service function and system mapping matrix

Based on the service function and system mapping matrix (SvcV-3a) in the service architecture, and according to the information attribute of the system resource, the sub service encapsulation of the system resource is carried out on the basis of the domain service template. The sub service is encapsulated by the system functional resources with the minimum granularity. According to its service performance parameters, it can meet the operational activity requirements to an extent. All the sub services form a sub service set, which provides a meta model for the next step of service composition and management. The implementation steps are as follows:

2.1   Design the resource service adapter, as shown in Fig. 3. Based on the service attributes of domain service template in step (1.3), and combined with the requirements of service composition and service management for service encapsulation, a standardized service adapter is designed.

2.2   Define a collection of system functional resources. Undertake the service architecture model, capture and refine all system functional resources involved in

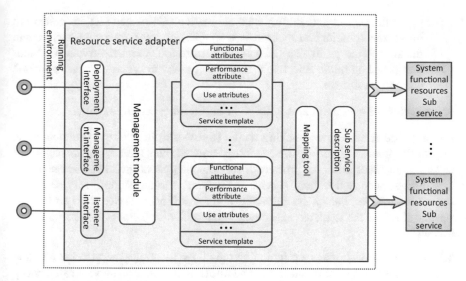

Fig. 3 Resource service adapter

avionics system functions from SvcV-3a view, merge them into categories, and generate system functional resources collection.

2.3   Generate subservice set. Take the system functional resources with their own information parameters in the set generated in step (2.2) as the input of the resource service adapter designed in step (2.1), bring the performance indicators and model parameters of all system functional resources into the corresponding type of domain service template, instantiate and encapsulate them to form the domain subservice set.

3.   Service composition based on service rule model

Taking the service rule model (SvcV-10a) in service architecture as one of the constraints of service composition, combining the sub service performance attributes and use attributes, the multi-objective optimization algorithm is used to combine the sub service sets and generate the service set, in which the service includes the operational performance index which can be achieved by the operational activities, and provides the input for the next priority policy of service invocation based on the combat task. The implementation steps are as follows:

3.1   Define service composition rules. Taking on the service architecture model, all service rules involved in refining avionics system functions are captured from the SvcV-10a view, and a set of service composition rules is generated.

3.2   Define how services are composed. According to the requirements of operational tasks, operational indexes are proposed for performance parameters, and the service combination mode is defined as meeting the operational indexes, spending the least and benefiting the most.

3.3   Build the service set. Taking the combination rule in step (3.1) as constraint input, the combination method in step (3.2) as optimization objective, and the subservice set in step (2.3) as optimization object, the service combination based on multi-objective optimization problem is carried out to generate domain service set.

4.   Service management and validation based on executable model

Based on the service event tracking model (SvcV-10c) in the service architecture, the whole process of service invocation is tracked, and the service management technologies involved in the process of service component invocation are realized, including service registration, service monitoring, service startup, service pause, service logout, etc. Finally, the rationality and feasibility of service management are verified through the dynamic executable model. The implementation steps are as follows:

4.1   Design service manager, including the above mentioned service list module, service management and service monitoring module, and service call priority policy module.

4.2   Define domain service management process. Take service architecture model, capture and refine all service management processes involved in avionics system function realization from svcv-10c view, and form standardized service management process as incentive of service manager.

4.3   Dynamic validation of executable models. According to the dynamic service invocation model generated by human–computer interface interaction, as shown in Fig. 4, verify the rationality and feasibility of avionics system service component design, and judge whether it meets the requirements of combat tasks, so as to carry out the next iteration.

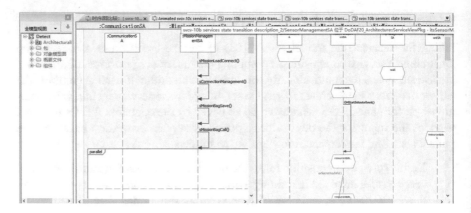

Fig. 4 Dynamic service invocation model

5 Conclusion

This paper first introduces the service architecture products for multi-platform avionics SoS, and elaborates the model products related to the development of underlying service components. Then, the basic theory of the key technologies in the process of service component development is studied, which lays the theoretical and technical foundation for the correlation mapping between the top-level architecture design and the bottom-level component development. Finally, this paper proposes a mapping method between service architecture design and service component development, which can connect the top-level service architecture and the bottom-level service components through model transfer, and realize seamless link and traceability of requirements, so as to reduce the cost of global maintenance caused by requirements or component changes. It provides the theoretical basis for enhancing system flexibility, improving software quality and reusability, and reducing software development cost.

References

1. Luo, X., Luo, A., Zhang, Y.: Architecture Technology of C⁴ISR System, pp. 4–6. National Defense Industry Press, Beijing (2010)
2. Zachman, J.A.: A framework for information systems architecture. IBM Syst. J. **26**(3), 276–292 (1987)
3. Integrated Architecture Panel of the C4ISR Integration Task Force: C4ISR Architecture Framework Version 1.0. The United States: Department of Defense (1996)
4. DoD Architecture Framework Working Group: DoD Architecture Framework Version 1.0. U.S, Department of Defense (2003)
5. Ministry of Defence: MOD Architectural Framework Viewpoint Overview Version 1.0. Ministry of Defence. https://www.modaf.com/files/MODAF%20Viewpoint%20Overview%20v1.0.pdf (2005)
6. Wang, L., Luo, X., Luo, A.: Research on service view products describing methods of C⁴ISR architecture. Sci. Technol. Eng. **10**(10), 2323–2328 (2010)
7. Wei, X., Chen, H., Zhang, M.: Service-system view raising and description in DODAF2.0. In: The 4th China Conference on Command and Control, Beijing, China (2016)
8. Xue, B., Zhan, Z., Wang, Y.: A new service-oriented architecture design method for avionics SOS. In: 32nd Congress of the International Council of the Aeronautical Sciences, Shanghai, China (2020) (Accepted)
9. Zhang, B.: Research an Application on Service Oriented Architecture. Taiyuan University of Technology, TaiYuan (2012)

Studies on Global Modeling Technology in Enterprise System Design and Governance

Yan Cheng, Xinguo Zhang, and Lefei Li

1 Introduction

Man-made systems basically fall into two categories based on their necessary components: engineering system such as equipment or bridge, constituted by parts or components and operating by physical laws, where man is not necessarily a component, and enterprise system constituted by man, such as combat command system, multinational companies, etc. Engineering system depends on continuous iteration and new product launch for optimization. While for enterprise system, its optimization depends on continuously benchmarking existing operational model and result against target, identifying deviations and correcting them in order to align operation results with the expected target. In such long-term governance, mature management models such as project management and total budget management as well as management rules such as check and balance between execution/decision/supervision, have evolved through incremental development of enterprise management science.

With growing complexity in system components and their interrelationships, especially by the further development in digital technologies, research and innovation of engineering system is increasingly dependent on digital system modeling and virtual simulation, which enables system's design and optimization at high cost-efficiency with high quality. Moreover, the application of modeling and simulation technologies have gradually expanded to use and maintenance phases of engineering system, and lead to various innovative business models, such as predictive maintenance,

Y. Cheng (✉) · X. Zhang · L. Li
Tsinghua University, Haidian District, Beijing 100084, People's Republic of China
e-mail: 13693329870@139.com

X. Zhang
e-mail: zhangxinguo@tsing.edu.cn

L. Li
e-mail: lilefei@tsinghua.edu.cn

© The Author(s), under exclusive license to Springer Nature Switzerland AG 2021
D. Krob et al. (eds.), *Complex Systems Design & Management*,
https://doi.org/10.1007/978-3-030-73539-5_32

product health management, etc. Yet such application in enterprise system principally remains in the phases of local modeling and structured analysis. For many, the art of enterprise system design and governance appears unsuitable for expression and analysis through full model representation of enterprise. As a result, the application of enterprise modeling and simulation techniques are mostly applied in management on local level and in a context of certainty, with proven effectiveness. For example, the generic domains of a corporation see a multitude of relatively mature theoretical methods, such as competitiveness model, equipment management model, knowledge management model, etc. as well as widely influential implementation methods including BMM, BSC, quality management model, etc. Yet few model theories and validation practice exist on the holistic level of enterprise.

In fact, without the global modeling specification as the foundation, the modeling and analysis of local business will be isolate from the holistic and dynamic changes of an enterprise system, where the analyzed model data tends to be one-time or time-specific and the analysis results are unlikely to be sustainably managed and upgraded into a continuously effective tool in enterprise governance. Global enterprise modeling means that the modeling elements can cover all the core requirements of enterprise management from strategy to execution, and can express the overall and various aspects of enterprise information Standardized global enterprise modeling provides a common language for people of all enterprise levels to create a transparent digital sand table for enterprise analysis, hence providing an environment for representation, analysis and decision of holistic strategic design; it may also provide a real and shared model data source for modeling in various domains, hence enabling accurate modeling and correlation analysis. Enabling continuous modeling and model transfer from a holistic perspective is indispensable for model-based enterprises.

Figure 1 illustrates the logic of this paper: different people in the enterprise have different concerns and are familiar with corresponding management tools and methodologies. Section 2 provides an analysis of the sequence of main management activities in an enterprise, stating that management activities of all types eventually and essentially span from architecture to process. As global management disciplines, architectural design, business model design and process development have become a **main thread** in enterprise management, providing methodologies and corresponding

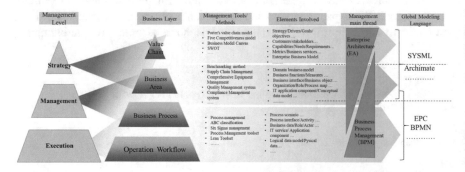

Fig. 1 Analysis path of global modeling method from management perspectives

modeling basis for the presentation, iterative verification and continuous implementation of validation of all management elements from strategy to implementation. Section 3 further analyzes the supporting modeling language supporting the management main thread methodology as illustrated in Fig. 1 on the right. Based on the existing international modeling standards and methodological theories, this paper provides approaches for building a uniform enterprise metamodel and forming enterprise-specific model classes by breakdown, and discusses the integration and collaboration of management methods through model presentation and interaction.

2 The "Management Thread" for Enterprise System Design and Governance

By fundamental changes have taken place in enterprise management objects and operational models. Single-domain, single-perspective management method applied in equipment, quality and cost management etc. plays a very limited role in leveraging the entire enterprise. Therefore, a new management system featuring horizontal collaboration and vertical breakdown in alignment is necessary in both holistic system governance and local business management. Here "horizontal" refers to the cross-domain, end-to-end process-execution layer that involves the eventual application of various management methods and tools. As the final value output is driven by process, integrated collaboration with process as carrier has become a consensus in orderly management. "Vertical" refers to the breakdown process from designing enterprise strategy to deployment execution elements. Essentially, the core of total enterprise governance is to build vertical breakdown in alignment while enabling the business design outcome of each level to be hierarchical verified against the upper layer. Based on the new scientific management theory proposed by Zhang [1], a system enabling further integrated collaboration in enterprise management is illustrated in Fig. 2, i.e. forming a methodological system and content system for forward design and governance of enterprise through vertical alignment and continuous optimization from mission to business architecture, business model and business process, and operational optimization featuring multi-element collaboration with process as a carrier, integrating local management methods and tools. In this system, holistic enterprise design and vertical element alignment are supported by architectural design; while execution on micro-level and implementation of various management methods are presented by process management. EA and BPM have constituted the main thread methodology of enterprise design and management.

Despite the fact that all enterprises are inherently in a certain architectural state, the task of architectural design is not to be taken for granted as it requires specific analysis, design and simulation verification by the enterprise manager. Architecture determines that an enterprise does what is right centering upon its strategic objective while process clarifies the correct way of doing it. The core of architectural method is to establish a big picture for the enterprise, to monitor its complexity and to describe

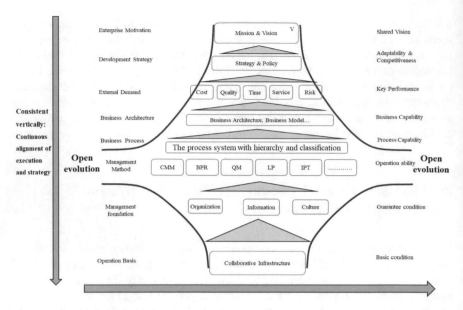

Fig. 2 Enterprise system design and governance featuring vertical alignment and horizontal collaboration

each element of the enterprise, their interrelationships as well as the relationship between as-is and to-be, focusing on enterprise design on a macro-level. In contrast, process management is a management technique with business process at its core, facilitating the collaboration between executive-level elements such as institution, responsibility, IT, etc., focusing on enterprise design on a micro-level.

In enterprise transformation and governance with long cycle and wide scope, the biggest challenge is continuous alignment of macro architectural design and micro process management, where business model plays a critical role. As a description of the logical relationships among business functions or business modules of the enterprise, business model is the fundamental logic of key activities, providing a mechanism and basis for further business rules design. Following the macro design of relationships between business domains and the deployment of strategic intention of the enterprise through architectural work, the internal operation rules to meet external demands will be described by the business model which tends to focus on a certain category of customers, business objects or domains, and the architectural performance in satisfying target customers and market demand and improving operational quality will be validated. The update and optimization of the business model will have an impact on the accomplishment of the intention of architectural design and furthermore on the re-definition of overall organization, institution and process in corresponding domains. There is an interactive relationship between the business architecture and the business model. In the TOGAF standard, Phase B: Business Architecture is where architects take the high-level business model artifact and develop a detailed set of architecture blueprints to enable more in-depth planning,

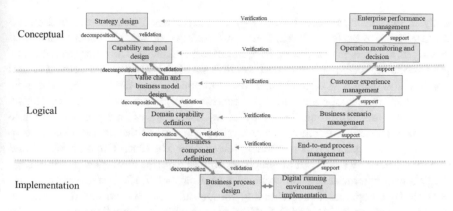

Fig. 3 The complete process in enterprise governance

investment, and option analysis [2]. For instance, when engine leasing service was first provided by an engine supplier, the content and approach of service received by customers went through a transformation within a short time, and architectural elements such as business structure, division of labor and skillset of personnel of engine suppliers were therefore adjusted. This is a typical case of macro business model transformation.

Various management mindsets, models and specialized management supporting tools are integrated in a layer- and domain-specific manner in enterprise "manage thread" constituted by architecture, business model and process. These management tools and methods continuously adjust the functions and operation behaviors of management elements. From a holistic view of the enterprise, a modern enterprise design process is eventually built, featuring hierarchical breakdown and verification and iterative validation from mission to process as illustrated in Fig. 3, as an enterprise management model characterized by integration and collaboration is shaped. In the decomposition of management elements on each level and in the management behaviors during enterprise operation in Fig. 2, the benchmark for verification and validation is a model system that fully demonstrates its design content and implementation requirement.

3   Multi-layer and Aligned Enterprise System Modeling Adapted to "Main Thread" Management Method

3.1   Approaches of Enterprise System Modeling Analysis

There are typically two types enterprise modeling approach: one aims at summarizing the key features of enterprise in the specific research domain, hence defining a multi-dimensional framework through structuralized content, such as Five Forces analysis and Business Model Canvas [3], etc. Yet it is usually used in elaboration and explanation of description, breakdown and initiative of specific issues rather than serving as the holistic description for enterprise governance. The other type of enterprise modeling shapes a common modeling standard through comprehensive and standardized definition of total enterprise components and their interrelationships, and enables wide-scope and multi-perspective enterprise modeling description by professional modeling tools for a faithful representation of enterprise status while supporting model-based enterprise analysis. Such capabilities are demonstrated in modeling languages such as Archimate released by The Open Group, SYSML and BMPN released by OMG, all of which are able to describe any object and interrelationship in an enterprise in a comprehensive way and their model symbols are comprehensible by computers, hence enabling computing, analysis and report of massive models. Besides, SYSML and BPMN are also capable of simulation and automated execution with the support of software tools. These modeling languages lie at the very core of implementing EA and BPM methodologies. Meanwhile, from the perspective of semantic composition of model elements, semantic elements on the process layer demonstrate obvious inheritability and scalability of those on the EA layer. The formation and mutual compatibility of corresponding modeling language standards and the expansion of supporting tools clearly demonstrate that the implementation system of enterprise governance methodologies from EA to BPM is maturing.

Modeling standards and tools for various types of information have always existed in the enterprise. In fact, model specifications developed in different domains including strategy, process, data, information system, infrastructure, etc. are creating new data silos, thus causing the correlation of enterprise model data to become unmanageable and making it difficult to ensure the authenticity and accuracy of single-domain models in enterprise due to the absence of continuous validation and interaction mechanism between models of different types. As illustrated in Fig. 4, EA model provides a normative definition of fundamental components of enterprise based on ontology on top of logical description of implementation objects, hence providing the basis for mapping and correlation between modeling languages in various domains. Process modeling, furthermore, elaborates the executive information in greater detail and, involving the steps and specific time and space data of implementation, is able to further integrate the subsystem modeling results on the implementation layer. Therefore, architectural model provides capabilities of expression, communication, simulation and decision on the layer of enterprise system design and sustained governance. In the collaborative enterprise management system from EA to BPM, the architectural modeling language and the process modeling language are the cornerstones of connectivity between models in different enterprise domains and implementation of an associated, model-based enterprise.

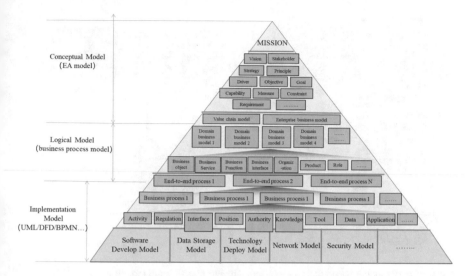

Fig. 4 An example of continuous modeling from concept to implementation in an enterprise system

3.2 Architecture Modeling in Enterprise System

As defined in ISO42010, the enterprise system architecture description will cover all fundamental elements the enterprise while providing a comprehensive description of the relationships between these elements, between the elements and their external environment as well as the principles of system design and evolution. In the latest Archimate standard [4], architecture description consists of seven interrelated parts including motivation, strategy, business, application, technology layers, etc. For an enterprise, a shortcut would be to adapt to and cut these modeling standards for application; however, for many enterprises it is necessary to translate their own management concepts and principles into logical expression of elements and their interrelationships and conduct revision and customization based on international standards in order to conduct model-based rule analysis and logical judgment after total modeling. For example, an enterprise advocating quantitative management may require all its objectives and business services to be measurable and all measurements to be obtained from business objects, which will produce new element rules and analysis requirements.

Before building a concrete model, model specifications need to be explicitly defined at two levels to meet the demand for flexible enterprise modeling, i.e. meta-model specification and specific model type specification. Where the former is a higher-level abstraction and specification of all enterprise elements and their relationships on the basis of the latter in order to provide a fundamental criterion for modeling in different domains and on different levels. More details about meta-model are provided in [5]. Metamodel is the holistic graphic expression of management laws as it reflects the composition of management elements and management

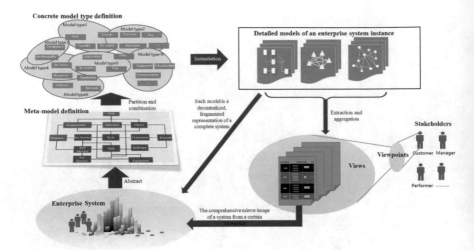

Fig. 5 Key components and their relationships in enterprise system modeling

logic of the enterprise; yet its complexity has limited its usability for modelers and analysts. Model design professionals, therefore, classify metamodels in different specific categories including business collaboration model, business composition relationship model, etc. with metamodel as a template. These specific model types interact with one another due to connected elements and their interrelationships while producing a multitude of examples. Managers on different levels of the enterprise with their respective concerns and requirements for information granularity, have different requirements for observation or analysis views which will efficiently and accurately emerge from a large number of dispersed models and faithfully reflect the status of enterprise system operation. The relationships between metamodel, specific model types and views are illustrated in Fig. 5, where the principle of enterprise modeling is comparable to engineering system modeling, of which the model-based, fully-defined specifications can be very complicated, while these specifications can be effectively simplified by classifying into modeling specifications based on different entity types. Once modeling is completed for each component respectively and a front view of the complete system is required, such a front view can be quickly formed by combining the model information of corresponding system components according to the rules. The advantage of doing this is that changes in model incurred by any component modification would be directly reflected in the view; while any change in bottom layer rules would be reflected on metamodel layer, thus triggering adjustment and re-evaluation of correlated model type. This provides support for the design, analysis, simulation and evaluation of model-based and model-driven system architecture.

In architecture-level modeling, effective metamodel design and model types partition based on metamodel are the key fundamental modeling techniques. In this article, a preliminary discussion is conducted on the rules of model types partition based on metamodel. Suppose the enterprise element involved in metamodel is E and the specific model type illustrated is M, the following basic principles are to be considered in defining the specific modeling specifications:

- E shall be attributed to different architectural levels or classified according to the roles determining element contents. For example, holistic elements such as principle and objective can be attributed to higher decision-making level such as strategic architecture or motivation architecture. Generally speaking, requirements and data are impacted by all architectural levels as they are cross-cutting architectural contents with specific and objective necessity.
- Generally speaking, the complexity of each M should ensure that its modeling work can be accomplished by the corresponding role independently. It should be noted that the modeling of M should not involve multiple riles or through collaboration by multiple positions of the same role, lest the difficulty of modeling should increase and the truthfulness of the basic model be compromised.
- Each M should include at least two different E, and such E should not fall into multiple architectural layers. The set of M must cover all E.
- Each E should appear in at least two model classes; in a given scenario, one class of model must used in defining such element while the other class validates whether the content of such element is appropriate (or meeting the expectation);
- Apart from requirement management models and governance models, other M should, whenever possible, avoid including elements across different architectural levels so that more flexibility can be generated by higher level architectural requirements as expressed on a lower level, and that "hard" definition of interrelationships between E may be avoided.

In summary, EA modeling requires not only the ontology-based definition of architectural elements, scientific definition of levels of architectural elements and supporting views for stakeholders for observation, analysis and decision-making, but also the finalization of specific model types in a systematic manner to support the expression of status and changes in various enterprise domains and the creation of corresponding views. Apart from M in diagram format as analyzed above, massive inventory models to ensure uniform data source and matrix model to regulate critical element relationships also exist. The definition of these EA components including architectural elements, architectural levels, model classes, perspectives/views, etc. constitute the core content of architecture-level modeling. Taking DODAF as an example, its metamodel includes 24 core elements, where 8 perspectives are defined and 52 specific model types are contained, most of which are directory or matrix models, and over 10 fall into M category.

3.3 Business Process Modeling in Enterprise System

Business process is the carrier for enterprise value realization as well as the endpoint of effects of all management and technical elements in an enterprise. Achievements in research and practice of business process management for the past 10 years are summarized and crystallized in Publication [6] which also offers best practice guidelines. In the phase of architectural design where process is but one element in business architecture, the result of business architectural design defines key process objectives and basic outline of the business architecture. According to publication [6], the business architecture design and optimization just is the input for the initial task of analysis phase in process management, After the definition of process value, it should be followed by the completion of the entire process map and 5 levels of process design in the phase of process design and construction, and eventually build a process deployment and operation environment in collaboration with other management elements. According to this standard process management methodology, a process model standard that enables effective transfer and conversion is needed from process analysis to implementation and monitoring of process automation. The key here is the model expression and transfer in three phases: first, model conformity transfer and conformance validation from architecture to process; second, level-by-level process modeling in conformance with guidelines; third, continuous transfer and conversion from business process model to workflow-defined model in information system construction. Through continuous model transfer throughout these three phases, a solution shall be found for whole-process conformity in implementing EA design and realizing business digital environment, and revision linkage of enterprise models and capabilities of rapid adjustment, adaptation and realization for endpoint IT execution model throughout different phases shall be built.

For model conformity transfer and conformance validation from architecture to process, latest process guidebooks and model standards such as publication [6] and BPMN2.0 [7] can be referred. In publication [6], 16 process element groups are defined, involving the oncology-based definition of over 60 process elements; while BMPN2.0 also includes definition of numerous process elements. As process models are an aggregate of various types of elements such as enterprise elements (enterprise organization, role, position), business elements (location, service, business object, etc.), technical elements (supporting application, technical tool), etc., some of these elements are inevitably architectural elements while others are on a more detailed level in process execution such as step or status. The overlapping part of architectural elements and process elements constitutes the foundation for conformance enterprise modeling, where architectural model provides the context for the description of process model while process model describes the correct business execution route.

There have widely applicable modeling methods such as EPC, Tunnel diagram, value stream diagram, etc. in process modeling. Attention should be paid to the bottom layer process description, i.e. detailed description of process activities. At this point, all management element requirements shall be broken down to the minimum and most detailed units. For example, regulations and specifications are clarified to the level of operations and rules of each activity; while the use of input and the production of output are clarified to data items, Such process description meets the requirements for automated process execution and simulation in BPMN2.0, hence can either be described in BPMN2.0 or translated into BPMN2.0 model by semantic mapping to provide input for automate execution in the next step.

For continuous transfer and conversion from business process model to workflow-defined model in information system construction, BPMN2.0 standard has provided sufficient support for the last mile in automated business process execution, hence becoming the standard process interface in many IT development platforms and is capable of receiving BPMN2.0 model and converting it into configurable work-flow engine for rapid and flexible realization of the process from being visible to being usable. Yet this conversion is less than impeccable as a significant gap tends to exist between the result of business description through process and the highly structuralized programming realization in information system. Therefore, clear model processing specifications need to be developed so as to gradually translate business-facing BPMN process model into one facing technical implementation; while access control, data rules, etc. shall be defined in detail. Based on Author's practice in large enterprises on an extensive scale, it is proven that process model with BPMN as eventual carrier provides an effective implementation approach for rapid digital implementation and flexible change for business process system. The next issue to be addressed by process model is the bilateral correlation and interaction between process model and business model in IT, which will enable model to become a "remote controller" for the management of holistic design and governance by enterprise.

4 Summary

This article offered a detailed discussion on full-model description from macro-design to micro-implementation of the enterprise through architectural modeling and process modeling, where the model description itself creates tremendous values for information sharing, business design and optimization of collaborative operation, providing a digital sand table of the AS-IS and the TO-BE for enterprise. Meanwhile, expression through visual model will have a major impact on the manner of enterprise management, analysis and decision-making, enabling simulation of enterprise operation under assumed conditions, hence helping managers in effective analysis and appropriate decision-making concerning all types of issues in enterprise operation and transformation design. At present, the enterprise system is far behind the engineering system in the application of modeling and simulation, which is, on one hand, attributable to insufficient skills of managers as the separation between management science and modeling theory and computer science has, to some extent, hindered managers from deep research and application of enterprise modeling technology; on the other hand, multitudes of hidden rules and fuzzy relationship between management elements in enterprise management have hindered the application of modeling technology. More time is still needed to build and adapt to new management modes in a full-fledged digital era to empower the collaboration of human intelligence and computer intelligence. Besides, it is equally important to build supporting tools and develop technologies, and more input from enterprise system researchers is needed in integrated application of successful modeling techniques and standards to accelerate evolution.

References

1. Zhang, X.: New scientific management: complexity-oriented modern management theory and method (2013)
2. Business Models: TOGAF® Series Guide. The Open Group (2018)
3. Osterwalder, A., Pigneur, Y., et al.: Business Model Generation, p. 14. Wiley, Hoboken (2010)
4. ArchiMate 3.0 Specification. The Open Group (2016)
5. Abu Bakar, N.A., Yaacob, S., Hussein, S.S.: Dynamic metamodel approach for government enterprise architecture model management. In: The Fifth Information Systems International Conference 2019. Procedia Comput. Sci. **161**, 894–902
6. von Rosing, M., Scheer, A.-W., von Scheel, H.: The Complete Business Process Handbook-Body of Knowledge from Process Modeling to BPM, vol. I. Elsevier Inc., Amsterdam (2015)
7. Allweyer, T.: BPMN 2.0 (2010)

Supporting Automotive Cooling and HVAC Systems Design Using a SysML-Modelica Transformation Approach

Junjie Yan, Biao Hu, Xin Wang, Minghui Yue, Xiaobing Liu,
Marco Forlingieri, and Richard Sun

1 Introduction

Nowadays the mechatronic products are getting more and more complex since a large amount of disciplines and domains are involved, so that the collaboration becomes of increasing importance. Systems engineering is the formal approach to manage the whole lifecycle of the product, i.e., requirements, functions, architectures, detailed designs, verifications and validations. Traditionally, documents are used to organize all the artifacts and activities, while nowadays Model-Based Systems Engineering (MBSE) was introduced by INCOSE and OMG for better management of systems engineering. System Modeling Language (SysML) provides syntax and standard to describe conceptual design. It is well-suited to model logical behavior with the help of behavior diagrams. On the other hand, simulation is also important on understanding how is the performance of the system. Modelica is one of the simulation languages that are widely used. These two modeling techniques should be well integrated and

J. Yan (✉) · B. Hu · X. Wang · M. Yue · X. Liu · R. Sun
Changan Auto Global R&D Center, Chongqing Changan Automobile Co., Ltd., Chongqing
401120, China
e-mail: yanjj5@changan.com.cn

B. Hu
e-mail: wxhubiao@126.com

X. Wang
e-mail: wangxinupc@163.com

X. Liu
e-mail: alberson@163.com

R. Sun
e-mail: sunld@changan.com.cn

M. Forlingieri
MBSE Consulting Ltd, 367-375 Queen's Road Central, Sheung Wan, Hong Kong, China
e-mail: marco.forlingieri@mbseconsulting.com

© The Author(s), under exclusive license to Springer Nature Switzerland AG 2021
D. Krob et al. (eds.), *Complex Systems Design & Management*,
https://doi.org/10.1007/978-3-030-73539-5_33

435

the traceability should be ensured. This will complete the closure of engineering process, since not only logical behaviors but also continuous dynamic behaviors need to be simulated, so that the requirements can be verified at performance level.

Johnson [1, 2] established syntax mapping between SysML and Modelica languages as they share similar structures of the existing semantics. This was further formalized and named as SysML4Modelica profile and continuously developed by OMG [3]. The method was applied as an attempt for electrical mobility design [4] and vehicle acceleration design [5], which are both prove-of-concepts. Besides Modelica tools such as MapleSim [6, 7], other continuous dynamic simulations tools were also investigated such as Simscape [8, 9], AMESim [10] and TRNSYS [11], to search for the paths of linking SysML models with simulation models. Kraus [12] conducted impact torque calculation for an automotive driveline system with the SysML Parametric Diagram (PAR), yet this calculation remains simple due to the limitation of modeling ability of PARs. Branscomb [5] and Bailey [13] explored the value of simply exporting simulation framework from SysML model [14] rather than a complete simulation since this reduces modeling effort in SysML, yet this kind of discussion is still rare.

Since system simulation is also one of the main activities of systems engineering that has been developed independently, it is worth reusing the existing system simulation models and link them with SysML models. However, the reports on this aspect are few. In this work the adoption of SysML-Modelica transformation in the enterprise is presented.

2 MBSE Framework at Changan Automobile

Changan CAE domain kicked off the MBSE adoption project in 2018 to improve collaboration with other departments. OpenModelica, GT-Suite, AMESim and AVL-Cruise are the main simulation tools that perform system simulations. The main idea is to develop a framework to integrate logical design and system simulation capabilities.

Figure 1 shows the MBSE framework. The systems engineer creates the SysML model using Cameo, where requirements are imported and modeled. IBD is also used to model the structure of the systems, such as cooling system and HVAC system, which will be discussed in details later.

Then the components behaviors are modeled either using SysML Parametric Diagram (PAR) or reusing simulation tool libraries. The first approach, UC1 as shown in Fig. 2a, PARs are applied to fully model the behaviors of the components. Here SysPhS [15] standard is applied so that later the SysML model could be exported as Modelica .mo file. In UC2, the second approach, the SysML model is created aligning with the architecture of Modelica model. By performing model transformation from SysML to Modelica, the structure of the system could be matched to the existing Modelica model. By any approach the systems engineer would be able to generate the .mo file to conduct system simulation, which means the consistency with SysML

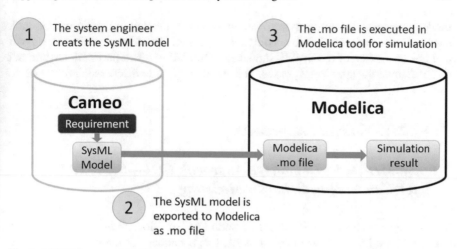

Fig. 1 MBSE framework in Changan automobile

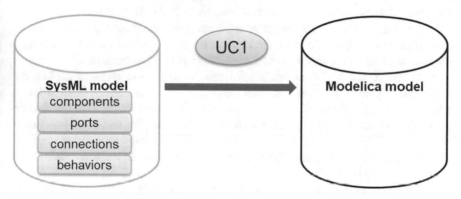

(a) UC1 approach – SysML model dominated

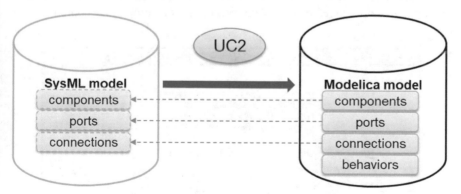

(b) UC2 approach – Modelica model dominated

Fig. 2 SysML-Modelica transformation approaches

model is ensured. Then the system simulation is conducted in the simulation tool, such as OpenModelica. We focus only on the SysML-Modelica transformation part and mention some of the SysML model and simulation result part in this work, but do not discuss the later V&V phases. That would be further discussed in future work.

3 SysML-Modelica Approaches

3.1 SysML Model Dominated Approach UC1—Using PAR for Components Behavior Modeling

The first approach applies PARs for components behavior modeling using SysPhS standard. SysPhS [15] was released in 2018 as a standard of the mapping between SysML and Modelica as well as MATLab Simulink. The conserved and non-conserved flow properties, flow rate and pressure in hydraulic problems for instance, are defined in SysPhS. Then the components and ports are modeled using these two properties, which forms the main structure of the system simulation since the main purpose of the system simulation is to predict the properties variations.

A hydraulic simulation of automotive cooling system is given to explain this approach. Figure 3 presents the decomposition of *VehicleForHydauSimulation*, which is later exported as .mo file for simulation. Figure 4 shows the IBD of *Vehicle-ForHydauSimulation*, and Fig. 5 shows the IBD of Cooling System about how pump, water jacket, thermostat, oil cooler, transmission cooler, high temperature radiator and tank are linked logically. All the ports are modeled as *VolumeFlowElement*, which is defined by SysPhS.

PARs were used to model the behavior of each component. Three types of constraints were defined as shown in Fig. 6, which are *ResistanceConstraint*, *PumpConstraint* and *TankConstraint*. *ResistanceConstraint* consists of four governing laws: (a) $pp - np = p$ refers to the definition of pressure drop of the component; (b)

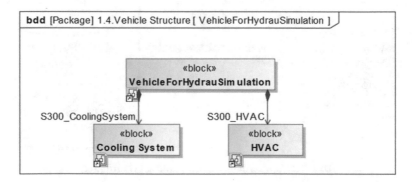

Fig. 3 Decomposition of *VehicleForHydauSimulation*

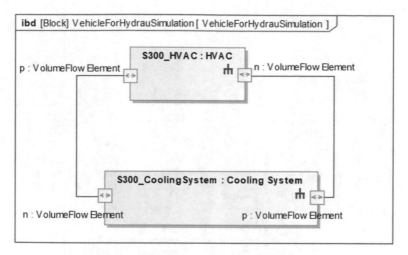

Fig. 4 Vehicle architecture focusing on thermal management domain

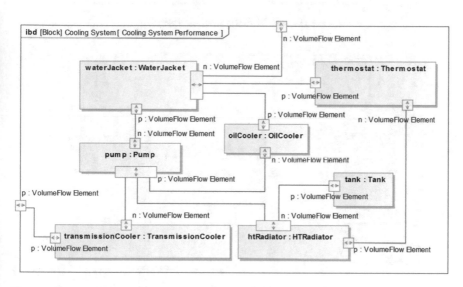

Fig. 5 IBD of cooling system

$pq + nq = 0$ refers to the conservation of mass; (c) $pq = q$ refers to the definition of the flowrate on the component; (d) $p = r * q$ refers to the linear resistance hypothesis. The resistance, r, is a parameter that need to be specified for each component that applying *ResistanceConstraint*. Figure 6 also presents the definition of other two constraints. Table 1 shows the summary of the usage of different constraints, and the parameters are specified.

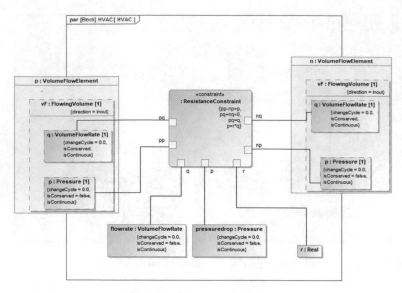

(a) PAR of *ResistanceConstraint*

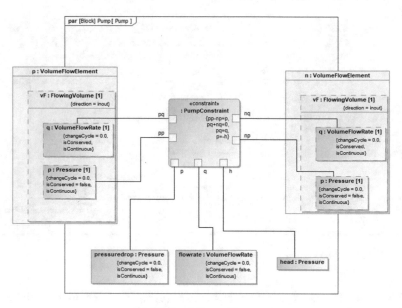

(b) PAR of *PumpConstraint*

Fig. 6 PARs of constraints

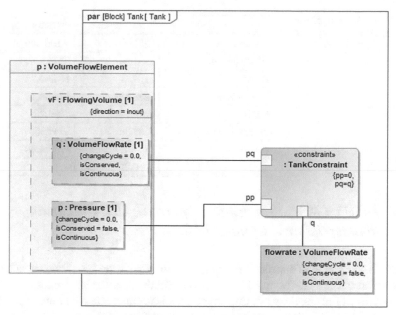

(c) PAR of *TankConstraint*

Fig. 6 (continued)

Table 1 Usages of constraints by components

	Constraint type	Parameters
Pump	Pump constraint	$h = 100,000$ Pa
Water jacket	Resistance constraint	$r = 64$ Pa s/m^3
HVAC	Resistance constraint	$r = 60$ Pa s/m^3
Thermostat	Resistance constraint	$r = 50$ Pa s/m^3
HT radiator	Resistance constraint	$r = 15$ Pa s/m^3
Transmission cooler	Resistance constraint	$r = 36$ Pa s/m^3
Oil cooler	Resistance constraint	$r = 22$ Pa s/m^3
Tank	Tank constraint	–

When SysML model is ready in Cameo, one can export .mo file from the model for simulation. Since this is a quite simple steady-state case, so that any Modelica-based simulation tool can be used to simulate the generated .mo file simply using default setting. OpenModelica is chosen as the tool, and Table 2 gives the simulation result for this case. The flowrate and pressure drop are reasonable as the mass conservation is achieved and the linear resistance law is followed.

Table 2 Simulation results of hydraulic case

	Flowrate/m^3/s	Pressure drop/Pa
Pump	1281.49	−100,000.00
Water jacket	1281.49	82,015.59
HVAC	187.34	11,240.25
Thermostat	276.68	13,834.16
HT radiator	276.68	4150.25
Transmission cooler	187.34	6744.15
Oil cooler	817.47	17,984.40
Tank	0.00	–

3.2 Modelica Model Dominated Approach UC2—Reusing Commercial or Customized Library

Although the previous approach is complete for simulation modeling, it takes too much effort of the systems engineers to model all the detailed information in PARs. For systems engineers, the more important work is to concentrate on functions, architectures and trade off. In order to know the performance of different design, they need to try different sets of parameters to find the best configuration leading to highest performance. Only key parameters are important to the systems engineers, which means the approach in UC1 is too heavy. Moreover, the system level simulation is to some extent mature in many industries including automotive, so that considering reusing the commercial and customized simulation ability is essential. In this approach, the responsibilities of the system engineer and the simulation engineer are clearly split so that they do not overlap each other.

A refrigeration simulation of automotive HVAC system is given for this Modelica model dominated approach UC2 because of the higher complexity. Note that we do not use the same simulation case as the cooling system because we want to highlight that the UC2 approach is more suitable to deal with complex case which UC1 cannot. Figure 7 shows the IBD of HVAC. The compressor, the condenser, the TXV and the evaporator form the basic refrigeration loop. Several boundary conditions are modeled including the compressor rotation speed and inlet air of condenser and evaporator.

Unlike simple hydraulic laws in Sect. 3.1, more governing laws, especially two-phase flow and phase-change heat transfer, should be included to model the refrigeration loop behaviors accurately. Thus, following Sect. 3.1 approach is annoying. Changan already has the Modelica modeling capability of refrigeration loop, so that a proper mapping is reasonable to reduce work and reuse existing simulation ability. To mapping SysML blocks to Modelica models, a stereotype *ModelicaBlock* is applied for all the components in this IBD. The relevant Modelica model path is specified so that later when transforming SysML model to Modelica model, it can be automatically matched. Figure 8 shows an instance

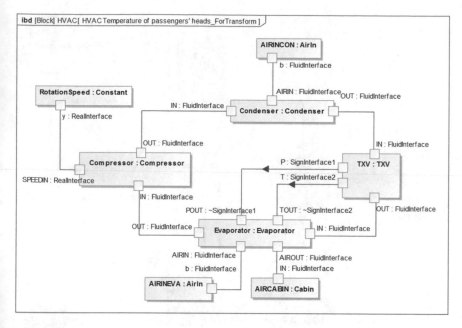

Fig. 7 IBD of HVAC system

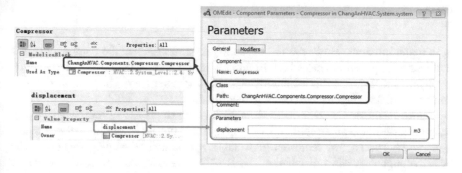

Fig. 8 Components and parameters mapping between SysML model and Modelica model

how this is applied for the block *Compressor*. The Modelica path, *Changan-HVAC.Components.Compressor.Compressor*, is defined as the *ModelicaBlock* name. When transforming model, this path will be read by OpenModelica to access the commercial or customized model, which is *ChanganHVAC* library here. Moreover, the parameter *displacement* is also transformed as the model parameter in Modelica. This allows the system engineers to concentrate on the key parameters and reuse existing libraries.

Figure 9 illustrates the transformed Modelica model of HVAC in OpenModelica. In Cameo SP2 only textual file can be generated, while in SP3 it can also come up

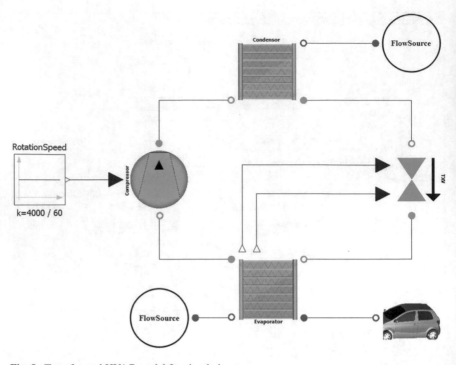

Fig. 9 Transformed HVAC model for simulation

with a graphic draft. However, the capability to well organize the components in a automatically generated model is still limited, so Fig. 9 is the final result after manual arrangement. Note that this arrangement is only for better view. The generated model can be simulated directly without any modification.

Figure 10 illustrates the typical simulation result of HVAC system, where T_evp represents the outlet air temperature of evaporator, T_cbn represents the cabin temperature, p_high and p_low represent high and low pressure of the refrigerant loop. The systems engineers can tell whether the HVAC system is well designed or not by monitoring the temperature decrease during a certain period.

3.3 Limitations and Future Works

The SysML-Modelica integrated framework and the two approaches could be used to integrate architectures and behaviors modeling while we avoid overlapping the capabilities of systems engineers and simulation engineers. SysML model is used as a generic model for requirements, functions, architectures, behaviors and parameters, so it covers more aspects of those in Modelica model. Thus, a configuration ability to choose which parts or parameters should be exported is required. In Cameo SP2,

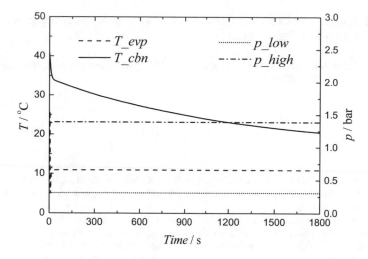

Fig. 10 Refrigeration simulation result

only the whole block could be exported with all the elements inside, while in Cameo SP3 it allows to export to Modelica only the content shown in the IBD and not the entire block. On one hand this feature can be used to reuse the SysML block for different simulation scopes (by selecting only the ports and parameters needed). On the other hand it has some limitations because only the IBD is simulated and not the block. This means that the simulation requirement linked to the block cannot be fully satisfied by the result of the simulation. This should be further discussed and developed in the future.

Two different examples, cooling system and HVAC system, have been used to show what is more suitable in term of complexity for UC1 and what for UC2. On the other hand, having two different example scopes for the two use cases does not allow us to have a proper comparison. Also the requirement traceability and V&V process should be further developed and discussed to form a complete MBSE framework in the future.

4 Conclusion

A MBSE framework that integrates SysML and Modelica modeling is presented. The requirements and IBDs are first modeled using SysML in Cameo. Later two approaches are developed for components behaviors modeling. The UC1 approach requires the systems engineer to model the equations using PARs, and SysPhS is applied as a standard. The UC2 approach makes full use of simulation capabilities of the enterprise, so that the mapping between SysML and Modelica models is carefully organized. The MBSE framework integrates conceptual design and system simulation as well as requirements.

Acknowledgements This work is supported by the Chongqing Science and Technology Project (cstc2019jscx-msxmX0026).

References

1. Johnson, T.A., Paredis, C.J.J., Burkhart, R., Jobe, J.M.: Modeling continuous system dynamics in SysML. In: ASME International Mechanical Engineering Congress and Exposition, Seattle, WA. ASME (2007)
2. Johnson, T., Kerzhner, A., Paredis, C.J.J., Burkhart, R.: Integrating models and simulations of continuous dynamics into SysML. J. Comput. Inf. Sci. Eng. **12**(1), 011002 (2012)
3. Paredis, C.J.J., Bernard, Y., Burkhart, R.M., de Koning, H.-P., Friedenthal, S., Fritzson, P., et al.: An overview of the SysML-Modelica transformation specification. In: 20th Anniversary International INCOSE Symposium, Chicago, IL, vol. 20, no. 1, pp. 709–722 (2010)
4. Votintseva, A., Witschel, P., Goedecke, A.: Analysis of a complex system for electrical mobility using a model-based engineering approach focusing on simulation. Procedia Comput. Sci. **6**, 57–62 (2011)
5. Branscomb, J.M., Paredis, C.J.J., Che, J., Jennings, M.J.: Supporting multidisciplinary vehicle analysis using a vehicle reference architecture model in SysML. Procedia Comput. Sci. **16**, 79–88 (2013)
6. Herzig, S.J.I., Rouquette, N.F., Forrest, S., Jenkins, J.S.: Integrating analytical models with descriptive system models: implementation of the OMG SyML standard for the tool-specific case of MapleSim and MagicDraw. Procedia Comput. Sci. **16**, 118–127 (2013)
7. Cao, Y., Liu, Y., Fan, H., Fan, B.: SysML-based uniform behavior modeling and automated mapping of design and simulation model for complex mechatronics. Comput. Aided Des. **45**(3), 764–776 (2013)
8. Cao, Y., Liu, Y., Paredis, C.J.J.: Integration of system-level design and analysis models of mechatronic system behavior based on SysML and simscape. In: ASME International Design Engineering Technical Conferences and Computers and Information in Engineering Conference, Montreal, Quebec, Canada. ASME, detc2010-28213 (2010)
9. Cao, Y., Liu, Y., Paredis, C.J.J.: System-level model integration of design and simulation for mechatronic systems based on SysML. Mechatronics **21**(6), 1063–1075 (2011)
10. Barbieri, G., Fantuzzi, C., Borsari, R.: A model-based design methodology for the development of mechatronic systems. Mechatronics **24**(7), 833–843 (2014)
11. Kim, S.H.: Automating building energy system modeling and analysis: an approach based on SysML and model transformations. Autom. Constr. **41**, 119–138 (2014)
12. Kraus, R., Papaioannou, G., Sivan, A.: Application of model based system engineering (MBSE) principles to an automotive driveline sub-system architecture. M.Sc. thesis, University of Detroit Mercy (2016)
13. Bailey, W.C.: Using model-based methods to support vehicle analysis planning. M.Sc. thesis, Georgia Institute of Technology (2013)
14. Friedenthal, S., Moore, A., Steiner, R.: A Practical Guide to SysML, 3rd edn. Morgan Kaufmann OMG Press, Waltham (2015)
15. SysML extension for physical interaction and signal flow simulation. Version 1.0. 2018. https://www.omg.org/spec/SysPhS/1.0/PDF

Towards Automated GUI Design of Display Control Systems Based on SysML and Ontologies

Jin Su, Jianjun Hu, Yue Cao, Yusheng Liu, Wang Chen, and Chao Wang

1 Introduction

As the complexity of the battlefield environment rising, the missions of armoured vehicles become more and more complex and flexible. The display control systems (DCSs) of the vehicles are responsible for managing and scheduling these complex missions [1]. Therefore, correspondingly, more advanced graphical user interface (GUI) of the DCSs is required to assist the passengers to accomplish the functions of vehicles [2]. However, the GUI of the display control systems is normally designed manually, which is error-prone and inefficient. Moreover, the GUI design cannot be traced back to the vehicle functions and hence is difficult to be verified. To this end, towards the automated GUI design, how to automatically identify the components of the GUI systems according to the vehicle functions is investigated in this study.

As observed in this study, intrinsic relationships exist between the functions of a vehicle and the GUI of its DCSs. On one hand, necessary instructions and parameters to realize the functions should be provided by the passengers via the GUI. On the other hand, execution status and results should be returned to the passengers via the GUI. This derives the basic idea to enable the automated retrieval of GUI components based on vehicle functions.

J. Su · J. Hu (✉) · W. Chen · C. Wang
China North Vehicle Research Institute, Beijing 100072, People's Republic of China

Y. Cao (✉)
Zhejiang University of Technology, Hangzhou 310014, People's Republic of China
e-mail: ycao@zju.edu.cn

Y. Liu
Zhejiang University, Hangzhou 310058, People's Republic of China
e-mail: ysliu@cad.zju.edu.cn

According to this idea, two key issues should be addressed. First, structured representations should be provided to facilitate designers to describe vehicle functions and GUI components. Second, reasoning mechanisms between the knowledge of vehicle functions and GUI components should be investigated.

In this study, an approach combining SysML and ontologies is proposed to address these two issues. First, a method to model the vehicle functions and GUI components is presented by extending SysML. Second, the semantics of the functions and GUI components is formally described in OWL2 such that the GUI components can be retrieved by functions according to their semantic similarity.

The rest of the paper is organized as follows. Section 2 reviews related work. Section 3 gives an overview of the proposed approach. Section 4 introduces the function and GUI component modeling method in SysML. Section 5 presents the formal descriptions in OWL2 and the automated retrieval procedure. Section 6 illustrates the proposed approach using the task planning function a case study. Section 7 concludes this paper and discusses future work.

2 Related Work

Related work of this study is reviewed from two aspects. First, literature about DCS design is discussed and corresponding challenges are summarized. Second, applications of SysML and ontologies in the early design of systems of different domains are illustrated to show the wide acceptance of these two enabling technologies.

Wu et al. [3] introduced a modular design approach of display control software based on the embedded operating system VxWorks and graphical interface development tool Tilcon. Xie et al. [4] proposed the future avionics system architecture, which includes five layers organized in a service-oriented manner. Shao and Zhang [5] designed a display control simulation system, whose design process includes both modular system design and detailed design for each subsystem. Liu et al. [6] introduced a hardware composition and software design process of the DCSs. Liu et al. [7] presented an interface design approach of DCSs and solved the problem of displaying Chinese characters. These works discussed the GUI design of DCSs more or less, but the GUI functions and components are manually identified based on designers' experience and not related to the functions of the vehicles. Moreover, no structured and computer-comprehensible modeling methods to the UI systems are provided.

SysML has been broadly adopted in the early design of complex systems. Many researchers extended SysML according to the domain-specific characteristics and proposed corresponding system design methodologies. Typical works are Cao et al. [8] for mechatronic system design, Thramboulidis [9] for cyber-physical system design, Vogel-Heuser et al. [10] for automation systems developed in IEC 61131, and Gao et al. [11] for satellite communication systems. These works prove the flexibility of SysML in various industrial domains. Besides that, some researchers combined SysML and semantic web technologies to support the model-driven design of complex systems. For example, Cao et al. [12] proposed an ontology-based approach to generate control software design from system design in SysML. Hästbacka and Kuikka [13] applied semantic web technologies to identify and analyze the complex structures of control application models. Feldmann et al. [14] managed consistency between system design and simulation models using a knowledge-based system. Therefore, ontologies and semantic web technologies are promising in making the model-driven system design more intelligent [15].

3 Method Overview

The overview of the proposed approach is shown in Fig. 1. It crosses both the model space described in SysML and the ontology space described in OWL2. In the model space, since the purpose of this study is to automatically generate GUI components constituting the display control interface from the functions of vehicles, the vehicle functions and GUI components, which are the source and target of the reasoning respectively should be specified in SysML. However, since SysML is a general modeling language, domain-specific stereotypes are defined. In addition, the intrinsic

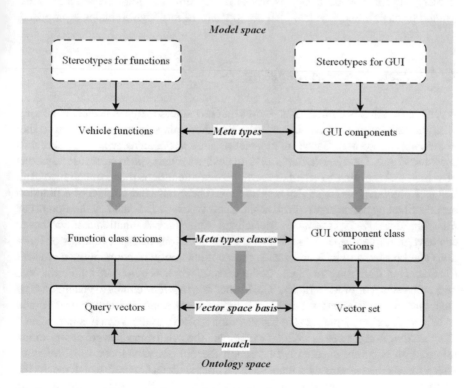

Fig. 1 Overview of the proposed approach combining SysML and ontology

correlations between functions and GUI components, which is that the parameters of functions should be displayed or operated via corresponding GUI components should be reflected in models. For this purpose, a set of common types, so-called meta types are defined to type the properties of both function parameters and GUI components.

The knowledge about vehicle functions and GUI components should be imported into the ontological space to enable the reasoning between them. To this end, the alignment between SysML and OWL2 is analyzed first, according to which the models can be transformed into OWL classes, properties, and axioms. With the formal descriptions, a retrieval algorithm is proposed by first mapping the query functions and GUI components to a vector space with the meta types as its basis and then calculating the semantic similarities between them.

4   Domain-Specific Modeling in SysML

SysML provides standard and graphical model elements that are capable to describe systems of different domains comprehensively. However, how to apply these elements to our specific design issues is not clarified. In this section, a concrete method to model functions and GUI components of armoured vehicles is proposed.

4.1   Vehicle Function Modeling

A vehicle function is a series of activities that are executed by the vehicle to accomplish the tasks issued by its supervisors. Therefore, its semantics is similar to that of *activities* in SysML. Based on this metaclass, a set of stereotypes are defined as shown in Fig. 2. The *VehicleFunction* is an abstract stereotype to represent functions of vehicles. According to whether it can be decomposed, a function can be classified as *ComplexFunction* or *AtomicFunction*. The inputs and outputs of a function are described by *parameters* of activities. Their types are indicated by the stereotype *FunctionParameterType*, which is specialized from *Block*. A function parameter type can have multiple properties such that the semantics of the inputs/outputs of functions can be specified in detail. Among these inputs/outputs, some of them are sent to or received from the passengers of the vehicle, whereas others are communicated with subsystems or other vehicles. The former ones are the sources to generated GUI components. To differentiated these two types of parameters, the stereotype *ActorPin* is defined. It can be applied to the inputs/outputs of a function when it is instantiated to an *action*. A tag *actor* is defined to indicate the role of the passenger whom the information is communicated with through the pin. Its values can be *Conductor*, which means the persons who manage the whole vehicle, *Gunner*, which means the persons who manage the gun, or *Driver*, which means the persons who drive the vehicle.

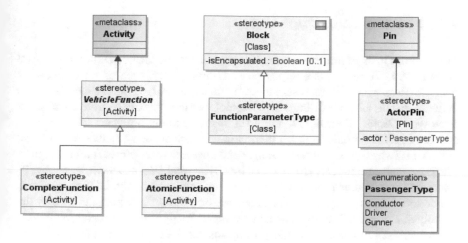

Fig. 2　Stereotypes for vehicle function modeling

These stereotypes can be used to describe individual functions. The model of an atomic function *ReceiveTask* is illustrated in Fig. 3. It means that the vehicle receives information of a task from its supervisor and then display it to the conductor. As shown in Fig. 3a, its input parameter is represented by *mainTask* whose type is *SupervisorTask*. Its output is the same but is set as an actor pin when it is instantiated. It means the task information is sent to the conductor to be processed further. Figure 3b shows the definition of *SupervisorTask*. Its attributes indicate the id, target location, performer and due time of the task.

It is similar to model a complex function except that besides that it contains multiple actions to indicate its sub-functions. They are instantiated from other functions and should be connected by *object/control flows* to represent the workflow of the complex function.

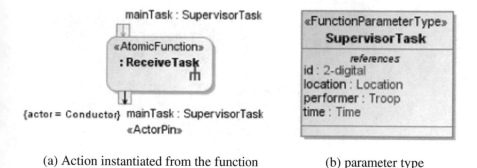

(a) Action instantiated from the function　　　　　(b) parameter type

Fig. 3　Model of the function *ReceiveTask*

4.2   DCS GUI Modeling

The GUI of DCSs of armoured vehicles is similar to the GUI of common software systems. A GUI system is composed of multiple views, each of which includes multiple components. To represent these elements, stereotypes are defined as shown in Fig. 4. All the GUI modeling related stereotypes are specialized from the abstract stereotype *DisplayControlGUI*. Different from common software GUIs, a GUI element of DCSs should indicate their target passengers and whether it should be always on. Two tags, i.e., *targetPassenger* and *alwaysOn* are defined for this purpose. The views and components are modeled by *UIView* and *UIComponent*, respectively. According to whether it is composed of other components, the GUI components can be divided into two classes, i.e., complex or atomic. They are indicated by the stereotype *ComplexUIComponent* and *AtomicUIComponent*. *AtomicUIComponent* has two sub-stereotypes. *UIComDisplay* means that the component is for displaying information and *UIComOperation* means that it is for editing information.

Using these stereotypes, the GUI elements designed by GUI designers can be translated into the system design space. They are modeled in SysML in an object-oriented way. Specifically, the widgets included in a GUI element such as textboxes,

Fig. 4 Stereotypes for GUI modeling

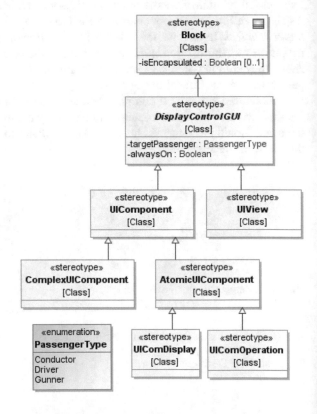

Fig. 5 Model of the GUI
component *Navigation*

«UIComDisplay»
«UIComOperation»
Navigation
references
duration : Duration
endPoint : Location
line : NavigationLine
performer : ExecutionUnit
startPoint : Location
waypoint : Location [0..*]

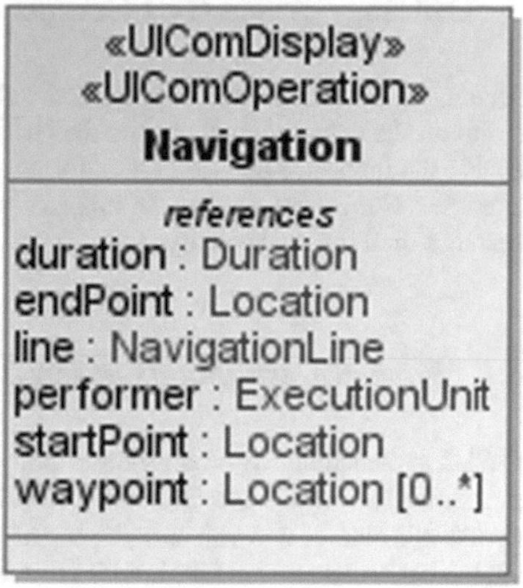

buttons, drop-down lists are abstracted as attributes of the GUI component in SysML according to the semantics of the information they display or manipulate. For example, Fig. 5 shows the model of *Navigation*. It is used to display and edit the route between two locations. The route is represented by its start, end, waypoints (*startPoint*, *endpoint*, *wayPoint*) and lines (*line*) between these points. It also shows the *performer* and expected *duration* of the route.

4.3 Meta Types

The above function and GUI models share some common basic types, which are used as the types of their attributes. For example, the function parameter type *SupervisorTask* has attributes whose types are *2-digital*, *Time*, and *Location* whereas the GUI component *Navigation* has attributes whose types are *Location* and *Duration*. Since these basic types are used to describe other types, they are named as *meta types*. Meta types are the bridge to link the functions and GUI components semantically. To differentiate meta types from others, a stereotype *MetaType* is defined based on *Block*. Typical meta types are *Time*, *Location*, *ID* which has sub-types such as *2-digital* and *3-digital*, and *ExecutionUnit* which has sub-types such as *Troop* and *Vehicle*.

5   Ontology-Based GUI Component Retrieval

Models in SysML provide a graphical and structured way for system designers to maintain the vehicle functions and the GUI components to be retrieved. However, to enable the inference between them, formal descriptions of their semantics should be extracted from these models. In this section, the knowledge is described in OWL2 first and then the retrieval algorithm is presented.

5.1   Formal Descriptions in OWL2

OWL2 is the standard ontology description language in the semantic web technologies stack. Its formal basis is a kind of description language (DL) SROIQ [16]. OWL2 uses OWL classes and object/data properties to represent concepts and roles, which are the basic elements in DL. Besides that, it also provides rich primitives to describe the semantics of concepts formally.

Similar to the hierarchical language architecture of SysML [17], the architecture of OWL2 can be divided into three levels, i.e., OWL primitives such as *owl:class* and *owl:dataproperty*, TBox of the knowledge base, which contains axioms to define concepts, and ABox of knowledge base constituted by individuals instantiated from the TBox. The alignment between the levels of SysML and OWL2 is the basis to extract knowledge from models to the ontological knowledge base. Adapted from [12], the alignment between SysML and OWL2 is shown in Fig. 6.

O3 of OWL2 has the same level of abstraction as M3 of SysML. Functions and GUI components are the focus of this study. Together with their stereotypes, they are mapped to the concepts and properties in OWL2 (O2). Instances of functions and GUI components, i.e., actions and part properties in SysML correspond to individuals in OWL2 (O1).

Since the reasoning of this study works on the O2 level, the mapping from SysML models to the TBox of the knowledge base is the focus. Models related to functions and GUI components can be transformed into the ontological knowledge base in 4 steps as follows.

1. **Upper-level class definition**. The domain-specific stereotypes are defined as OWL classes, i.e., *VehicleFunction*, *FunctionParameterType*, *DisplayControlUI*, *MetaType*, and their sub-stereotypes.
2. **Lower-level class definition**. Functions, function parameter types, GUI components and meta types defined by system designers are mapped to sub-classes of the above upper-level classes. For example, *ReceiveTask* is defined as a sub-class *AtomicFunction*.
3. **Property definition**. Properties of the above objects are mapped to OWL object properties. For example, the attribute *time* of *SupervisorTask* is defined as an object property *hasTime*, whose domain is set as *SupervisorTask*.

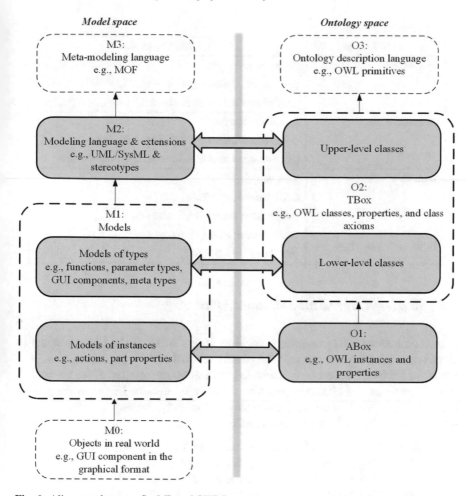

Fig. 6 Alignment between SysML and OWL2

4. ***Class axioms definition***. Class axioms are defined to each of the OWL classes by declaring their properties and restrictions on these properties.

For example, Table 1 shows the formal descriptions of two functions *ReceiveTask* and *DecomposeTask* and a function parameter type *SupervisorTask*. By declaring the equivalent relations between the classes and their class expressions, class axioms can be defined.

These axioms are the core for reasoning on the ontological base. They depict the semantics of the concepts by their properties so that the concepts in the knowledge base can be related. Besides that, since the instances of the classes must obey the axioms, inferences such as instance classification and consistency check can be conducted.

Table 1 Sample class axioms

OWL Class	Class Expression
ReceiveTask	AtomicFunction and (hasParameterIn **exactly** 1 SupervisorTask) and (hasParameterConductorOut **exactly** 1 SupervisorTask)
DecomposeTask	AtomicFunction and (hasParameterIn **exactly** 1 SupervisorTask) and (hasParameterConductorOut **min** 2 DecomposedTask)
SupervisorTask	Task and (hasTime **exactly** 1 Time) and (hasLocation **exactly** 1 Location) and (hasPerformer **exactly** 1 Troop) and (hasId **exactly** 1 2-digital)

5.2 Retrieval Based on Semantic Similarity

The basic idea of GUI component retrieval is that some of the inputs/outputs of functions require interactions with the passengers so that these parameters should be implemented by GUI components with the most similar properties. Since the semantics of functions and GUI components are formally described by their properties, such similarity is called *semantic similarity* in this study. According to this idea, the retrieval algorithm includes 4 steps as follows.

1. *Filtering*. The input and output parameters of functions should be implemented by GUI components for operating and displaying, respectively. It works as a basic condition to filter out the candidate dataset of GUI components to be retrieved. Specifically, for each function $f = \langle P_{in}, P_{out}, P_{in}^a, P_{out}^a \rangle$, the parameters P_{in}^a and P_{out}^a that require interactions with passengers include a set of parameters $\{p_1, p_2, \cdots, p_n\}$. For each parameter p_i, if it is an input parameter, the candidate GUI components to be retrieved in the ontological knowledge base are all the sub-classes of *UIComOperation*, otherwise, the dataset is all the sub-classes of *UIComDisplay*.

2. *Property alignment*. A parameter p_i and a GUI component c_j are both described by their properties, i.e., $p_i = \{a_1, a_2, \ldots, a_n\}$ and $c_j = \{a_1, a_2, \ldots, a_m\}$. Since each property can be denoted as $a_k = \langle n_k, t_k \rangle$ and $t \in MT$ where MT denotes the set of meta types, p_i and c_j can be transformed to a unified form $p_i = \{t_1, t_2, \ldots, t_n\}$ and $c_j = \{t_1, t_2, \ldots, t_m\}$.

3. *Vector generation*. Since p_i and c_j are both described by the common meta types, they can be mapped to a vector space whose basis is the top-level meta types $\langle t_1, t_2, \ldots, t_h \rangle$. The value v_k of the coordinate for t_k can be set as follows. For a parameter p or GUI component c, if it has at least one property whose

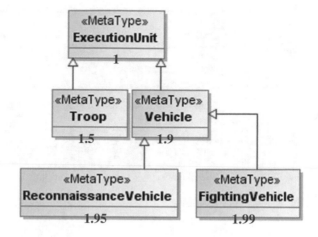

Fig. 7 Sample coordinate values for a meta type hierarchy

type is meta type t, then if t is the top-level meta type t_k, v_k is set to 1. If t is a sub-type of t_k, then v_k should be determined empirically to reflect the semantic distances among these sub-meta types. For example, Fig. 7 shows the sample values of the sub-types of *ExecutionUnit*. For other meta types that none of the properties of p and c is typed as, the corresponding coordinates are set to 0.

4. ***Similarity calculation.*** Since the parameters and GUI components are currently depicted by quantitative vectors, their similarity can be determined by the normalized Euclidean distances. The GUI components are then ordered according to the similarities with the query function parameter and the most similar components can be retrieved for designers to choose.

6 Case Study

The proposed approach is implemented using MagicDraw 16.8 as the SysML modeling platform and Protégé 5.1 as the ontology maintenance tool. The knowledge extraction and retrieval algorithms are implemented in Java using JENA APIs. The task planning function of an armoured vehicle is used as a case study to illustrate the proposed approach.

The function is defined as a stereotyped activity named *PlanTask*. Its internal work-flow is shown in the activity diagram in Fig. 8. This complex function is composed of 6 atomic functions. In the beginning, the vehicle receives a task (denoted as the parameter *mainTask*) from its supervisor and displays the task to the conductor of the vehicle. Then, the conductor decomposes the main task to several sub-tasks through the GUI. The conductor can edit some of the decomposed tasks and generate routes for accomplishing these tasks. These routes can be edited and then issued to corresponding units to execute. Please note that each function parameter is described by

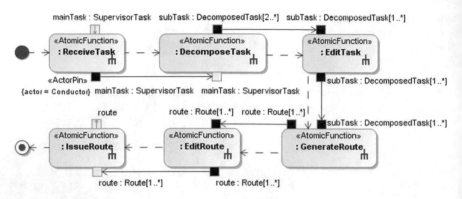

Fig. 8 Model of *PlanTask*

its type and multiplicity. The parameters requiring interactions with the conductor is set as actor pins (marked by black squares in the figure for conciseness).

The models are then imported into the ontological knowledge base as shown in Fig. 9. Besides knowledge about vehicle functions, the GUI components are pre-stored in the knowledge base. The formal descriptions of certain concepts are also shown in this figure.

By querying the knowledge base with each of the function parameters of the sub-functions, the GUI components to implement the function *PlanTask* can be retrieved. For example, Table 2 shows the similarities between the query parameter

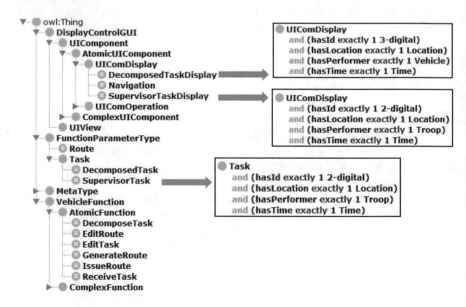

Fig. 9 Ontology after functions are imported

Table 2 Similarities between *SupervisorTask* and GUI display components

	SupervisorTask (query)	SupervisorTaskdisplay	DecomposedTaskDisplay	Navigation
Time	1	1	1	0
Location	1	1	1	1
Duration	0	0	0	1
ID	1.5	1.5	1.9	0
Execution unit	1.5	1.5	1.9	1
Navigation line	0	0	0	1
Similarity	/	1	0.639	0.299

mainTask:SupervisorTask of the function *ReceiveTask* and the 3 GUI components, which are all components for displaying because this parameter is an output. The query and the GUI components shown in the columns of the table are mapped to the vector space whose basis is the meta types shown in the rows. According to the similarities, *SupervisorTaskDisplay* if the best match of this parameter. The similarity of *DecomposeTask* is much higher than *Navigation*.

7 Conclusion and Future Work

As a necessary step towards automated GUI design of armoured vehicles, in this study, an approach to generating the GUI components of DCSs from the vehicle functions is proposed. It combines SysML and ontologies to support both the modeling and reasoning between the involved objects. The contributions of this study can be summarized as follows.

1. A modeling method of functions and display control GUI components of armoured vehicles is proposed based on stereotypes and meta types.
2. The alignment between SysML and OWL2 is analyzed, according to which knowledge contained in SysML models can be transformed into formal descriptions in OWL2.
3. Based on the formal descriptions, GUI components for displaying or operating the parameters of vehicle functions can be automatically retrieved.

The retrieval process in this study is essentially an application of information retrieval technologies in the model-based system design domain. However, it is currently in a rather raw form. Therefore, it should be improved from multiple aspects in the future. For example, systematic methods to quantify the query and dataset should be provided. The precision and recall ratios should be analyzed to evaluate different methods.

As a subsequent step, how to group and arrange the retrieved GUI components to generate the layout of each view of the GUI system should be investigated in the future.

References

1. Zhang, J., Gao, X., Wu, Y., Zhang, A., Shi, G.: Design and realization of display control management system in an avionics integrated simulation system. Fire Control Command Control **31**, 40–47 (2006)
2. Mao, M., Liu, Y., Hu, J.J.: Research on the overall design of integrated electronic information system for tanks and armored vehicles. Acta Armamentarii **38**, 1192–1202 (2017)
3. Wu, Y., Xuejun, Z., Huaxian, L., Xiangmin, G.: Design and realization of display control software in integrated avionic system for general aviation based on the VxWorks. In: Proceedings of the World Congress on Intelligent Control and Automation, July 2018, pp. 1295–1299 (2019)
4. Xie, Y., Hu, X., Duan, H., Wang, S., Zhou, Q.: Research on display control system and cross platform communication of open avionics system. In: Proceedings of 2017 12th IEEE Conference on Industrial Electronics and Applications, ICIEA 2017 2018–February, pp. 1796–1801 (2018)
5. Shao, F., Zhang, A.: simulation of display control management in integrated avionic system. Electron. Opt. Control **17**, 89–92 (2010)
6. Liu, Q., Liu, C., Chu, H.: Design of display and control terminal for electric drive armored vehicle. Autom. Instrum. 50–53 (2011)
7. Liu, B., Li, J., Zhao, A., Chen, Y.: Design of graphic user interface of the fire control system based on tilcon. Comput. Meas. Control **16**, 1153–1160 (2008)
8. Cao, Y., Liu, Y., Fan, H., Fan, B.: SysML-based uniform behavior modeling and automated mapping of design and simulation model for complex mechatronics. Comput. Aided Des. **45**, 764–776 (2013)
9. Thramboulidis, K.: A cyber-physical system-based approach for industrial automation systems. Comput. Ind. **72**, 92–102 (2015)
10. Vogel-Heuser, B., Schütz, D., Frank, T., Legat, C.: Model-driven engineering of manufacturing automation software projects—a SysML-based approach. Mechatronics **24**, 883–897 (2014)
11. Gao, S., Liu, Y., Fan, L., Cao, Y., Long, G.: Model-based semi-physical simulation platform architecting for satellite communication system. In: 2018 13th System of Systems Engineering Conference on SoSE 2018, pp. 379–386 (2018)
12. Cao, Y., Liu, Y., Wang, H., Zhao, J., Ye, X.: Ontology-based model-driven design of distributed control applications in manufacturing systems. J. Eng. Des. 1–40 (2019)
13. Hästbacka, D., Kuikka, S.: Semantics enhanced engineering and model reasoning for control application development. Multimed. Tools Appl. **65**, 47–62 (2013)
14. Feldmann, S., et al.: Towards effective management of inconsistencies in model-based engineering of automated production systems. IFAC-PapersOnLine **48**, 916–923 (2015)
15. Vyatkin, V.: Software engineering in industrial automation: state-of-the-art review. IEEE Trans. Ind. Inform. **9**, 1234–1249 (2013)
16. Horrocks, I., Kutz, O., Sattler, U.: The even more irresistible SROIQ. In: Proceedings of the 10th International Conference on Principles of Knowledge Representation and Reasoning (KR2006), pp. 57–67 (2006)
17. OMG. Model-Driven Architecture: (2014) Available at: https://www.omg.org/mda/. Accessed 10 Oct 2019

Posters

A Detailed Process for Aviation Systems Requirements Analysis and Definition via Model-Based Approach

Teng Li

Abstract Due to the system complexity, it is difficult to define system requirements clearly and correctly by natural languages. Although engineers spend plenty of time on system requirements analysis, system requirements still suffer some drawbacks. In the paper, a new and powerful approach is proposed to analyze and define system requirements. The model-based approach was utilized to analyze system characteristics and then define system requirements of a helicopter air conditioner system. In this process, System Modeling Language was used to build the model to analyze and describe the system characteristics through different views. And qualified system requirements of the helicopter air conditioner system, could be derived effectively by the model.

Keywords System requirements · Model-based approach · System modeling language · Helicopter air conditioner system

T. Li (✉)
Avic-Digital CO., Ltd., Jingshunlu No. 7, Chaoyang District, Beijing, China

© The Author(s), under exclusive license to Springer Nature Switzerland AG 2021
D. Krob et al. (eds.), *Complex Systems Design & Management*,
https://doi.org/10.1007/978-3-030-73539-5_35

463

A Prediction Method for Gas Turbine Overhaul Makespan with Uncertainties

Tao Li, Zhen Chen, and Yusong Liu

Abstract The prediction of a gas turbine overhaul makespan benefits much from the real-time supervision of overhaul process and fast response to abnormal conditions, which ensure the product delivery in time for customers. Previous research indicated that many uncertainties contribute to the inaccurate of overhaul process and thus lead to delivery delay. A novel prediction method compromised with Program Evaluation and Review Technique (PERT) and Monte Carlo simulation is proposed to solve this issue. The uncertain factors affecting the overhaul makespan are investigated first, then the generic overhaul process modelling is produced to understand the relationship and constraints between detailed processes. Based on that, the duration time distribution of the overhaul makespan plans can be generated. It is possible to finalize the prediction method considering the distribution of overhaul time probability and critical work path, and the process dependences. A gas turbine is used as a case study to examine this method. The results show it is advanced in the efficiency compared with traditional methods, and has the potential to meet the requirements of high demand overhaul service for airliners.

Keywords Gas turbine overhaul · Uncertainty · PERT · Makespan prediction · Simulation

T. Li (✉) · Z. Chen · Y. Liu
AVIC Chengdu Aircraft Industrial (Group) Co., Ltd., Chengdu 610092, China

© The Author(s), under exclusive license to Springer Nature Switzerland AG 2021
D. Krob et al. (eds.), *Complex Systems Design & Management*,
https://doi.org/10.1007/978-3-030-73539-5_36

465

A Preliminary Research on Performance Prediction Model of Catapult Launched Take-Off for a Large Wingspan Unmanned Aerial Vehicle

Hongrui Xiong, Tao Li, Heyu Li, and Changgui Yu

Abstract The design of Unmanned Aerial Vehicle (UAV) is a typical Systems Engineering (SE) process, which consists of design uncertainties and trade-offs. In predicting the catapult launch of a large wingspan UAV, there are several design drivers and constraints that contribute to the take-off performance. This research investigates the distance estimation methods for UAV take-off in previous research first. The conventional methods, which mostly used are for the distance estimation of the land-based takeoff plane or studying the model of the catapult landing gear only. Thus a gap is found that there are no existing catapult launch performance prediction methods for a large wingspan UAV. It is proposed a fast prediction method based on dynamic and kinematical equations model to deal with this issue. A case study is used to examine the proposed method, and the results show the feasibility in catapult launch performance prediction, especially in the case of short taxing in runway. The new method has the advantage of fast parameter configuration re-build, more reliable performance prediction, and ensures the safety in UAV flight test.

H. Xiong (✉) · T. Li · H. Li · C. Yu
AVIC Cheng Du Aircraft Industrial (Group) Co., Ltd., No. 105, Chengfei street, Qingyang District, Chengdu, Sichuan, China

© The Author(s), under exclusive license to Springer Nature Switzerland AG 2021   467
D. Krob et al. (eds.), *Complex Systems Design & Management*,
https://doi.org/10.1007/978-3-030-73539-5_37

Application of Data Bus-Based Software Architecture in Wind Turbine Control Software

Meiyu Cui and Yongjun Qie

Abstract In this paper, a data bus-based software architecture is designed to provide standardized, transparent software interfaces, to reduce coupling between software modules, and shield the complexity of operating system for synchronous reactive system software design. First of all, based on synchronous theory, a data bus-based software architecture is designed. And then, common data space is designated, it is open to all application layer software module, each application layer software module can directly access the common data space. At the same time, a data protection mechanism has been established to ensure that every data has single source and all data are updated uniformly. Finally, the software architecture is applied in wind turbine control software, and the wind turbine control software architecture is reconstructed.

M. Cui · Y. Qie (✉)
Tsinghua University, Haidian District, Beijing, China
e-mail: xiyj19@mail.tsinghua.edu.cn

© The Author(s), under exclusive license to Springer Nature Switzerland AG 2021
D. Krob et al. (eds.), *Complex Systems Design & Management*,
https://doi.org/10.1007/978-3-030-73539-5_38

Application of Model Based System Engineering in Hydraulic Energy System Design

Quanrun Mou, Xiaolong Tong, Zhenghong Li, and Liangliang Liu

Abstract In this paper, the system engineering method based on the model is applied to the design of hydraulic energy system. Based on Doors software, the system level and product level requirements are allocated from top to bottom and retrospected back up. Based on Rhapsody software, function logic modeling of hydraulic energy system and correlation modeling between requirement and activity are carried out. Based on Amesim software, system performance simulation modeling and verifying is implemented.

Q. Mou (✉)
Shenyang Aerospace University, Shenbei, Shenyang, China

X. Tong · Z. Li · L. Liu
Shenyang Aircraft Design Institute, Tawan, Shenyang, China

471

Architecture Design of High Safety Helicopter Flight Control System with Direct Control Mode

Liqiang Teng, Wenshan Wang, and Yining Liu

Abstract The complex Electronic Control Unit (ECU) is the core of helicopter Flight Control System (FCS), which can cause flight accident when the FCS redundancy is degraded quickly due to its failure, or when ECU fails simultaneously due to common mode failure. Accordingly, a 3×2 redundant helicopter FCS architecture with direct control mode is designed, by comparing and analyzing the FCS architectural characteristics of typical international advanced NH-90 and S-92 helicopters. In primary control mode, this FCS can fail-operate twice, and ensure level 1 handling qualities. After three failures, the FCS switches to direct control mode and can still tolerate once failure, and the handling qualities changed to level 2. Meanwhile, non-similar design is used to avoid ECU's common mode failure. Finally, the safety of primary control mode is analyzed by fault tree analysis. This paper can provide reference for the FCS architecture design of medium and large helicopters.

L. Teng (✉) · W. Wang · Y. Liu
AVIC Qing'an Group Co.Ltd., Xi'an 710077, China
e-mail: liqiangteng@126.com

© The Author(s), under exclusive license to Springer Nature Switzerland AG 2021
D. Krob et al. (eds.), *Complex Systems Design & Management*,
https://doi.org/10.1007/978-3-030-73539-5_40

473

Design Method of Aviation Architecture Based on Model Base

Zhijuan Zhan, Qing Zhou, Bingfei Li, and Bing Xue

Abstract Drawing on dodaf ideas and summarizing aviation examples, the paper proposes an aviation architecture design method based on model base. The input content and form of aviation architecture design are defined, the organization form and metadata mapping relationship of meta model in the architecture model base are determined, the template data organization is designed. It realizes the automatic generation method of aviation system task architecture, functional architecture and requirements, and improves the efficiency of aviation architecture design.

Z. Zhan (✉) · Q. Zhou · B. Li · B. Xue
China National Aeronautical Radio Electronics Research Institute, Shanghai 200233, China
e-mail: 13916905819@163.com

Q. Zhou
e-mail: Zhouqing0@163.com

B. Li
e-mail: Libingfei612@126.com

B. Xue
e-mail: xueicb16@126.com

Experiences with Applying Scenario-Based Approach to Refine Aircraft Stakeholder Requirements

Wenhao Zhu and Fuxing Tao

Abstract The development of aircraft often involves much effort of requirements elicitation, definition, and review, and the initial version of these requirements are provided by aircraft stakeholders. In practice, as the initial stakeholders' requirements do not often meet the characteristics and attributes of INCOSE as good requirements, they might cause errors and difficulties to requirements verification and validation. Although the Easy Approach to Requirements Syntax (EARS) has been introduced to provide an easy method to define natural language requirements for years, the engineers and stakeholders still need to reach a shared comprehension of the requirements before defining them. The approach reported in this paper is to combine a scenario-based approach with templates as an elementary guidance for system engineers to refine requirements at the early stages of aircraft development project, so as to eliminate possible errors and problems caused by natural language and insufficient communication.

W. Zhu (✉) · F. Tao
AVIC Digital Corporation Ltd., Building 19E, Compound A5, Shuguangxili, Chaoyang District, Beijing 100028, China
e-mail: zhuwh@avic-digital.com

F. Tao
e-mail: taofx@avic-digital.com

Formal Modeling and Correctness Proof of Spatial Partition Algorithm

Liping Zhu, Fangfang Wu, Pei Yang, Jixun Yan, and Li Ma

Abstract In the development of the embedded real-time operating system which follows ARINC653 (Standard Interface of Avionics Application Software), the partition protection and address translation algorithm of the memory management unit based on PowerPC E200 processor is proposed and formally verified. Using the interactive theorem proving tool Coq, firstly, address translation is formally modeled. Secondly, the three requirements that the algorithm must meet are expressed in the form of theorems. Finally, the strategies and construction methods of the tool are used to prove the correctness of the algorithm. The algorithm verification results show that the formal method theoretically guarantees the correctness of the key algorithm of the operating system, overcomes the incompleteness of the traditional test methods, and provides a strong guarantee for the development of high-quality safety critical software.

L. Zhu (✉) · F. Wu · P. Yang · J. Yan · L. Ma
AVIC Xi'an Flight Automatic Control Research Institute, Jinye Road 129, Xi'an, China
e-mail: zlponline@163.com

© The Author(s), under exclusive license to Springer Nature Switzerland AG 2021 479
D. Krob et al. (eds.), *Complex Systems Design & Management*,
https://doi.org/10.1007/978-3-030-73539-5_43

Integrated Configuration Management Based on System Engineering

Qiu Xi, Zhong Jin, Hui Wei, Wang Lixin, Yang Yi, and Wang Jw

Abstract The design and manufacturing process of an aircraft is a huge and complex SE-System Engineering involving a large group of specialized expertise. In the aviation design and manufacturing company (units), the production mode of small batch and multi modified production mode has become a trend. In this paper, configuration management is examined from the perspective of system engineering, the engineering domain involved in aircraft products is divided into many specialized fields, and according to the professional characteristics of merger into the order-cluster, requirement-cluster, design-cluster, manufacture-cluster, repair-cluster, finance-cluster, manpower-cluster, airworthiness-cluster etc., for these clusters of unified configuration management system called "Configuration management based on system engineering". Need is a macro perspective to control product requirements, product design, product manufacturing, supply chain, trial inspection, maintenance, human property supply and a series of processes. The management of

Q. Xi (✉) · W. Lixin
AVIC Xi'An Aircraft Industry (Group) Co., Ltd., Xi'an, China
e-mail: qx143@163.com

W. Lixin
e-mail: wlx662@163.com

Z. Jin · H. Wei · Y. Yi
AVIC Digital Co., Ltd., Shenzhen, China
e-mail: zhongjin70@126.com

H. Wei
e-mail: huiwei1314@139.com

Y. Yi
e-mail: yangyi2424@126.com

W. Jw
Shanghai Aircraft Design and Research Institute, Shanghai, China
e-mail: hswangjunwen@126.com

multi professional coupling data based on system engineering can be realized by integrating configuration management, improving data consistency to remove barriers between and within the enterprise.

Model Driven Verification of Airplane Scenarios, Requirements and Functions

Chao Zhan, Dongsheng Chen, Yanbo Zhang, and Meixiang Peng

Abstract The correctness and completeness of requirements is critical for a successful project. Minor mistakes will be significantly amplified later in the development phase and will require significant financial and personnel resources to modify them. It is therefore crucial to rigorously validate the operational scenarios and functions as early as possible in the aircraft design cycle. Due to the disadvantages of natural language for describing increasingly complex systems, models have been adopted to produce a standardized means of visualization and simulation. This has become known as Model Based Systems Engineering (MBSE). In this paper, the scenarios of systems operation will be described, utilizing a top-down design process based on MBSE. The system functional analysis will focus on the translation of the higher-level requirements into a coherent description the system functions. Every function is then refined by decomposition within each relevant scenario. These functions are then allocated to corresponding systems, sub-systems, or items, to capture the functional architecture. Finally, the correctness and completeness of the system scenarios and functional model is verified through state machine execution to complete the MBSE process. Through this method, we can fully define the scenarios, functions, and requirements early in the design phase.

C. Zhan (✉) · D. Chen
Shanghai Aircraft Design and Research Institute, COMAC, Shanghai, China
e-mail: zhanchao@comac.cc

D. Chen
e-mail: chendongsheng@comac.cc

Y. Zhang · M. Peng
Beijing RUNKE TONGYONG Technologies CO,. Ltd., Beijing, China
e-mail: jkrescuezhang@outlook.com

M. Peng
e-mail: pengmeixiang@outlook.com

© The Author(s), under exclusive license to Springer Nature Switzerland AG 2021
D. Krob et al. (eds.), *Complex Systems Design & Management*,
https://doi.org/10.1007/978-3-030-73539-5_45

Optimal Design of Airborne Test System Based on Model Analysis

Xiao-Lin Li, Shuai Xie, and Wei Yuan

Abstract Model-based systems engineering design ideas play an increasing significant role in the development and testing of new aircraft, the finite element method as an important means of model analysis can provide a theoretical basis for the design of airborne test systems. First, propose an optimized design scheme for model-based airborne test system, design a multi-source data fusion real-time processing system for UAV, make the real-time processing and analysis speed of the original data, link data and multi-band test data reach a subtle level; Then, according to the complex coupling heat transfer mechanism of the UAV, based on the structure characteristics, a method of calculating the thermal environment of the UAV in different regions was proposed, realize the rational design of temperature test structure layout. Finally, based on the theory of modal analysis and simulation experiment, the optimal selection of vibration auxiliary equipment is realized to solve the vibration data overrun problem caused by resonance. The test results show that the airborne test system has a reasonable layout and measurement design, which can effectively reduce about 32.5% of test flight resources, and can be applied to other flight tests.

X.-L. Li (✉) · S. Xie · W. Yuan
Aviation Key Laboratory of Science and Technology on Flight Test, Chinese Flight Test Establishment, Xi'an 710089, Shanxi, China

485

Practice of ARCADIA and Capella in Civil Radar Design

Renfei Xu, Wenhua Fang, and Wei Yin

Abstract ARChitecture Analysis and Design Integrated Approach (ARCADIA) is a kind of Model-Based System Engineering (MBSE) methodology developed by Thales, and Capella is a kind of modeling tool dedicated to ARCADIA. In this paper, we will introduce our practice of ARCADIA/Capella in civil radar design, including the reason we choose this solution and how we use it in civil radar design. We will also briefly introduce our extension of Capella in parametric modeling, dynamic execution and simulation, system and subsystem collaboration, interface detailed design, and document generation.

R. Xu (✉) · W. Fang
Shanghai PGM Technology Co., Ltd., No. 2966 on Jinke Road, Shanghai, China

W. Yin
No. 1th Institute of CETC, No. 8th of Guorui Street, Yuhuatai District, Nanjing, China

© The Author(s), under exclusive license to Springer Nature Switzerland AG 2021   487
D. Krob et al. (eds.), *Complex Systems Design & Management*,
https://doi.org/10.1007/978-3-030-73539-5_47

Research on Distributed Harmony-SE Model Integration Method for Complex UAV System

Guang Zhan, Miao Wang, Zhixiao Sun, Yuanjie Lu, and Yang Bai

Abstract The distributed modeling of complex unmanned aerial vehicle (UAV) system is faced with the problem of multiple use cases and multiple models integration. After studying the purpose and principles of model integration, we proposed a Harmony-SE based tight coupling integration method for the white box functional logic model of UAV system, model integration process was described in detail. The whole system model integration of complex UAV and the whole system functional logic verification based on state machine diagram are realized, which is of great value to realize the collaborative design and parallel modeling of multiple use cases of complex UAV system.

G. Zhan (✉) · M. Wang · Z. Sun · Y. Lu · Y. Bai
Shenyang Aircraft Design and Research Institute (SADRI), AVIC No. 40 Tawan Road, Shenyang, Liaoning, China

© The Author(s), under exclusive license to Springer Nature Switzerland AG 2021   489
D. Krob et al. (eds.), *Complex Systems Design & Management*,
https://doi.org/10.1007/978-3-030-73539-5_48

Research on Multi-disciplinary Integrated Design Method of Remote Sensing Satellites

Yongsheng Wu, Guangyuan Wang, and Jiaguo Zu

Abstract The improvement of mission capability of remote sensing satellites has made it more demanding for the in-depth, detailed, and intensive design. It is required that the ultimate goal of system performance should be carried out around the mission capability, which results in significant changes in product integration and verification pattern. This paper firstly introduces the concept and requirements of multi-disciplinary integrated design method, and then presents the key issues of the method in worldwide developments and applications. Finally, the practice in integrated collaborative analysis platform establishment is demonstrated, and the simulation verification is conducted and summarized.

Y. Wu (✉) · G. Wang · J. Zu
Institute of Remote Sensing Satellite, CAST, Beijing 100094, China
e-mail: winforever19890727@126.com

G. Wang
e-mail: zhuichilun@126.com

© The Author(s), under exclusive license to Springer Nature Switzerland AG 2021   491
D. Krob et al. (eds.), *Complex Systems Design & Management*,
https://doi.org/10.1007/978-3-030-73539-5_49

Research on Multi-physical Modeling and Co-simulation of Aircraft

Dangdang Zheng, Liqiang Ren, Ying Wu, and Juntang Liu

Abstract Advanced aircraft system is more and more integrated, which makes the problem of multiple physical coupling between different physical systems more and more complex. Multi-physical modeling and co-simulation is a kind of important means for the early design study of multi-physical coupling, supporting systems integrated design and verification. Combined with the methodology of model-based system engineering, the key technologies of multi-physical system modeling and co-simulation are systematically analyzed, and the engineering application mode of multi-physical co-simulation for different stages of aircraft development is proposed.

D. Zheng (✉)
Northwestern Polytechnical University, Xi'an, People's Republic of China
e-mail: zhengdangdang@126.com

D. Zheng · L. Ren · Y. Wu · J. Liu
The First Aircraft Institute of AVIC, Xi'an, People's Republic of China
e-mail: 18717393589@163.com

Y. Wu
e-mail: wuying@163.com

J. Liu
e-mail: juntangliu@163.com

Research on Practice Methods of Complex Aircraft Requirement Management

Zhixiao Sun, Dong Kan, Yang Bai, and Yuanjie Lu

Abstract The development of complex aircraft is complicated system engineering. Facing to the requirements exponential growth of complex aircraft, manufacturers put forward stringent standard to requirement quality control, therefore, it is very urgent to explore a suitable method for requirement management of complex aircraft development. The paper uses the development of complex aircraft as an example based on system engineering method, discusses the engineering application practice of requirement management technology from requirement acquisition, attribute and traceability, verification and validation, change control. Research shows when implementing requirement management method in this paper on complex aircraft, it can effectively improve the integrity, systemic, validity, coherence of requirement, further enhance the quality of requirement and efficiency of management, and the development cycle is shortened and the cost is reduced in the future.

Z. Sun (✉) · D. Kan · Y. Bai · Y. Lu
INCOSE CSEP, Shenyang Aircraft Design and Research Institute (SADRI), AVIC, No. 40 Tawan Road, Shenyang, Liaoning, China

Y. Bai
e-mail: baijordon@163.com

Y. Lu
e-mail: plough5221@sina.com

© The Author(s), under exclusive license to Springer Nature Switzerland AG 2021
D. Krob et al. (eds.), *Complex Systems Design & Management*,
https://doi.org/10.1007/978-3-030-73539-5_51

Retrospect and Prospect of Aircraft Comfort Design

Peng Li and Kaixiang Li

Abstract Passenger comfort has not been completely integrated into the entire aircraft design process yet. Sometimes, the cabin vibration environment is intolerant during flight, causing seriously adverse reactions such as discomfort, fatigue, and reduced task-performing ability. Here, the aircraft digital FEM model is used to predict the cabin environment, and the acceleration response results can be used as the input to evaluate the human comfort level. Four methods for evaluating human comfort level (the absorbing power method, NASA ride quality method, ISO and BSI standard) are commonly used. The four methods are coded into a plug-in program, which can be used in aircraft FEM model, to facilitate the evaluation process. Finally, the procedures and suggestions of cabin comfort design are given, for example, engine damping installation, seat cushion updating, seat vibration isolation treatment, etc. In this way, the aircraft structure can be modified and improved according to the passenger comfort feedback.

P. Li (✉) · K. Li
Aircraft Strength Research Institute, Yanta District, 86 Dianzierlu Road, Xi'an Cty, China

D. Krob et al. (eds.), *Complex Systems Design & Management*,
https://doi.org/10.1007/978-3-030-73539-5_52
497

SoS Architecture Models Transformation for Mission Simulation in Aircraft Top-Level Demonstration

Zang Jing

Abstract The continuous modeling is necessary to achieve the virtual verification and validation in aircraft top-level demonstration. To fill in the gaps between SoS architecture models and mission simulation models, a model transforming method for mission simulation is proposed. The operational/ system activity sequences models can be mapped to the parts of the task flows of scenario script in mission simulation system. The operational/ system state transitions models can be transformed to the behavior models though format conversion of the SoS architecture outputs. A sample case is provided to illustrate a specific scenario with architecture models transformation. The result preliminary verifies the consistency of model transformation from SoS architecture to mission simulation.

Z. Jing (✉)
Aviation Industry Development Research Center of China, Xiaoguandongli No. 14, Beijing 100029, People's Republic of China
e-mail: zangjing2006@163.com

The Application Research of System Cooperative Design Engineering Based on MSFC Architecture

Hongjie Xu, Jinhu Ren, and Yingru Wang

Abstract Many research institutions are exploring the best practice of model-based system engineering in order to improve the development quality of complex system, reduce the design cost and shorten the development cycle. The collaborative development of complex systems by MBSE is the focus of system engineering research and the foothold of engineering application. This paper proposes a new modeling idea-the MBSE system collaborative design idea based on MSFC data architecture, which used a unify data source and effective transfer of the model for system analysis and multi-professional collaborative design and used MSFC architecture to implement collaborative design for a certain aileron system to verified the feasibility of the theory.This paper verified the relevant requirements in the aileron system analysis process through the system modeling and simulation of the control subsystem, the hydraulic drive subsystem, the actuator subsystem and the rudder surface. The research shows that MSFC architecture was suitable for the collaborative design process of complex system that can be used as the modeling idea of MBSE engineering application.

Keywords MBSE · MSFC · System modeling · Collaborative development · Requirements closed-loop verification

H. Xu (✉)
Northwestern Polytechnica University, Xi'an 710000, China
e-mail: xuhj@avic-digital.com

H. Xu · J. Ren · Y. Wang
AVIC Digital Technology Co., Ltd., No. 7 Jingshun road, Chaoyang district, Beijing 100028, China

501

Printed in the United States
by Baker & Taylor Publisher Services